AF721005

Virology

SCIENCES

Biology, Field Director – Marie-Christine Maurel

Virology, Subject Head – Félix Augusto Rey and María Carla Saleh

Virology

Coordinated by
María Carla Saleh
Félix Augusto Rey

ISTE

WILEY

First published 2021 in Great Britain and the United States by ISTE Ltd and John Wiley & Sons, Inc.

Apart from any fair dealing for the purposes of research or private study, or criticism or review, as permitted under the Copyright, Designs and Patents Act 1988, this publication may only be reproduced, stored or transmitted, in any form or by any means, with the prior permission in writing of the publishers, or in the case of reprographic reproduction in accordance with the terms and licenses issued by the CLA. Enquiries concerning reproduction outside these terms should be sent to the publishers at the undermentioned address:

ISTE Ltd
27-37 St George's Road
London SW19 4EU
UK

www.iste.co.uk

John Wiley & Sons, Inc.
111 River Street
Hoboken, NJ 07030
USA

www.wiley.com

© ISTE Ltd 2021
The rights of María Carla Saleh and Félix Augusto Rey to be identified as the authors of this work have been asserted by them in accordance with the Copyright, Designs and Patents Act 1988.

Library of Congress Control Number: 2020949883

British Library Cataloguing-in-Publication Data
A CIP record for this book is available from the British Library
ISBN 978-1-78945-023-1

ERC code:
LS1 Molecular Biology, Biochemistry, Structural Biology and Molecular Biophysics
LS1_4 RNA synthesis, processing, modification, degradation
LS6 Immunity and Infection
LS6_5 Biology of pathogens (e.g. bacteria, viruses, parasites, fungi)
LS6_8 Infectious diseases in animals and plants

MIX
Paper from responsible sources
FSC® C013604

Contents

Introduction

María-Carla SALEH[1] and Félix AUGUSTO REY[2]

[1] *Viruses and RNAi Unit, Institut Pasteur and CNRS UMR 3569, Paris, France*

[2] *Structural Virology Unit, Institut Pasteur and CNRS UMR 3569, Paris, France*

Viruses are ubiquitous and have been a major driver in the evolution of life on Earth. Without viruses, life would certainly not be as we know it. Viruses are intimately linked to cells: invading a cell is mandatory for their reproduction and perpetuation in nature. The origin of viruses goes back to the origin of life, which emerged from primitive self-replicating nucleic acid molecules. The viruses that are mostly studied are those that cause disease in humans, animals or plants; however, disease-causing viruses are only an epiphenomenon in the global picture of the origin and evolution of life. Most viruses have evolved with their host and do not cause disease. Virus pathogenicity is most often the result of a change of context, such as the infection of new hosts in which they have not evolved and adapted. A good example is constituted by zoonotic viruses, such as the Ebola virus or the current coronavirus that is responsible for the COVID-19 pandemic, which are bat viruses and do not cause disease in their natural host.

Viruses have genome coding for all of the information required for their perpetuation in nature. Some viruses have very small genomes, while others have very large genomes. Baltimore introduced a classification of viruses into different groups, according to the way the viral genome produces messenger RNA once inside the cell, so that they can be translated into viral proteins (Baltimore 1971). There are seven virus groups according to this criterion, the simplest one being those in which the genome is a positive-sense, single-stranded RNA (+ssRNA) molecule (group IV in Baltimore's classification), which can be directly translated by the ribosomes in a cell. There is no transcription step required for group IV viruses, and the naked

Virology,
coordinated by María-Carla SALEH and Félix AUGUSTO REY. © ISTE Ltd 2021

genome is infectious, contrary to the RNA viruses in the other groups. One of the virally encoded enzymes is polymerase, which is capable of replicating the +ssRNA genome to make more progeny within the cell, as cells normally lack RNA-directed RNA polymerase activity. The replication of viruses in group IV takes place in the cytoplasm of the cell. Viruses in the other groups require the transcription of their genome in order to produce messenger RNA. Most of them encode viral polymerase, which needs to first transcribe the genome into messenger RNA to synthesize the viral proteins. In this case, the viral RNA is only infectious in association with the viral polymerase, and it is this complex that is delivered into the cytoplasm of the cell for a productive infection. After the initial transcription round, replication ensues – i.e. the synthesis of more genomic RNA – often using the same virally encoded polymerase, which has dual transcriptase and replicase activities – to be incorporated into new viral particles. For viruses in Baltimore groups VI and VII, the viral polymerase is a "reverse transcriptase" (RT). In group VI, the genome is a +ssRNA molecule that is not used as a messenger, but is transformed by RT into double-stranded DNA (dsDNA) that gains the cell nucleus and is integrated into the host cell genome to be transcribed by cellular polymerases. In the viruses in group VII, the genome is a DNA molecule that is maintained as an episome in the nucleus of the infected cell (i.e. it is not integrated into the host genome), in the form of a covalently closed circular DNA (cccDNA), like in the hepatitis B virus (HBV). This DNA genome is transcribed by cellular polymerase, and the RNA of full-length transcripts, called pregenomic RNA (prgRNA), is incorporated, along with RT, into newly formed viral particles. Once inside the particle, RT transforms the prgRNA into an incomplete double-stranded linear DNA genome that will be used to infect other cells.

This book provides several snapshots into the universe of viruses. They provide more detail on the replication of viruses in the most complex groups in the Baltimore classification. In Chapter 1, Lindsey M. Costantini and Blossom Damania discuss the replication of DNA viruses, which belong to Baltimore groups I (dsDNA viruses) and II (ssDNA viruses), as well as HBV (group VII). Most of these viruses replicate in the nucleus of the cell, but there are exceptions, such as the poxviruses. This chapter also highlights the specific features of some of the DNA viruses causing serious diseases in humans, notably the so-called oncogenic viruses, showing that some become integrated into the host genome, but others remain as episomes. In Chapter 2, Michelle M. Arnold, Albie van Dijk and Susana López analyze the life cycle of the viruses in group III, which have a double-stranded RNA genome that is always contained in an icosahedral proteinaceous capsid. This protein shell protects these viruses from the detection of their genomes by the cell. Indeed, dsRNA is not normally present in a cell, and therefore constitutes a pathogen-associated molecular pattern (PAMP) that is recognized by pattern recognition receptors (PRRs), which trigger the innate immune system of the host. For this reason, all

RNA viruses have developed the means to isolate their replication compartments, in which transient dsRNA molecules form without being exposed to PAMPs.

In Chapter 3, Rachel Fearns describes the replication of viruses in group V, which have a negative-sense ssRNA genome in complex with a protein (termed "N" for nucleoprotein) that polymerizes all along the genome, and is associated with a large viral polymerase that is first used for transcription and then for replication. Some of these viruses have a segmented genome. While most of them replicate in the cytoplasm, some gain the nucleus for replication, for example the influenza virus. This chapter also describes the replication of one of the smallest viruses that infects humans, the hepatitis delta virus (HDV). Although it has a negative-sense ssRNA genome, it is so small that it does not code for a polymerase enzyme, nor for a capsid protein: it instead uses the capsid and envelope protein of a co-infecting HBV to propagate to other cells. For its replication, the HDV genome enters the nucleus and co-opts the cellular RNA polymerase II, which normally only transcribes DNA, for its own transcription and replication. This is a remarkable ability of this virus, to make cellular enzymes work in a different way for its own benefit.

Viruses can also hijack the metabolic pathways of a cell in order to make progeny and convert the cell into a virus-producing factory. During evolution, however, there is a constant arms race in which cells fight back – or more precisely, the whole organism fights back, sacrificing infected cells so that the infected organism survives as a whole. The PRRs detect foreign motifs (PAMPs) in the cell by using different mechanisms, in order to trigger a strong innate immune response that leads to the death and elimination of the infected cell. Viruses, in turn, have developed ways to counter this detection. In Chapter 4, Rachel Netzband and Cara T. Pager describe the extensive epigenetic modifications of viral nucleic acids that make them appear like cellular RNA and avoid or limit their detection by PRRs. On the contrary, it is in the virus' benefit to avoid killing its host. In Chapter 5, Carolina B. López describes the defective interfering particles generated by deletions, or other defects, in the replicated viral genome, which trigger the innate immunity of the cell. This delicate balance allows the virus to become persistent, with replication only at a basal level, protecting its host but ensuring its viability. This is another illustration of the constant interplay between viruses and cells during evolution, both mutually adapting in a race that is reminiscent of the Red Queen in Lewis Carrol's fiction Through the Looking-Glass (1871), constantly running to stay in the same place.

Evolution has been such that viruses have developed highly specific and sophisticated ways of interacting with their host. In Chapter 6, Matthew Phillips, Bria F. Dunlap, Megan T. Baldridge and Stephanie M. Karst illustrate the way enteric viruses co-opt the intestinal flora in order to augment their infectiveness. They bind to bacterial polysaccharides or other structures with high affinity and specificity, in order to use them as vehicles to efficiently reach their target cells in the intestinal epithelium. Similarly, viruses that are transmitted by arthropod vectors – notably insects or ticks – have evolved very efficient and specific ways of interacting with their vector in order to be transported and maximize their chances of finding new, uninfected hosts. Such an example is described in Chapter 7 by Swapna Priya Rajarapu, Diane E. Ullman, Marilyne Uzest, Dorith Rotenberg, Norma A. Ordaz and Anna E. Whitfield, who analyzed the interactions of plant viruses with their insect vectors. Some viruses have developed highly efficient ways to bind to an insect's mouth while it feeds on an infected plant, so that they can be transported to other susceptible plants. Some plant viruses replicate in their vector, and change insect behavior so that it is more attracted to uninfected plant hosts, thereby spreading the infection more easily. They further discuss the role of bacterial endosymbionts in the vertical transmission of viruses in the vector species, as a way of maintaining the virus even when no plant hosts are available. The last two chapters are perfect examples of the evolution of viruses as part of the global evolution of life, as viruses evolve in response to changes in their hosts and vectors. Finally, in Chapter 8, Rachele Cagliani, Alessandra Mozzi, Chiara Pontremoli and Manuela Sironi conclude this book with an analysis of the evolution of human viruses known for their medical relevance. They consider many of the viruses mentioned in the previous chapters from a different perspective, describing the sources of viral genetic diversity, their evolution compared to their host range, and the zoonotic viruses responsible for public health crises such as influenza viruses, coronaviruses and several retroviruses, including HIV-1.

In their vast majority, the chapters presented in this book were written by female authors. In a way, this book is an homage to those talented women who have contributed to the study of viruses in the last century or so. The study of virology and the impact of these studies on the advancement of contemporary biology have been marked by women who had to fight the stereotypes that claim women and men are not equal in their thirst for knowledge. In order to have the career they wanted, to have access to education and to walk uncommon paths, they broke rules, they often worked alone (with no funding and no recognition) and they believed in themselves when others doubted them. Creativity, persistence and a love of discovery were the greatest tools these women used in order to make significant contributions.

Did you know that it was a female scientist who first succeeded in culturing a virus *ex vivo* and studying it under laboratory conditions? The work of the amazing Edna Steinhardt (1874–1941) changed virology. In 1913, she published a study

(Steinhardt *et al.* 1913) focusing on a culture of the vaccinia virus (belonging to the Poxviridae family), present in the cornea and plasma tissue of a guinea pig, in order to produce new virions and study the virus biology outside of the human body. Her legacy is one of the most important for modern virology.

And what about June Almeida (1930–2007)? The woman who discovered the first coronavirus! She left school at the age of 16, with little formal education, but got a job as a laboratory technician in histopathology at the Glasgow Royal Infirmary. It was in Canada, at the Ontario Cancer Institute, that Dr. Almeida developed her outstanding skills with an electron microscope. In 1964, Almeida was recruited by St Thomas's Hospital Medical School in London. She pioneered a method to better visualize viruses, by using antibodies to agglutinate them. In 1966, she identified a group of viruses (Almeida and Tyrell 1967) that were named coronavirus, because of the crown – or halo – surrounding them on the viral image, in nasal washings of volunteers with the common cold.

Even today, despite massive campaigns of awareness and education, women in science still represent just 29% of researchers globally (according to the United Nations), and their work rarely gains the recognition it deserves. True gender equality and women's access to science is a permanent pursuit and involves us all. When the editorial group approached us to prepare yet another book in virology, we chose to give the voice to the women virologists of our time. It is a small contribution and a tiny step towards equality. Through these eight chapters, we want to applaud, recognize and shout out the everyday battles of women scientists around the world.

This book has been written and edited during the SARS coronavirus-2 pandemic. Originally, we had contributors for each group of viruses from the Baltimore classification, as well as special chapters dedicated to exploring new frontiers in virology (zoonotic outbreaks, phages, virome, etc.). These chapters are not represented because these women virologists are the same principal investigators, department chairs, mothers, daughters, partners, siblings and cancer survivors, who saw their world turned upside down by COVID-19. Priorities were reset for all of us. Dices were thrown again. It couldn't be timelier to present you this book, certainly incomplete but definitely outstanding by the quality of its contributors, writing about viruses when one virus has stopped the world, reminding us all that there is still much more to be done to achieve equality in science and in all aspects of life.

References

Almeida, J.D. and Tyrrell, D.A. (1967). The morphology of three previously uncharacterized human respiratory viruses that grow in organ culture. *J. Gen. Virol.*, 1(2), 175–178.

Baltimore, D. (1971). Expression of animal virus genomes. *Bacteriol. Rev.*, 35, 235–241.

Steinhardt, E., Israeli, C., and Lambert, R.A. (1913). Studies on the cultivation of the virus of vaccinia. *J. Infect. Dis.*, 13(2), 294–300.

1

DNA Viruses

Lindsey M. COSTANTINI and Blossom DAMANIA
North Carolina Central University, Chapel Hill, USA

1.1. Introduction to DNA viruses

Our fundamental understanding of cellular biology has accelerated due to decades of research investigating the diverse mechanisms that DNA viruses have adapted to, in order to survive and exploit their host cells for survival. DNA viruses exhibit immense diversity, from the types of cells that DNA viruses infect to the DNA virions' size and shape. DNA viruses infect every domain of life including bacteria, archaea and eukaryotes. In the early stages of virology as a field, research on DNA viruses that infect bacteria (bacteriophages) profoundly contributed to establishing modern molecular biology. Specifically, the small DNA genomes that encode a limited number of genes made DNA bacteriophages ideal tools to examine the processes that govern DNA replication and transcription. In fact, the 1969 Nobel Prize in Physiology of Medicine honored researchers whose phage research contributed to the discovery that protein expression is mediated by messenger RNA and mRNA (Cobb 2015), as well as identifying the essential steps required for phage and viral reproduction: the separation of capsid and genome, replication of viral genomes, expression of viral proteins and formation of new viral progeny (NobelPrize.org 2019). To date, viral research has been recognized by the Nobel Committee on six different occasions. Notably, the 2008, 1975 and 1969 recognized research and breakthroughs studying DNA viruses, human papilloma virus, simian virus 40 (SV40) and DNA bacteriophage, respectively.

For a color version of all figures in this book, see www.iste.co.uk/saleh/virology.zip.

Virology,
coordinated by María-Carla SALEH and Félix AUGUSTO REY. © ISTE Ltd 2021

The first DNA bacteriophage was discovered at the turn of the 20th Century. Thirty-five years later, in 1940, the first image visualizing a bacteriophage was captured (Ruska 1940; Ackermann 2011) using the then newly developed electron microscope. Even today, electron microscopy along with sequencing methods are routinely used to discover and characterize new viruses. The most recently cited DNA virus discovery is from the newest class of so-called mega-viruses, viruses whose particles are larger than 250 nanometers in diameter and are known to infect amoeba, a unicellular eukaryote (Aherfi *et al.* 2016). The Medusavirus was isolated from some hot springs in Japan and like other megaviruses, infects *Acanthamoeba castellanii* (Yoshikawa *et al.* 2019). A single-particle cryo-electron microscopy revealed that the surface of the icosahedral capsid includes unique spike-like projections that extend approximately 14 nanometers from the surface. This newly characterized virus measures 260 nanometers in diameter, with a 381-kilobase DNA genome, which encodes 461 proteins. Of particular interest, the Medusavirus genome encodes genes for five histones and various DNA synthesis and transcriptional proteins, providing a clearer evolutionary connection between viruses and eukaryotic cells (Yoshikawa *et al.* 2019).

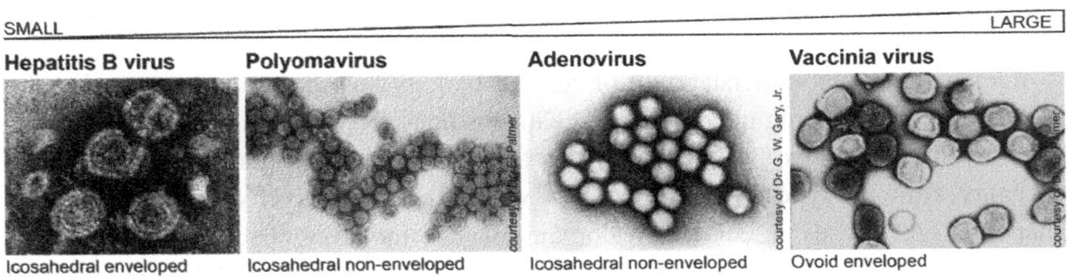

Figure 1.1. *Representative electron micrographs of small, medium and large DNA viruses. Hepatitis B virus (HBV) and Polyomavirus are relatively small viruses, approximately 40 nanometers in diameter, Adenoviruses are 90 nanometers and Vaccinia virus ranges between 220 and 400 nanometers in diameter. Images are at different magnifications. Images courtesy of Dr. Erskine Palmer and Dr. G. Williams Gary, Jr. available at the CDC Public Health Library*

1.1.1. *What are the most abundant DNA viruses?*

This is a challenging question. DNA viruses are everywhere, on animals, plants, soil, and fresh and salt water. However, in the last decade, significant effort has been focused on the human microbiome, with the expectation that by understanding the communities of microorganisms that call the human body their home, the scientific and medical communities may better address and potentially treat diseases (Human Microbiome Project Consortium 2012). The initial efforts were heavily focused on microbial communities, specifically those organisms that possess 16S ribosomal genes. As a result, viruses were excluded from the primary analysis. However, this was only temporary and the importance and abundance of viruses has since been an area of growing research interest.

Commonly known as the human virome, which is the viral component of the human microbiome, many researchers are invested in identifying the viruses that not only infect human cells, but also the microorganisms that live in and on our cells. The human virome includes viruses that infect human cells as well as accounting for viruses that infect non-human cells found associated with humans. It also includes viruses that infect bacteria, eukaryotes, and the sequences of viruses that have integrated into the human genome (Wylie *et al.* 2012).

To evaluate the viral diversity of the human body, researchers currently use high-throughput, deep sequencing approaches to identify unique non-human sequences from various samples taken from the digestive and respiratory tracts, blood, stool, skin and many other body sites (Rascovan *et al.* 2016). Viruses that infect both eukaryotes and prokaryotes have been identified. Statistically, the most abundant viruses are bacteriophages (Rascovan *et al.* 2016), which infect the estimated 4 trillion bacterial cells (Sender *et al.* 2016) that colonize the human body. One of the first studies examining the DNA viral communities of the respiratory tract analyzed airway samples from cystic fibrosis (CF) patients compared to non-diseased individuals. Interestingly, the DNA phage communities from the CF patient samples were less varied than those of control non-diseased individuals (Willner *et al.* 2009). These and similar studies underscore the ever-changing viewpoint that the complex interplay between eukaryotic cells, bacterial cells and viruses has significant implications on human health.

Generally, there are numerous challenges that emerge when using sequencing methods to identify the presence of viruses (Wylie *et al.* 2012). Chiefly, the high variability in genome size, confirmation and the abundance of viral genomes in tested samples may lead to an inherent bias, or underrepresent the number of viruses detected. The greatest challenge is the number of viruses that are yet to be discovered, which have no available reference sequences to compare new viral sequences with. Regardless of these and other challenges, significant discoveries have been made in the attempts to characterize the complete human virome.

One of the abundant DNA viruses detected in samples from the gut are Anelloviruses (Rascovan *et al.* 2016). The *Anelloviridae* family of viruses includes the torque teno virus (TTV), which is ubiquitous in all humans. At present, TTV is not known to cause any specific disease. Anelloviruses are relatively small, 30–32 nanometers diameter, non-enveloped viruses that encode a 3.8-kilobase negative-sense single-stranded DNA (ssDNA) circular genome. TTV was the first ssDNA virus identified to infect humans (Brajão de Oliveira 2015).

Human virome analysis of the eukaryotic double-stranded DNA (dsDNA) sequences (and ssDNA virus sequences that have a known dsDNA intermediate) from 102 individuals at a five body site samples identified the presence of DNA viruses from nine different families: *Geminiviridae*, *Herpesviridae*, *Nanoviridae*, *Papillomaviridae*, *Poxviridae*, *Parvoviridae*, *Polyomaviridae*, *Adenoviridae* and *Circoviridae* (Rascovan *et al.* 2016) and specifically revealed a high abundance of herpesviruses as well as numerous papillomaviruses (Wylie *et al.* 2014). Moreover, in individuals who were sampled over two time points, there was a persistent detection of both herpesviruses and papillomaviruses. In a recent study analyzing over 8,000 human blood samples, 42% of the samples included DNA-virus DNA (Moustafa *et al.* 2017). Of the DNA viruses, both human herpesviruses and papilloma viruses were detected. These measurements represent underestimates of these viruses in the human virome due to the nature and viral tropism, and the fact that latent (dormant, not actively replicating) and non-productive infections would not be detectable; however, these results indicate the prevalence of the two viruses featured in this chapter.

Although our focus is primarily on DNA viruses infecting humans, if we consider the Earth's proportion of land to ocean, it is unsurprising that most viruses reside in the oceans and infect aquatic prokaryotic, archaeal and eukaryotic cells. It has been estimated that 10^{31} virions are found in the oceans (Suttle 2007). A majority of the discovered marine bacteriophage (viruses that infect bacteria that reside in oceans) contain double-stranded DNA genomes in a tailed phage structure (Ackermann 2007). In the non-human DNA viruses, there are also viruses that infect archaea, prokaryotic cells that live in extreme environments that may have high temperatures, salinity and pressure. These represent an interesting subtype of viruses and could provide clues to the evolution of viruses. For example, studies comparing common protein folds have identified conserved structures in major capsid proteins (the viral protein that forms the protein shell around the viral genome), and discovered a possible evolutionary link between viruses known to infect archaea, prokaryotes or eukaryotes (Krupovic and Bamford 2011).

1.1.2. *Human DNA viruses*

Even for the subset of DNA viruses that specifically infect humans, the virion structures, genome architectures and infectious cycles are highly variable.

Consequently, DNA viruses evolved distinct tactics for viral reproduction and pathogenesis in human cells. This chapter primarily focuses on two human viruses as a means to explore the conserved similarities and unique differences of DNA viruses. Human papilloma viruses (HPV) and Kaposi's sarcoma-associated herpesvirus (KSHV), also known as human herpesvirus-8 (HHV-8), are of particular interest because both viruses are causative agents of several human cancers, and therefore represent a subset of DNA viruses that are highly relevant to human disease (Walboomers *et al.* 1999).

Current estimates predict that viruses cause 15–20% of worldwide human cancers. It is important to note that viral infections do not always lead to cancer development; in fact, the majority of individuals infected with an oncogenic virus will not develop cancer. However, for the cancer-associated DNA viruses discussed, viral infection is a necessary step in cancer progression. Of the seven oncogenic human viruses currently identified, five are DNA viruses. HPV, Hepatitis B (HBV), Merkel cell polyomavirus (MCV), Epstein–Barr virus (EBV) and KSHV are all oncogenic viruses. In populations that are at the highest risk of developing HPV, HBV, EBV and KSHV-associated cancers, a combination of chronic infection and added risk factors (co-infection, inflammation, smoking, alcohol drinking, and exposure to a high concentration of chemical pollutants) contributes to the onset of cancer. In addition, infected individuals in developing countries are at a higher risk for viral-associated cancers. HPV strains are classified as "high risk" or "low risk". Several types of high-risk HPVs are associated with cervical, anal, penile, vaginal and oropharyngeal cancers. Persistent HBV infection is linked to liver cancer. EBV is most commonly associated with Hodgkin and Burkitt lymphomas. KSHV is the causative agent of Kaposi's sarcoma, multicentric Castleman's disease and primary effusion lymphoma (Chang *et al.* 1994).

Although this chapter focuses primarily on two human DNA viruses (HPV and KSHV), the common viral themes of all DNA viruses will be highlighted, including: members of *Geminiviridae*, *Herpesviridae*, *Nanoviridae*, *Papillomaviridae*, *Poxviridae*, *Parvoviridae*, *Polyomaviridae*, *Adenoviridae* and *Circoviridae*, specifically T4 bacteriophage, various large nucleocytoplasmic large DNA viruses, torque teno virus, HBV, EBV, pox viruses, Simian virus-40 and variola virus. Numerous books and reviews have also focused on the world of DNA viruses and oncogenic DNA viruses. For additional resources the authors recommend (IARC 2006; Lukac and Yuan 2007; Damania and Pipas 2009; Mesri *et al.* 2014; Winther *et al.* 2018; Rohrmann 2019).

1.2. Taxonomy and structure

Historically, scientists have used many criteria to classify viruses. They have been grouped into families by virion size, capsid geometry, nucleocapsid conformation (whether nucleocapsids are naked or enveloped by a lipid membrane), type of viral genome (RNA or DNA), genome nucleotide conformation (linear, circular, gapped, segmented) as well as via the Baltimore system (I–VII), which classifies viruses according to the steps required to generate mRNA from different viral genome types. By grouping viruses on the basis of these standards, the DNA viruses discovered to date have virions with icosahedral, helical or other complex symmetries, and include both naked or non-enveloped and enveloped (host cell-derived lipid bilayer acquired during viral assemble or egress) capsids (metastable viral protein structure that protects and transports the viral genome), with either linear or circular single-stranded DNA (ssDNA) or double-stranded DNA (dsDNA) genomes. In the Baltimore classification, DNA viruses appear as class I, II or VII viruses. HBV is classified as a class VII, reverse transcribing virus that contains a partial dsDNA circular genome encased in an enveloped spherical capsid.

Of the currently identified DNA viruses, the dimensions of DNA virus particles can range from the smallest *Porcine circovirus*-1, with a 17 nanometer diameter, to the largest *Pandoravirus salinus* DNA virus, measuring 1 micron in length by 0.5 micron in width. By comparison, this size difference roughly equals the diameter of a soccer ball (approximately 22 cm) compared to an inflated hot-air balloon (15 meters). The size of the genome is linked to the size and shape of the particle: large genomes are packaged into large particles, while small genomes are packaged into smaller particles. During new particle formation, newly synthesized genomes must be compressed in order to be packaged into viral capsids. DNA viral capsids often have icosahedral geometry. Others do not. For example, the largest DNA virus particle, pandoravirus (Philippe *et al.* 2013) is a group I DNA virus, with a genome that is approximately 2,500 kilobases in length, encoding an estimated 2,556 open reading frames (ORFs) contained in an enveloped ovoid particle. The largest virus particles were initially thought to be bacteria. The mimivirus (La Scola *et al.* 2003) (Microbe Mimicking virus) was the first giant eukaryote-infecting virus to be discovered. Since then, many more viruses with large genomes packaged into large particles have been discovered. These viruses include *Marselleviridae*, *Pandoraviridae*, *Pithovirus*, Faustovirus and *Mollivirus* (Aherfi *et al.* 2016). The distinctive size of these virus particles is large enough that they are visible with standard light microscopy techniques. By contrast, most virus particles (with diameters less than 200 nanometers) can only be viewed using electron microscopy.

Therefore, the classification of virion geometry and structure is primarily done using electron microscopy (Figure 1.1). Alternatively, if the purified virus is non-enveloped, small, and forms crystals, the virus geometry can be examined using X-ray crystallography. The HPV and KSHV particles are representative of relatively small and large human DNA viruses with dsDNA genomes and icosahedral capsid geometry. The high-resolution structure of the KSHV capsid was captured using Cryo-electron microscopy (Cryo-EM) (Dai *et al.* 2018), while multiple HPV-16 capsid structures were captured using time-lapse Cryo-EM, to determine capsid structural differences during HPV virion maturation (Cardone *et al.* 2014). Data from structural studies can promote greater understanding of viral nucleocapsid (viral genome in the capsid structure) assembly and disassembly, thereby enhancing the likelihood of identifying mechanisms to inhibit assembly during the later steps of the viral infectious cycle to prevent spread or disassembly during viral entry, to prevent initial infection.

1.2.1. *Small DNA tumor virus, e.g. human papillomavirus*

The HPV virion is 60 nanometers in diameter assembled with icosahedral symmetry and has no membrane envelope (i.e. it is "naked"). It contains an 8 kb double-stranded, circular DNA genome. HPV is a member of the *Papillomaviridae* family. The *Papillomaviridae* also include viruses that infect birds and reptiles, as well as other mammals. This family is subdivided, based on the sequence similarity of the major capsid protein, L1 ORF, into five genera: alpha-, beta-, gamma-, mupa- and nupapapillomaviruses (de Villiers 2013; Van Doorslaer *et al.* 2018). The papillomavirus genera are categorized further into species (i.e. Alpha-1-14), which are in turn subdivided into types, denoted by numbers, for example, HPV-16, where the numbers follow the historical order in which the genome of each HPV type was fully sequenced (Bzhalava *et al.* 2015). HPV types can also be considered as high-risk or low-risk based on their association with aggressive and less aggressive cancers. The World Health Organization (WHO) International Agency for Research on Cancer (IARC) has categorized 12 HPV types as group 1 carcinogens (causing human cancers), which have been cited to cause 90% of cervical cancers. To date, more than 150 species of *Papillomaviridae* have been described (Van Doorslaer *et al.* 2018).

1.2.2. *Large DNA tumor virus, e.g. Kaposi's sarcoma-associated herpesvirus*

KSHV is the most recent of the eight human *Herpesviridae* known family members to be identified (Chang *et al.* 1994). The family *Herpesviridae* is divided into sub-families, named alpha-, beta-, and gamma-, based on the amino acid sequence alignment of highly conserved herpesvirus genes (McGeoch *et al.* 1995). KSHV is a gamma-herpesvirus, and more specifically, a gamma-2-lymphotropic virus. KSHV virions are enveloped, and the capsids are relatively large with icosahedral symmetry. The KSHV genome packaged in the capsid is linear dsDNA. A complex layer of proteins and RNA called the tegument lies just under the membrane and outside of the capsid. The tegument is formed during new virus particle assembly and egress. Several of the KSHV tegument proteins have functional roles in viral assembly, primary infection, inhibiting host antiviral responses and inducing cell signaling (Sathish *et al.* 2012). These features are shared by the 100 known herpesviruses that infect invertebrates or vertebrates, eight infecting humans and two, EBV and KSHV, are considered group 1 carcinogens by the WHO. Worldwide EBV prevalence is high, greater than 90%, with the majority of infections being asymptomatic.

1.3. Genomes

Viral genomes encode the genes necessary for viral replication, expression, the assembly of new virions, proteins that regulate the timing of viral processes, as well as proteins that modulate host cells and promote the spread to new cells, and evasion of host cell surveillance. In addition to being single- or double-stranded, DNA virus genomes can be linear or circular, or may have more complex configurations, similar to the gapped DNA genomes of the hepadnaviruses (like HBV). Of the known viruses with linear ssDNA genomes, most are small, with lengths ranging from 4 to 12.5 kilobases, encoding less than five gene products. Those with circular ssDNA genomes range from 1.8 to 24.9 kilobases and encode less than five gene products. Those with circular dsDNA genomes can be from 4.5 to 610 kilobases in size (Hulo *et al.* 2011). Those with linear dsDNA viral genomes fall between 14.5 and 2,500 kilobases in size. Overall, linear dsDNA genomes are usually longer than ssDNA genomes and as such, encode a greater number of genes.

Viral particles that carry smaller viral genomes tend to encode fewer genes and rely heavily on the host cell machinery for the expression and replication of their genomes. DNA viruses with large genomes are often less reliant on the host cell for replication. But all viral genomes, large or small, must produce mRNA that can be translated by host ribosomes.

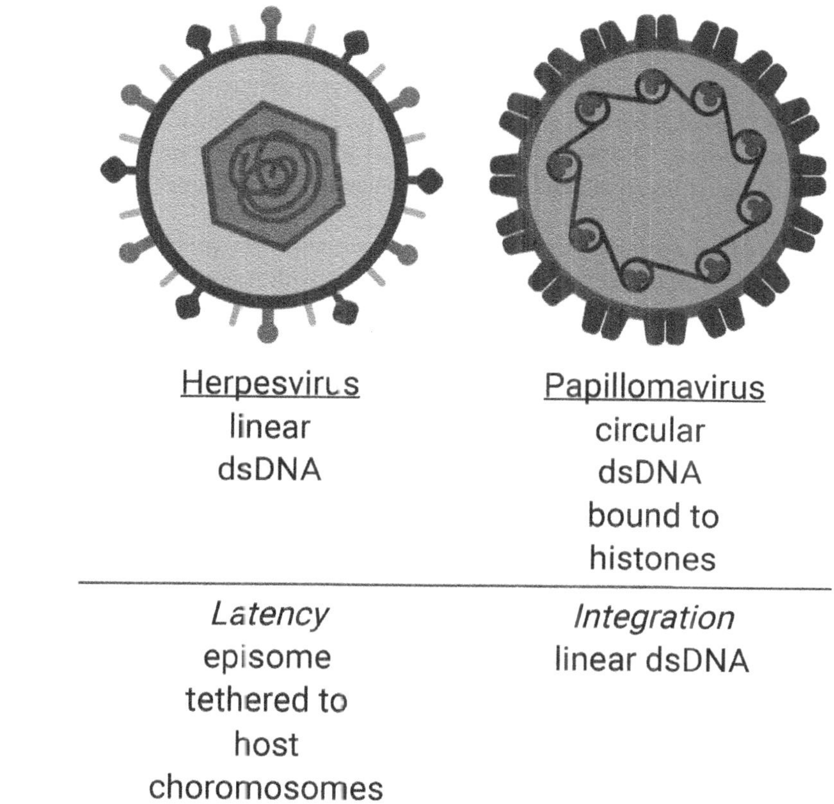

Figure 1.2. *DNA viral genomes can be linear, circular, single- and double-stranded. In virus particles, the small double-stranded HPV genome is decorated with host histones. Certain HPV strains are known to integrate into the host genome, converting from a circular to linear integrated state. The KSHV genome in virus particles is linear; during latency, the KSHV genome becomes circularized and decorated with viral proteins and cellular histone proteins to form the KSHV episome. Created with BioRender*

1.3.1. *HPV, a small DNA tumor virus genome*

The HPV circular dsDNA genome is approximately 8 kilobases in size. The genome encodes eight viral proteins. Ninety percent of the entire HPV genome encodes the eight ORFs. The remaining 10% includes two polyadenylation sites (the early, pA_E and late, pA_L) and the so-called long control region (LCR). The LCR contains the origin of replication and binding sites for transcriptional regulators. It does not contain protein-coding sequences. During the assembly of new virions, the HPV genome is associated with cellular histones to create a circular chromatin-like structure in the virions. The position of the cellular histones bound to viral genomes plays a role in regulating viral gene expression in newly infected cells.

In rare events, the HPV genome can integrate into the cellular genome of infected cells via a non-specific recombination event with a preference for fragile sites (Smith *et al.* 1992). HeLa cells, one of the most widely used immortalized cell culture cell lines, harbors an integrated partial copy of the HPV genome (Adey *et al.* 2013). Papers published citing HeLa cells, immortalized human cervical cancer cell line, isolated from Henrietta Lacks in the 1950s, number close to 110,000 PubMed citations. Over 70 years later, the widespread usage of this cell culture cell line continues, which is in part due to the presence of the HPV genome encoded in these cells.

1.3.2. *KSHV, a large DNA tumor virus genome*

The linear KSHV dsDNA genome is approximately 160–175 kilobases, composed of 145 kilobases, encoding viral genes plus terminal sequences of 20–35 kilobases of GC-rich DNA (85% GC-rich). These sequences are often called terminal repeats. The KSHV genome forms a chromatinized episome (circularized genome covered with nucleosomes) tethered to the host chromatin by viral proteins in the nucleus of infected cells during the latent or dormant phase. The terminal repeat regions are joined together via viral proteins to form the viral episome. To date, the KSHV genome is not known to integrate into the host cell chromosomes. In addition to protein coding genes, the KSHV genome also encodes microRNAs and long non-coding RNAs (lncRNA). The KSHV genome codes for more than 90 proteins (20 ORFs are unique to KSHV and are not conserved among other herpesviruses) and 12 microRNAs. The KSHV genome also contains two origins of replication denoted as left and right, each more than 1 kilobase in length, that dictate the starting location for the viral replication machinery to initiate viral DNA replication.

1.4. Gene expression and regulation

Transcription of DNA viral genes is regulated at multiple levels, including: (1) the location of transcription in the cell; (2) usage of cellular or viral transcriptional machinery; (3) open (transcriptionally active) or closed (repressive) DNA conformations; and (4) transcriptional inhibitors and activators that restrict or promote the expression of viral proteins. Transcription may ensue as soon as the transcriptional machinery and viral genome co-localize. When transcription relies on cellular RNA polymerases like polymerase II, the viral DNA must be double-stranded. For viruses with ssDNA or gapped DNA genomes, repair and synthesis of the complementary strand must precede transcription. Parvoviruses are one of the two potentially known ssDNA viruses that infect humans and contain a linear ssDNA genome, with DNA hairpin structures at both ends of the genome. These hairpin structures contain self-complementary DNA sequences, which serve as primers and sites where DNA synthesis is initiated to generate the fully dsDNA genome. After the parvovirus genome is fully synthesized into dsDNA, viral gene expression proceeds. Therefore, the presence and localization of cellular DNA synthesis proteins are a pre-requirement to viral protein transcription.

DNA viruses that rely on the cellular RNA polymerases for mRNA synthesis have evolved mechanisms to successfully transport the viral genomes to the nucleus of the infected eukaryotic cells. However, a particular subset of DNA virus genomes replicates in the cytoplasm and encode a viral DNA-dependent RNA polymerase; these viruses are often called "nucleocytoplasmic large DNA viruses" or NCLDV. They include the poxviruses, mimivirus, pandoravirus, etc. In the example of the pox virus, vaccinia virus, when new virus particles assemble, the viral components required for transcription and RNA processing of viral mRNAs are packaged in virions, and are therefore present when a virus infects a new cell. After entry, the vaccinia virus DNA-dependent RNA polymerase begins viral transcription in the cytoplasm, which eliminates the requirement for the viral genome to traffic into the nucleus.

The expression of genes encoded by DNA viruses occurs in a stepwise, temporal manner, whereby the expression of a subset of initial genes occurs first, and once viral DNA replication ensues, the expression of the majority of viral genes occurs. This latter phase of gene expression typically corresponds to the viral proteins required for viral assembly, which includes capsid subunit proteins. Viral DNA expression can also be subdivided into phases. The initial phase typically ensures the expression of viral activators, whose gene products ensure viral reproduction. This phase is followed by the expression of the genes needed to replicate the viral genome, and finally, the gene products that are required in the highest abundances will be expressed later. The temporal regulation of gene expression is one mechanism that ensures successful viral reproduction. For example, the genes that encode the viral capsid subunits must be made in quantity in order to produce the subunits required to assemble new virus particles. Therefore, if newly replicated viral genomes act as transcriptional templates prior to packaging, greater amounts of mRNA encoding viral structural proteins will be translated. The strict temporal control needed to achieve both the high abundance of viral capsids and structural proteins and the copies of the newly replicated viral genomes is essential for efficient replication and transmission.

1.4.1. *Small DNA tumor virus gene expression, the HPV example*

Successful viral reproduction is also controlled by the absence or presence of certain cellular transcription factors and, for certain viruses, the presence or absence of certain epigenetic modifications involving nucleosomes. The HPV genome is covered with cellular histones and is also methylated on CpG residues in the genome, the functional consequences of which – as well as the correlation to the viral stage of replication – are not well characterized. However, the fact remains that activating (opened) and repressive (closed) histone modifications, as well as epigenetic changes such as methylation to the viral genomes, play an important role in governing when viral genes may be expressed.

Viruses with small DNA genomes usually encode multiple proteins in a single mRNA. Accordingly, these viruses rely in part on the cellular RNA processing and splicing machinery to express the proteins. This reliance provides tissue and cell-based control of viral gene expression. For example, HPV protein expression is dependent on the infected cell type and the localization in the epithelial layers of the skin. The temporal expression of HPV proteins is subdivided into early and late episodes. Early (E) genes are composed of six non-structural genes (E1, E2, E4, E5, E6 and E7), while late genes (L1 and L2) encode two structural genes. The E genes are further categorized by protein functions; proteins involved in (1) replication of viral genomes, (2) transcription of viral genes and (3) modulation of host cell processes, such as the enhancement of cellular proliferation and inhibition of apoptosis. E1 and E2 are involved in viral DNA replication and the regulation of early transcriptional events. E4, E5, E6 and E7 are viral oncogenes that affect the cell cycle. More specifically, E1 and E2 ensure the HPV genome segregates into progeny daughter cells during cellular division. E2 accomplishes this by tethering the HPV mini-chromosome to the host chromatin. E1 is a DNA helicase, and E4 enhances DNA replication. E5 is a transmembrane protein that inhibits apoptosis and stimulates proliferation. HPV E6 and E7 are the major drivers of cellular transformation and are highly associated with carcinogenesis. The L genes encode the two capsid proteins that assemble to form the HPV capsid. L1 is the major capsid protein, and L2 is the minor capsid protein. The HPV capsid is composed of 360 L1 copies arranged into pentameric capsomeres that associate with approximately 12 L2 monomers. L2 is important for both the entry and egress of HPV.

Figure 1.3. *Time line of viral gene expression for HPV and KSHV. The HPV and KSHV genomes have been color-coded to indicate the proteins that are expressed during early (insert color) or late (insert color) gene expression. The KSHV genes are further subdivided into latent and lytic. The lytic genes of KSHV are coded into immediate early (IE), early (E) and late (L) genes (insert colors associated)*

1.4.2. *Large DNA tumor virus gene expression, the KSHV example*

Like HPV, epigenetic modification of the viral genomes dictates viral gene expression and the stage of the viral life cycle of KSHV. The genome of latent KSHV is an episome covered with histones and is also heavily CpG methylated. During latency, the majority of the KSHV genome is maintained in a closed chromatin confirmation, thereby restricting gene expression to the latency proteins, by maintaining nucleosome-free promoters or histones with activating modifications.

The approximately 165-kilobase dsDNA genomes from many strains of KSHV have been completely sequenced. Nearly 80% of the KSHV genome is in the coding region, which corresponds to about 140 kilobases. In the coding region there are 75 highly conserved genes found in closely related alpha- and beta-herpesviruses and 15 genes unique to KSHV, denoted with a K as part of the gene name. The KSHV transcripts detected during latency include K1, vIL6 (K2), Kaposin (K12), LANA (ORF73), v-Cyclin (ORF72) and vFLIP (ORF71). In addition to the latency proteins, 25 microRNAs (miRNAs) encoded by 12 precursor miRNA, known as miR-K1-12, can be detected during the KSHV latent phase. During the productive phase of the KSHV replication cycle, the remaining viral genes are expressed in a temporally regulated manner and are categorized into immediate early (IE), early (E) and late (L) proteins (Zhong *et al.* 1997). IE mRNAs include transcripts encoding RTA (ORF50), ORF45 and ORFK4.2. RTA is an essential KSHV protein that is required for the shift from latency into the productive phase (Lukac *et al.* 1998). Once expressed and in the presence of other IE proteins, the E genes include transcripts encoding the polyadenylated nuclear RNA (PAN RNA), Kaposin, ORF57, k-bZIP (K8), K5, K9, K14, K15, ORF6, ORF21 and ORF74. Finally, the L gene transcripts, such as the major capsid protein (MCP, ORF25) and small viral capsid (ORF65), are produced. KSHV encodes viral DNA replication proteins (ORF6, ORF9, ORF40/41, ORF44, ORF56, ORF59). The transcriptional cascade produces the KSHV-encoded DNA replication proteins that replicate and produce the KSHV genome copies, which are assembled into new virions in the nucleus. Before being packaged, these copies may act as transcriptional templates for the generation of high levels of structural proteins.

The KSHV gene expression profile of *de novo* infection includes transcripts of latent and lytic genes. The lytic genes aid in establishing the initial infection; then, shortly after primary infection, the latency program is initiated. At this phase of the KSHV life cycle, viral proteins create a favorable environment to evade host surveillance that ensures the KSHV genome is maintained throughout cellular division. Over the course of a lifelong infection, KSHV will sporadically reactivate to begin productive replication. During reactivation from latency, the temporal transcription program results in the expression of all KSHV proteins required to produce new virions. Recent reexamination of the KSHV genome, transcriptome and proteome has confirmed previously identified traditional cis-regulatory elements and has revealed novel strategies, including alterative transcriptional start sites and novel small open reading frames (less than 100 amino acids long) (Arias *et al.* 2014).

1.4.3. *DNA virus inhibition of cellular gene expression*

Viral genomes have evolved unique strategies that favor viral gene expression over cellular gene expression. DNA viral genomes may encode genes that preferentially eliminate or reduce cellular mRNAs via exo- and endonuclease activity. This action enables viral transcripts to outcompete host mRNAs so that cellular translation machinery will translate larger amounts of viral protein. KSHV, ORF37 (SOX), a viral exonuclease, shuts off host cell gene expression in order to promote viral protein translation (Glaunsinger and Ganem 2004). DNA viruses have evolved many tactics to accomplish the initial induction of viral transcription during primary infection to ensure the establishment of viral genomes.

The next section expands on the functions of DNA virus proteins, including not only the viral proteins necessary for viral reproduction, but also the proteins that alter host cell functions and function to evade or modulate host cell defenses. Viral protein functions are divided into three major categories: (1) viral proteins required for the formation of new virus particles; (2) the proteins necessary to modulate cellular pathways; and (3) a unique class of viral proteins that mimic cellular proteins.

The last category has important implications about viral–host cell co-evolution. The discussion of viral proteins and their functions will be framed according to how they function during viral infection. Viral glycoproteins on the surface of virions and capsid proteins are the first to engage the host cell. Viral proteins that function immediately after entry are packaged in virions, and no transcription or translation is required, as they are released to function in the host cell. In addition to promoting the entry and delivery of the viral genome into the nucleus, viral proteins present and functioning at the time of primary infection promote the initiation of the transcription of additional viral genes. The viral proteins that promote transcription can do so by directly binding DNA sequences and serve to coactivate viral gene transcription by host RNA polymerase II. Alternatively, these viral proteins may function indirectly by promoting cellular proteins to coactivate the transcription of viral genes. Subsequent viral proteins promote viral genome replication, translation and the assembly of new virions. Simultaneously, with these essential replication proteins, viral genomes also express proteins that modulate host processes to evade host detection and establish an environment conducive to the virus infection.

1.5. Infectious cycle

The complete infectious cycle for a virus includes attachment, entry, replication, assembly and egress. Together, the viral reproduction cycle promotes the assembly of new virions and the consequent infection of new cells. The first step is attachment and entry into permissive cells, i.e. those expressing the necessary cellular surface proteins needed for entry. Following the initial steps of entry, new copies of the viral genomes are synthesized, while new viral proteins are produced, together these and additional viral proteins unite to assemble into new virus particles. The subcellular localization of the steps in the viral replication cycle is virus dependent. Typically, for DNA viruses that infect eukaryotic cells, the viral replication process begins in the nucleus and is followed by intracellular trafficking and the release of new virions from the infected cell.

The infectious cycle of the human hepatitis B virus (HBV) is known to produce structurally different viral particles; only 5% of the viral replication cycle successfully results in a mature virus particle including a viral genome, while the other 95% of particles lack a copy of the HBV genome. In addition, the assembly steps of the HBV reproductive cycle represent one of the more diverse ways DNA viruses may replicate in host cells. HBV is a relatively small DNA virus with a circular genome that is not completely double-stranded. During new particle assembly, the viral pregenomic RNA is packaged with the viral reverse transcriptase in the newly formed capsid. Once inside the capsid, the pregenomic RNA is reverse-transcribed into an ssDNA copy, and subsequently, the second incomplete DNA strand is synthesized. This viral strategy requires a high abundance of capsid proteins prior to DNA replication. Both HPV and KSHV initiate genome amplification prior to capsid protein expression.

1.5.1. *Small DNA tumor virus life cycle, the HPV example*

HPV is primarily transmitted though skin-to-skin and sexual contact. HPV infects basal epithelial cells by penetrating through the stratified epidermis to the basal cells, usually at the site of a damaged epithelium, which enables the virus particles to reach the lower level of the epithelium (Kines *et al.* 2009). At the cell surface, HPV L1 and L2 proteins interact with heparan sulfate proteoglycans at the basal membrane and subsequently, with receptors found on the surface of keratinocytes. The alpha-6 integrin has been implicated as a potential cellular receptor in studies examining HPV-16 virus-like particles' entry *in vitro* (Yoon *et al.* 2001). A proteolytic event in the L2 capsid protein via furin, a cellular protease, is required for virus attachment (Richards *et al.* 2006). Viral entry via endocytosis, capsid uncoating and transport of the HPV genome into the nucleus, occurs in basal cells.

After acute primary infection, persistent infection of HPV occurs in the basal cells, during which episomal copies of the genome are maintained in the nucleus of infected cells and low levels of early proteins (E1 and E2) are expressed. Genome maintenance in the basal cells is essential to the HPV life cycle (McBride *et al.* 2006). Infected basal epithelial cells divide and give rise to new daughter cells harboring the HPV genome. HPV viral genome replication initiation is governed by E1/E2 binding to the HPV origin of replication. The helicase activity of E1 is essential to initiate viral DNA replication. The function of E2, the viral transcriptional regulator, varies in the different layers of the stratified epithelium, due to wide-ranging levels of the E2 protein. In the basal epithelium, E2 together with bromodomain containing protein 4 (Brd4), tethers the HPV genome to the cellular chromatin in order to maintain viral genomes during cell division (McBride *et al.* 2006). Competition between HPV E2 and cellular Sp1 controls the expression of E6 and E7 proteins. When E2 abundance is low, Sp1 binding promotes transcriptional activation and the expression of E6 and E7, leading to the inactivation of p53 and pRb and promoting cellular DNA replication and activity of the cellular DNA polymerase. Enhanced polymerase activity drives HPV genome amplification. The balance between the abundance of E6 and E7 with E1 and E2 is pivotal to determining whether infected cells remain undifferentiated or continuously divide. Once cells switch and maintain a lower abundance of E6 and E7, the expression of the late (L1 and L2) proteins promotes the epithelial cells at the granular layer to terminally differentiate into cornified epithelium, stimulating the final steps of the viral replication cycle of HPV.

During transcription of the HPV genome, the activity of either the early or late promoter produces the primary transcriptions encoding multiple viral genes. Subsequent RNA processing yields mRNA for each viral gene. The abundance of RNA processing factors governs early or late mRNA processing with undifferentiated or differentiated infected cells. High E2 expression promotes early promoter gene expression and E1, E2, E4, E5, E6 and E7, while lowering the E6 and E7 levels drives proteins expressed from the late promoter, L1 and L2. L2, via the E2 protein, associates with newly synthesized copies of the HPV genome. New virion assembly occurs in the nuclei of cells, in the topmost layer of the stratified epithelium. It is thought that the oxidizing environment of terminally differentiated keratinocytes is required for this process. Terminal keratinocytes have decreased mitochondrial oxidative phosphorylation, causing a switch from reducing to oxidizing the cellular environment. The now oxidizing environment is conducive to the formation of the L1 disulfide bonds necessary for the stabilization of newly formed icosahedral capsids. New virus particles are released when the surface keratinocytes are shed.

HPV may also induce a non-productive pathway (where no new virus particles are produced), that promotes cellular transformation and potentially, cancer progression. This pathway does not produce infectious HPV particles, but rather changes the infected cell's expression profile via the upregulation and overexpression of HPV proteins E6 and E7. HPV E6 and E7 are examples of some of the first viral oncogenes discovered and studied. The HPV E6 and E7 proteins enhance cell division, inhibit apoptosis and increase chromosomal instability, primarily by the inactivation of two tumor suppressor proteins, p53 and retinoblastoma protein (pRb) (Moody and Laimins 2010). The E6 protein is a ubiquitin ligase and has been shown to ubiquitinate p53 (Scheffner *et al.*, 1990). Degradation and low levels of p53 promote progression through cell division without repairing potential DNA damage. E6 has also been shown to activate telomerase, which promotes the lengthening and continued division of HPV-infected cells. E7 binds pRb, which causes pRb to release E2F, and promotes the signals for the cell cycle to progress and uncontrolled proliferation to proceed (Dyson *et al.* 1989). These virally induced cellular changes are examples of viral proteins that function to alter normal cellular processes and, in the case of HPV oncogenes, promote cellular transformation.

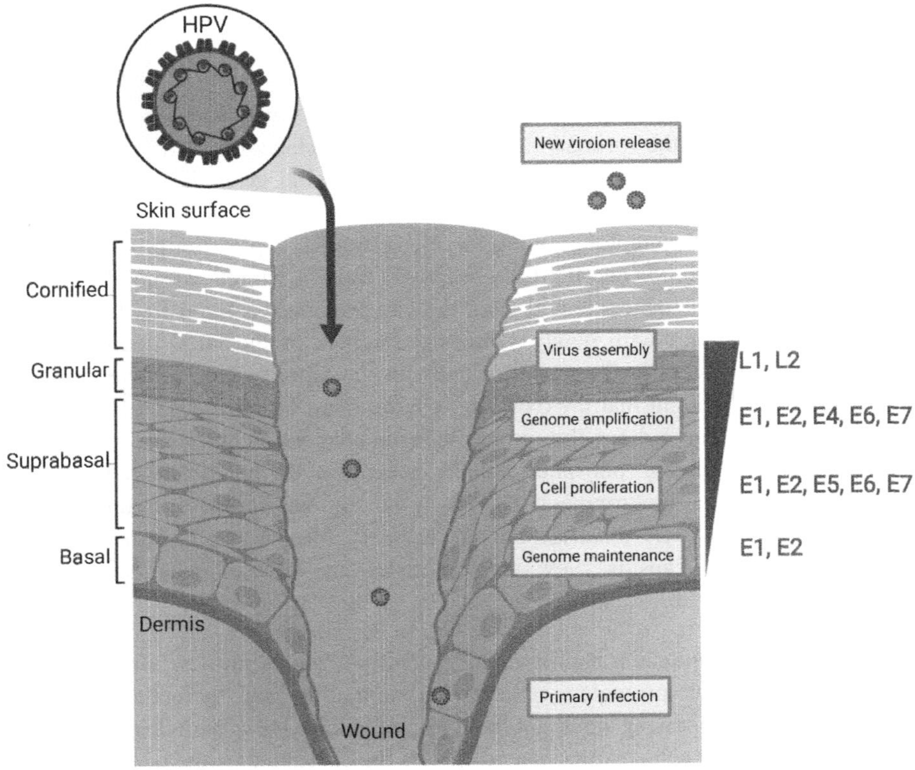

Figure 1.4. *The life cycle of HPV. HPV infects human stratified epithelium. In the stratified epithelium are layers, at the base is the basal membrane on which the basal cells are attached, on top of the basal cells and moving towards the tissue surface are the spinous layer, granular layer and finally the cornified layer (top most layer). Each layer of cells is characterized by a specific gene expression pattern. The distinct cellular layers of the stratified epithelium promote the early viral protein expression in the spinous layer, late protein gene expression in the granular layer and finally virion assembly and shedding to occur in the upper most layer of the cornified epithelium. Created with BioRender*

1.5.2. *Large DNA tumor virus life cycle, the KSHV example*

KSHV is known to infect monocytes, macrophages, dendritic cells, epithelial cells, endothelial cells and B lymphocytes cells. B cells serve as the KSHV latency reservoir, harboring the KSHV viral episome and promoting virally infected B cells to proliferate by inhibiting apoptosis and promoting cell division.

KSHV is primarily transmitted via saliva, but sexual transmission and infection via blood transfusions and organ transplants have been documented (Martin *et al.* 1998). The primary step of viral replication involves a subset of the KSHV glycoproteins initiating attachment. KSHV expresses five herpesvirus-conserved glycoproteins (gB, gH, gL, gM and gN) and five unique KSHV glycoproteins (gpK8.1A, gpK8.1B, K1, K14 and K15). Many human herpesviruses receptor–viral protein interactions involve the binding of viral glycoprotein B (gB), interaction with heparin sulfate to target lymphocytes, epithelial, endothelial and monocytes. Heparan sulfate is ubiquitously expressed on the surface of many cells and may aggregate KSHV particles on the surface of cells (Akula *et al.* 2001). Based on the presence of conserved heparin binding domains, gpK8.1A and gB likely mediate this initial interaction, and gH and ORF4 bind heparan sulfate. Following heparin sulfate binding, gB–integrin interactions suggest additional co-receptors involved in KSHV entry into lymphocytes, endothelial and epithelial cells (Akula *et al.* 2002). Cells that express the dendritic cell-specific intracellular adhesion molecule 3-grabbing non-integrin, DC-SIGN, such as macrophages and activated B cells on their surface, will bind KSHV (Rappocciolo *et al.* 2006). Furthermore, KSHV attachment and infection into adherent cells can be mediated by xCT (Kaleeba and Berger 2006), a cellular transmembrane transporter protein and ephrin receptor, receptor protein–tyrosine kinase, EphA2 (Hahn *et al.* 2012). The multiple cell surface proteins identified to be involved in KSHV attachment and entry indicate that the cell-specific expression of different cell surface proteins may enable viral attachment of different cell types. The general consensus is that heparan sulfate binding (since ubiquitously found on cells) may serve to increase the concentration of viral particles on the surface of cells, in order to enhance the likelihood of encountering the actual entry receptors. Future investigations comparing the similarities and differences in different KSHV-permissible cell types will likely reveal additional details about the mechanisms that control KSHV entry. After attachment, KSHV enter cells via endocytosis into clathrin-coated vesicles. Accordingly, the inhibition of clathrin mediated endocytosis and acidification block KSHV entry. In the cytoplasm, the KSHV-nucleocapsid traffics to the nucleus along microtubules to reach the nuclear pores and gain the cell nucleus. From here, the KSHV genome can be actively transcribed, enable the expression of KSHV proteins or be circularized to form the viral episome. The former indicates the lytic or active phase of KSHV infection, while the latter is indicative of viral latency.

The KSHV reproduction cycle begins with primary infection, and cycles between latency and lytic phases. KSHV, like all herpesviruses, establishes a lifelong infection in an infected host. During primary infection, it follows a typical viral life cycle, beginning with attachment, entry, and then followed by the KSHV genome localizing to the nucleus. The two-stage life cycle is largely controlled by two key KSHV proteins, Latency-associated nuclear antigen (LANA) and replication and transcription activator (ORF50/RTA). Shortly after infection, the primary default pathway for KSHV is latency. Latency is the dormant phase in which no new virions are produced, and the genome is maintained with each cell division. Over the course of KSHV infection, RTA-driven reactivation and re-entry into the lytic cycle will result in the release of new infectious particles. In the lytic phase and during reactivation, the IE proteins promote the activation of E genes necessary for the synthesis of new copies of the genome. As the L proteins become abundant, including the capsid proteins and the glycoproteins, new KSHV nucleocapsids are assembled and exit the nucleus. The major components of the KSHV capsid are the major and minor capsid proteins (ORF25 and ORF26), and ORF62, which is required for capsid assembly and binds viral DNA.

The KSHV infectious particle has a complex structure: the assembled nucleocapsid must exit the infected cell nucleus to acquire the necessary lipid bilayer with viral glycoproteins, as well as the tegument-associated proteins from cytoplasmic localized vesicles. There are multiple budding events during the maturation of the virus particle. First, the assembled nucleocapsid buds through the nuclear envelope, after which the nucleocapsid interacts with vesicles that are likely derived from the Golgi complex. At this stage, the nucleocapsid acquires a primary membrane with the viral glycoproteins and tegument layer. The virus particle is trafficked to the cellular membrane and released via a lipid raft-mediated process (Wang *et al.* 2015).

1.6. Viral-induced cellular survival

DNA viruses have evolved mechanisms to alter normal cell functions in order to maintain viral genomes and produce new virus particles. The Simian virus 40 (SV40), a small DNA virus, encodes nine proteins including the large and small T-antigen proteins – arguably one of the most well-known viral proteins that alter cellular behavior. Results of early SV40 research identified the cellular transformation capabilities of large T antigens. This discovery has advanced cell biology research by enabling primary mammalian fibroblast immortalization (i.e. mouse embryotic fibroblasts, MEFs). In SV40-infected cells, large and small T-antigen proteins drive cell cycle progression into S-phase. Once in S-phase, SV-40 replication proceeds, producing new progeny. Productive infection of SV-40 normally results in cell death. However, in a low percentage of infected cells, the integration of T-antigen genes into host cell chromosomes causes unregulated cell division via blocking cellular pRB and p53, in order to promote proliferation and evasion of apoptosis (Ahuja *et al.* 2005). Today, commercially available kits and vectors encoding large T-antigens are sold for the purpose of immortalizing primary mammalian cells. In order to orchestrate both viral survival and oncogenesis, both KSHV and HPV encode proteins that specifically target cellular signaling cascades that promote cell proliferation and inhibit cell death.

1.6.1. *Small DNA tumor virus enhancement of cell survival, e.g. HPV*

Pivotal to the HPV replication cycle and viral oncogenesis, are the activities of E6 and E7. As previously described (section 1.5.1), these viral proteins make direct interactions with cellular proteins such as pRb and p53 to enhance cell cycle progression and pro-proliferation signaling (Hoppe-Seyler *et al.* 2018). A third oncogenic viral protein, HPV E5, is a membrane protein that associates with Golgi, endoplasmic reticulum and nuclear envelope membranes. The E5 protein is reported to transform cells by modulating epidermal growth factor receptor (EGFR) signaling, thereby enhancing cell proliferation, angiogenesis and cell survival (Kim *et al.* 2010).

1.6.2. *Large DNA tumor virus enhancement of cell survival, e.g. KSHV*

Several KSHV proteins are homologous to cellular host proteins. The expression of these viral proteins is known to modulate immune pathways, regulate cell cycles and inhibit apoptotic signaling. Additionally, a growing body of evidence indicates that the KSHV miRNAs not only regulate the viral life cycle, but also affect the cell cycle, apoptosis, angiogenesis and host immune surveillance. Over 2,000 cellular targets of the KSHV miRNAs have been identified (Gottwein and Cullen 2008), and continuing studies seek to validate these targets.

The KSHV viral interleukin-6 protein, vIL-6, a homolog of cellular IL-6, effectively blocks the antiviral interferon pathway that would ordinarily cause cell cycle arrest and thereby inhibit viral replication. In response to increases in IFN-α, vIL-6 expression is induced, which binds gp130 to the interleukin transducer, to promote proliferation and cell survival signals (Chatterjee *et al.* 2002). The KSHV homologue to cellular cyclin D, vCyclin, has been shown to interact with cellular cyclin-dependent kinases and cyclins and phosphorylate pRb to promote cell cycle progression (Chang *et al.* 1996). LANA, the major KSHV latency protein, has no cellular homolog. The major function of LANA is to ensure the viral episome and latency is maintained. LANA also interacts with many cellular proteins and has been shown to inhibit p53 and inactivate pRb, blocking cell death and enhancing cell division (Uppal *et al.* 2014). This section highlights a few of the KSHV protein activities and downstream cellular impacts. Overall, the activities of these and other

KSHV proteins ensures KSHV survival and contributes to the development of KSHV-associated diseases (Giffin and Damania 2014).

1.7. Disease prevalence and prevention

The pathogenesis of DNA virus infections can range greatly in severity. Many of them cause no obvious disease symptoms, while others induce severe disease and potentially death. To prevent the most severe diseases and mortality, identifying effective methods to prevent and treat viral infections is a high priority. To date, one of the most frequently cited examples of vaccination-based eradication of a virus is the variola virus, the DNA virus that causes small pox. In 1980, small pox was said to be eradicated on the basis of a worldwide vaccination (WHO 1980). A major factor that contributed to this victory against an infectious agent was due to the fact that variola virus is transmitted directly between humans and has no animal reservoir; therefore, with the successful administration of an effective vaccine, small pox is the first human virus to be eradicated (Parrino and Graham 2006). The strategies to both preemptively prevent and treat HPV and KSHV infections have had varying levels of success.

1.7.1. *HPV, a small tumor DNA virus and disease associations*

HPV infections are common and are most often cleared within two years of primary infection. The majority of HPV infections go undetected because most result in the development of benign warts in infected cutaneous or mucosal stratified epithelium that are resolved by the host's normal immune system. Recurrent respiratory papillomatosis (RRP) is associated with low-risk HPV types 6 and 11. RRP can occur in adults and children and results in benign squamous papilloma in the respiratory tract (Mounts *et al.* 1982). However, infections with high-risk types, HPV-16 and HPV-18, are associated with squamous cell carcinomas; additionally, HPV-18 is associated with adenocarcinoma in the cervix. A second striking difference observed between HPV types, in the majority of HPV-18-infected individuals that develop cancer, is that the HPV-18 genome is found to integrate into the host genome; however, this is not observed in all HPV-induced cancers. Cervical cancer accounts for an estimated 5% of worldwide cancers, according to the International Agency for Research on Cancer (IARC) and the World Health Organization (WHO). HPV is known to cause squamous cell carcinoma in the cells of the anogenital tract, specifically cervical, penial, anus, vulva and vagina. In addition, HPV is associated with head and neck carcinomas, including cancers in the oropharynx, oral cavity and larynx. HPV cancer progression is associated with a persistent infection and constitutive expression of the HPV oncogenes. Cellular transformation induced by HPV infection and the subsequent expression of viral oncoproteins promotes cellular genome instability and the increased likelihood of compounding mutations that promote cell proliferation and inhibit cell death.

1.7.1.1. *Vaccinations and recommended treatments for HPV*

Currently, there are three vaccines available for HPV. They target a combination of papilloma viral types, including a quadrivalent HPV (effective against types: HPV-6, -11, -16 and -18), a bivalent HPV (HPV-16 and -18) and most recently, a vaccine protecting against nine types (HPV-6, -11, -16, -18, -31, -33, -45, -52 and -58). All three vaccines, which were approved by the FDA and are recommended by the CDC, target HPV-16 and HPV-18, the two types predicted to cause 70% of cervical cancers. The first HPV vaccines began to be administered in 2006 and although highly effective in individuals who received the vaccine, those who did not receive the vaccine prior to exposure remained at risk of HPV infection. Unfortunately, the highest at-risk populations are in developing countries and in 2018, accounted for approximately 570,000 new cases of cervical cancer and 311,000 deaths, making cervical cancer the second most common cancer in underdeveloped countries (Ferlay *et al.* 2019; Okunade 2019). The availability of the current HPV vaccines is the lowest in underdeveloped countries; therefore, the risks of HPV infection and carcinogenic risk are still high in these areas.

In addition to efforts to advance prophylactic treatments, the CDC and WHO recommendations include effective screening methods to detect infection and monitor disease progression. Even in cases when consistent monitoring is in place, the frequency and consistency of testing can be sporadic between different patient cohorts and thus produces differing levels of successful early detection.

There are many open questions that remain about HPV infection and the onset of subsequent diseases (Woodman *et al.* 2007). For example, how the persistence of the infection and varying levels of viral load influences disease progression, the requirement of viral genome integration prior to disease onset, the timing of epigenetic changes (i e. methylation of the viral genome) varies throughout HPV infection and how these modifications contribute to cancer progression. Addressing these and other areas will provide insights into the HPV life cycle and could potentially lead to the identification of improved diagnostics to differentiate the onset of benign infections from the instances that have a high risk of developing into cancer.

1.7.2. *KSHV, a large DNA tumor virus and disease associations*

Herpesvirus infections are most severe in immunocompromised individuals. KSHV infection can be detected in peripheral blood mononuclear cells, saliva, oropharyngeal mucosa, semen, vaginal secretions and prostate glands. KSHV is endemic to African and Mediterranean regions, ranging from 20% to 80% seropositive in adult populations, while in North America and Northern Europe, seroprevalence is relatively low with estimates predicting less than 10% (Minhas and Wood 2014). In China, KSHV prevalence ranges from 7% to 16% (Minhas and Wood 2014). In non-endemic areas, such as in the United States, KSHV infections are highest among men who have sex with men.

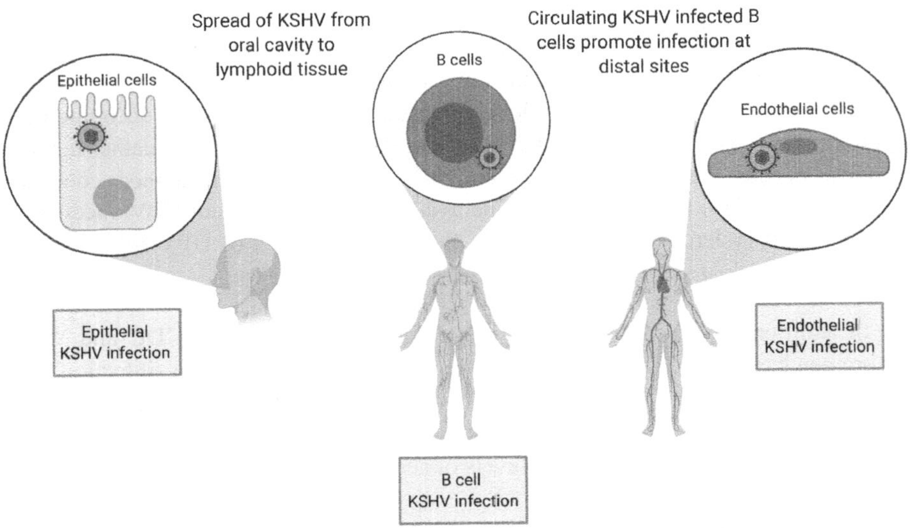

Figure 1.5. *KSHV permissive cells and potential mechanism for KSHV infection to distal body sites. KSHV is transmitted primarily through the saliva and will encounter oral epithelial cells, and the infected epithelial cells will produce new virions that disseminate throughout the lymphatic tissues to infect B cells. KSHV-infected B cells and KSHV will circulate throughout the body and infect endothelial cells. Created with BioRender*

KSHV is associated with three human diseases, Kaposi's sarcoma (KS), primary effusion lymphoma (PEL) and Multicentric Castleman's disease (MCD). KS lesions develop from KSHV-infected endothelial cells. Infected cells undergo an endothelial-to-mesenchymal transition into spindle cells. There are four classes of KS: (1) classical, (2) endemic or African, (3) iatrogenic (transplant-associated) and

(4) AIDS-associated. KS lesions present as dark patches and can be present on the skin and internal organs (lungs, liver and gastrointestinal organs). PEL is characterized by the proliferation of infected B cells into the pericardium, pleura or peritoneum spaces. Frequently, PEL is observed in patients co-infected with KSHV and EBV. Like AID-associated KS, PEL frequency is higher in HIV-positive individuals. MCD is a rare, polyclonal lymphoproliferative disease.

1.7.2.1. *Vaccinations and recommended treatments for KSHV*

There are neither vaccines nor treatments that can prevent KSHV infection. Standard KS treatments include anti-cancer drugs that inhibit topoisomerase (doxorubicin) and inhibit microtubule function (taxol). Much effort has been applied to deciphering the most potent antiviral targets specific to KSHV protein activities (Dittmer *et al.* 2012; Coen *et al.* 2014). Current strategies to treat KSHV-induced cancers include attempts to inhibit the upregulation of the cellular angiogenic, inflammatory, proliferation pathways (Dittmer and Damania 2016). Alternatively, in AIDS-associated cases of KS, treatments with anti-retroviral therapies have had mixed outcomes. Efforts seek to identify the most effective strategy and timing of treatments to ensure that combinational therapeutic success is ongoing.

As with any virus without a definitive cure, a targeted therapeutic approach to prevent viral replication and the spread of infection in host cells is the primary goal. In large DNA tumor viruses that cycle into a latent state, this approach is further complicated when the virus persists in a dormant state at levels below detection. Therefore, identifying a preventative, such as a vaccine would be highly efficacious in preventing KSHV infection and associated cancers.

1.8. Conclusion

Historically, DNA virus research has laid the foundations for the basic biology of cellular processes. As more DNA viruses are identified, there are increasing opportunities to ascertain the evolutionarily conserved mechanisms that DNA viruses have evolved in order to modulate or bypass host cell pathways, as well as the molecular processes and protein activities that pre-date the first cell. Presently, there is a growing appreciation for the contributions of viruses to the human microbiome, many of which are bacteriophages that infect our gut bacteria. Finally, as increasingly more viruses are identified as contributing factors to the development of some human cancers, the mechanisms to prevent and inhibit DNA viral infections will likely garner additional focus.

KSHV and HPV were emphasized here to compare two major families of DNA viruses, *Herpesviridae* and *Papillomaviridae*, that infect humans. Herpes and papilloma viruses infect the majority of the worldwide population. The KSHV- and HPV-specific sections serve to provide the reader with details about these two viruses and emphasize the molecular differences in viral reproduction cycles and modes of pathogenesis. Other DNA virus families are not featured as predominantly, but are rather included as examples to highlight key differences and discoveries in virology.

In closing, it is imperative to reiterate that the oncogenic viruses described in this chapter have cancer as a possible pathogenic outcome, and by no means an absolute outcome of viral infection.

1.9. References

Ackermann, H.-W. (2007). 5500 phages examined in the electron microscope. *Archives of Virology*, 152, 227–243. doi: 10.1007/s00705-006-0849-1.

Ackermann, H.-W. (2011). The first phage electron micrographs. *Bacteriophage* 1(4), 225–227. doi: 10.4161/bact.1.4.17280.

Adey, A., Burton, J.N., Kitzman, J.O., Hiatt, J.B., Lewis, A.P., Martin, B.K., Qiu, R., Lee, C. and Shendure, J. (2013). The haplotype-resolved genome and epigenome of the aneuploid HeLa cancer cell line. *Nature*, 500(7461), 207–211. doi: 10.1038/nature12064.

Aherfi, S. *et al.* (2016). Giant viruses of amoebas: An update. *Frontiers in Microbiology*, 7, 349. doi: 10.3389/fmicb.2016.00349.

Ahuja, D., Sáenz-Robles, M.T., and Pipas, J.M. (2005). SV40 large T antigen targets multiple cellular pathways to elicit cellular transformation. *Oncogene*, 24(52), 7729–7745. doi: 10.1038/sj.onc.1209046.

Akula, S.M. *et al.* (2001). Human herpesvirus 8 interaction with target cells involves heparan sulfate. *Virology*, 282(2), 245–55. doi: 10.1006/viro.2000.0851.

Akula, S.M. *et al.* (2002). Integrin α3β1 (CD 49c/29) is a cellular receptor for Kaposi's sarcoma-associated herpesvirus (KSHV/HHV-8) entry into the target cells. *Cell*, 108(3), 407–419. doi: 10.1016/S0092-8674(02)00628-1.

Arias, C. *et al.* (2014). KSHV 2.0: A comprehensive annotation of the Kaposi's sarcoma-associated herpesvirus genome using next-generation sequencing reveals novel genomic and functional features. *PLoS Pathogens*, 10(1), e1003847. doi: 10.1371/journal.ppat.1003847.

Brajão de Oliveira, K. (2015). Torque teno virus: A ubiquitous virus. *Revista brasileira de hematologia e hemoterapia*, 37(6), 357–358. doi: 10.1016/j.bjhh.2015.07.009.

Bzhalava, D., Eklund, C., and Dillner, J. (2015). International standardization and classification of human papillomavirus types. *Virology*, 476, 341–344. doi: 10.1016/j.virol.2014.12.028.

Cardone, G. *et al.* (2014). Maturation of the human papillomavirus 16 capsid. *mBio*, 5(4), e01104-14. doi: 10.1128/mBio.01104-14.

Chang, Y. *et al.* (1994). Identification of herpesvirus-like DNA sequences in AIDS-associated Kaposi's sarcoma. *Science*, 266(5192), 1865–1869. doi: 10.1126/science.7997879.

Chang, Y. *et al.* (1996). Cyclin encoded by KS herpesvirus. *Nature*, 382(6590), 410–410. doi: 10.1038/382410a0.

Chatterjee, M. *et al.* (2002). Viral IL-6-induced cell proliferation and immune evasion of interferon activity. *Science*, 298(5597), 1432–1435. doi: 10.1126/science.1074883.

Cobb, M. (2015). Who discovered messenger RNA? *Current Biology*, 25(13), R526–R532. doi: 10.1016/J.CUB.2015.05.032.

Coen, N. *et al.* (2014). KSHV targeted therapy: An update on inhibitors of viral lytic replication. *Viruses*, 6(11), 4731–4759. doi: 10.3390/v6114731.

Dai, X. *et al.* (2018). Structure and mutagenesis reveal essential capsid protein interactions for KSHV replication. *Nature*, 553(7689), 521–525. doi: 10.1038/nature25438.

Damania, B. and Pipas, J.M. (eds) (2009). *DNA Tumor Viruses*. Springer-Verlag, New York. doi: 10.1007/978-0-387-68945-6.

Dittmer, D.P. and Damania, B. (2016). Kaposi sarcoma-associated herpesvirus: Immunobiology, oncogenesis, and therapy. *Journal of Clinical Investigation*, 126(9), 3165–3175. doi: 10.1172/JCI84418.

Dittmer, D.P., Richards, K.L., and Damania, B. (2012). Treatment of Kaposi sarcoma-associated herpesvirus-associated cancers. *Frontiers in Microbiology*, 3, 141. doi: 10.3389/fmicb.2012.00141.

Dyson, N. *et al.* (1989). The human papilloma virus-16 E7 oncoprotein is able to bind to the retinoblastoma gene product. *Science*, 243(4893), 934–937. doi: 10.1126/science.2537532.

Ferlay, J. *et al.* (2019). Estimating the global cancer incidence and mortality in 2018: GLOBOCAN sources and methods. *International Journal of Cancer* 1941–1953. doi: 10.1002/ijc.31937.

Giffin, L. and Damania, B. (2014). KSHV: Pathways to tumorigenesis and persistent infection. *Advances in Virus Research*, 88, 111–159. doi: 10.1016/B978-0-12-800098-4.00002-7.

Glaunsinger, B. and Ganem, D. (2004). Lytic KSHV infection inhibits host gene expression by accelerating global mRNA turnover. *Molecular Cell*, 13(5), 713–723. doi: 10.1016/S1097-2765(04)00091-7.

Gottwein, E. and Cullen, B.R. (2008). Viral and cellular microRNAs as determinants of viral pathogenesis and immunity. *Cell Host and Microbe*, 3(6), 375–387. doi: 10.1016/j.chom.2008.05.002.

Hahn, A.S. *et al.* (2012). The ephrin receptor tyrosine kinase A2 is a cellular receptor for Kaposi's sarcoma-associated herpesvirus. *Nature Medicine*, 18(6), 961–966. doi: 10.1038/nm.2805.

Hoppe-Seyler, K. *et al.* (2018). The HPV E6/E7 oncogenes: Key factors for viral carcinogenesis and therapeutic targets. *Trends in Microbiology*, 26(2), 158–168. doi: 10.1016/J.TIM.2017.07.007.

Hulo, C. *et al.* (2011). ViralZone: A knowledge resource to understand virus diversity. *Nucleic Acids Research*, 39(SUPPL. 1), D576–D582. doi: 10.1093/nar/gkq901.

Human Microbiome Project Consortium (2012). A framework for human microbiome research. *Nature*, 486(7402), 215–221. doi: 10.1038/nature11209.

IARC (2006). *Human Papillomaviruses*. World Health Organization [Online]. Available at: https://www.ncbi.nlm.nih.gov/books/NBK321760/ [Accessed: 23 September 2019].

Kaleeba, J.A.R. and Berger, E.A. (2006). Kaposi's sarcoma-associated herpesvirus fusion-entry receptor: Cystine transporter xCT. *Science*, 311(5769), 1921–1924. doi: 10.1126/science.1120878.

Kim, M.-K. *et al.* (2010). Human papillomavirus type 16 E5 oncoprotein as a new target for cervical cancer treatment. *Biochemical Pharmacology*, 80(12), 1930–1935. doi: 10.1016/J.BCP.2010.07.013.

Kines, R.C. *et al.* (2009). The initial steps leading to papillomavirus infection occur on the basement membrane prior to cell surface binding. *Proceedings of the National Academy of Sciences*, 106(48), 20458–20463. doi: 10.1073/pnas.0908502106.

Krupovic, M. and Bamford, D.H. (2011). Double-stranded DNA viruses: 20 families and only five different architectural principles for virion assembly. *Current Opinion in Virology*, 118–124. doi: 10.1016/j.coviro.2011.06.001.

La Scola, B. *et al.* (2003). A giant virus in amoebae. *Science*, 299(5615), 2033. doi: 10.1126/science.1081867.

Lukac, D.M. and Yuan, Y. (2007). *Reactivation and Lytic Replication of KSHV*. Cambridge University Press [Online]. Available at: http://www.ncbi.nlm.nih.gov/books/NBK47395/ [Accessed 21 April 2015].

Lukac, D.M. *et al.* (1998). Reactivation of Kaposi's sarcoma-associated herpesvirus infection from latency by expression of the ORF 50 transactivator, a homolog of the EBV R protein. *Virology*, 252(2), 304–312. doi: 10.1006/viro.1998.9486.

Martin, J.N. *et al.* (1998). Sexual transmission and the natural history of human herpesvirus 8 infection. *New England Journal of Medicine*, 338(14), 948–954. doi: 10.1056/NEJM199804023381403.

McBride, A.A., Oliveira, J.G., and McPhillips, M.G. (2006). Partitioning viral genomes in mitosis: Same idea, different targets. *Cell Cycle*, 5(14), 1499–1502. doi: 10.4161/cc.5.14.3094.

McGeoch, D.J. *et al.* (1995). Molecular phylogeny and evolutionary timescale for the family of mammalian herpesviruses. *Journal of Molecular Biology*, 247(3), 443–458. doi: 10.1006/jmbi.1995.0152.

Mesri, E.A., Feitelson, M.A., and Munger, K. (2014). Human viral oncogenesis: A cancer hallmarks analysis. *Cell Host & Microbe*, 15(3), 266–282. doi: 10.1016/j.chom.2014.02.011.

Minhas, V. and Wood, C. (2014). Epidemiology and transmission of Kaposi's sarcoma-associated herpesvirus. *Viruses*, 6(11), 4178–4194. doi: 10.3390/v6114178.

Moody, C.A. and Laimins, L.A. (2010). Human papillomavirus oncoproteins: Pathways to transformation. *Nature Reviews Cancer*, 10(8), 550–60. doi: 10.1038/nrc2886.

Mounts, P., Shah, K.V., and Kashima, H. (1982). Viral etiology of juvenile- and adult-onset squamous papilloma of the larynx. *Proceedings of the National Academy of Sciences*, 79(17), 5425–5429. doi: 10.1073/pnas.79.17.5425.

Moustafa, A. *et al.* (2017). The blood DNA virome in 8,000 humans. *PLoS Pathogens*, 13(3), e1006292. doi: 10.1371/journal.ppat.1006292.

NobelPrize.org (2019) The Nobel Prize in Physiology or Medicine 1969, Press release [Online]. Available at: https://www.nobelprize.org/prizes/medicine/1969/press-release/ [Accessed: 16 August 2019].

Okunade, K.S. (2019). Human papillomavirus and cervical cancer. *Journal of Obstetrics and Gynaecology*, 40(5), 602–608. doi: 10.1080/01443615.2019.1634030.

Parrino, J. and Graham, B.S. (2006). Smallpox vaccines: Past, present, and future. *Journal of Allergy and Clinical Immunology*, 118(6), 1320–1326. doi: 10.1016/j.jaci.2006.09.037.

Philippe, N. *et al.* (2013). Pandoraviruses: Amoeba viruses with genomes up to 2.5 Mb reaching that of parasitic eukaryotes. *Science*, 341(6143), 281–286. doi: 10.1126/science.1239181.

Rappocciolo, G. *et al.* (2006). DC-SIGN is a receptor for human herpesvirus 8 on dendritic cells and macrophages. *Journal of Immunology*, 176(3), 1741–1749. doi: 10.4049/jimmunol.176.3.1741.

Rascovan, N., Duraisamy, R., and Desnues, C. (2016). Metagenomics and the human virome in asymptomatic individuals. *Annual Review of Microbiology*, 70(1), 125–141. doi: 10.1146/annurev-micro-102215-095431.

Richards, R.M. *et al.* (2006). Cleavage of the papillomavirus minor capsid protein, L2, at a furin consensus site is necessary for infection. *Proceedings of the National Academy of Sciences of the United States of America*, 103(5), 1522–1527. doi: 10.1073/pnas.0508815103.

Rohrmann, G.F. (2019) *Baculovirus Molecular Biology*. National Center for Biotechnology Information, USA [Online]. Available at: https://www.ncbi.nlm.nih.gov/books/NBK543458/ [Accessed 23 September 2019].

Ruska, H. (1940). Die Sichtbarmachung der bakteriophagen Lyse im Übermikroskop. *Die Naturwissenschaften*, 28(3), 45–46. doi: 10.1007/BF01486931.

Sathish, N., Wang, X., and Yuan, Y. (2012). Tegument proteins of Kaposi's Sarcoma-associated herpesvirus and related gamma-herpesviruses. *Frontiers in Microbiology*, 3, 98. doi: 10.3389/fmicb.2012.00098.

Scheffner, M. *et al.* (1990). The E6 oncoprotein encoded by human papillomavirus types 16 and 18 promotes the degradation of p53. *Cell*, 63(6), 1129–1136. doi: 10.1016/0092-8674(90)90409-8.

Sender, R., Fuchs, S., and Milo, R. (2016). Revised estimates for the number of human and bacteria cells in the body. *PLoS Biology*, 14(8), e1002533. doi: 10.1371/journal.pbio.1002533.

Smith, P.P. *et al.* (1992). Viral integration and fragile sites in human papillomavirus-lmmortalized human keratinocyte cell lines. *Genes, Chromosomes and Cancer*, 5(2), 150–157. doi: 10.1002/gcc.2870050209.

Suttle, C.A. (2007). Marine viruses – Major players in the global ecosystem. *Nature Reviews Microbiology*, 5(10), 801–812. doi: 10.1038/nrmicro1750.

Tzellos, S. and Farrell, P.J. (2012). Epstein–Barr virus sequence variation-biology and disease. *Pathogens*, 1(2), 156–175. doi: 10.3390/pathogens1020156.

Uppal, T. *et al.* (2014). KSHV LANA – The Master Regulator of KSHV Latency. *Viruses*, 6(12), 4961–4998. doi: 10.3390/v6124961.

Van Doorslaer, K. *et al.* (2018). ICTV virus taxonomy profile: Papillomaviridae. *Journal of General Virology*, 99(8), 989–990. doi: 10.1099/jgv.0.001105.

de Villiers, E.-M. (2013). Cross-roads in the classification of papillomaviruses. *Virology*, 445(1–2), 2–10. doi: 10.1016/J.VIROL.2013.04.023.

Walboomers, J.M.M. *et al.* (1999). Human papillomavirus is a necessary cause of invasive cervical cancer worldwide. *The Journal of Pathology*, 189(1), 12–19. doi: 10.1002/(SICI)1096-9896(199909)189:1<12::AID-PATH431>3.0.CO;2-F.

Wang, X. *et al.* (2015). Mono-ubiquitylated ORF45 mediates association of KSHV particles with internal lipid rafts for viral assembly and egress. *PLOS Pathogens*, 11(12), e1005332. doi: 10.1371/journal.ppat.1005332.

WHO (1980). The Global Eradication of Smallpox. Final Report, Global Commission for the Certification of Smallpox Eradication, Geneva. doi: 10.1136/pgmj.58.685.734-c.

Willner, D. *et al.* (2009). Metagenomic analysis of respiratory tract DNA viral communities in cystic fibrosis and non-cystic fibrosis individuals. *PloS O*, 4(10), e7370. doi: 10.1371/journal.pone.0007370.

Winther, J.F. *et al.* (2018). Aetiological agents. In *Rapid Infectious Diseases and Tropical Medicine*, R. Isba (ed.). Blackwell Publishing Ltd, Massachusetts. doi: 10.1002/9781119548430.ch2.

Woodman, C.B.J., Collins, S.I., and Young, L.S. (2007). The natural history of cervical HPV infection: Unresolved issues. *Nature Reviews Cancer*, 7(1), 11–22. doi: 10.1038/nrc2050.

Wylie, K.M., Weinstock, G.M., and Storch, G.A. (2012). Emerging view of the human virome. *Translational Research*, 160(4), 283–290. doi: 10.1016/j.trsl.2012.03.006.

Wylie, K.M. *et al.* (2014). Metagenomic analysis of double-stranded DNA viruses in healthy adults. *BMC Biology*, 12(1), 71. doi: 10.1186/s12915-014-0071-7.

Yoon, C.-S. *et al.* (2001). α6 integrin is the main receptor of human papillomavirus type 16 VLP. *Biochemical and Biophysical Research Communications*, 283(3), 668–673. doi: 10.1006/BBRC.2001.4838.

Yoshikawa, G. *et al.* (2019). Medusavirus, a novel large DNA virus discovered from hot spring water. *Journal of Virology*, 93(8), e02130-18. doi: 10.1128/jvi.02130-18.

Zhong, W. *et al.* (1997). Characterization of ribonucleoprotein complexes containing an abundant polyadenylated nuclear RNA encoded by Kaposi's sarcoma-associated herpesvirus (human herpesvirus 8). *Journal of Virology*, 71(2), 1207–1212. [Online]. Available at: http://www.ncbi.nlm.nih.gov/pubmed/8995643 [Accessed 28 September 2019].

2

Double-stranded RNA Viruses

Michelle M. ARNOLD[1], Albie VAN DIJK[2] and Susana LÓPEZ[3]

[1] *Department of Microbiology and Immunology, Louisiana State University Health Sciences Center, Shreveport, USA*

[2] *Department of Biochemistry, North-West University, Potchefstroom, South Africa*

[3] *Department of Developmental Genetics, Instituto de Biotecnología UNAM, Cuernavaca, Mexico*

2.1. Introduction

Double-stranded RNA (dsRNA) viruses represent a large group of viruses that infect a wide range of vertebrates, invertebrates, plants, bacteria and fungi demonstrating their excellent ability to adapt to different ecological environments. In contrast to the infectious nature of naked RNA of the positive-sense RNA viruses, the genome of dsRNA viruses is inert; the infectious unit of these viruses is a ribonucleoprotein complex containing the genomic dsRNA associated with the viral polymerase components required for the synthesis of mRNA (Whelan 2013). Although dsRNA viruses share the need to encode and package an RNA-dependent RNA polymerase (RdRp), there are distinct differences among these viruses in terms of the number of genome segments, replication strategies and genome packaging. The dsRNA viruses vary in the number of segments that compose their genome (from 1 to 12). In replicating the genome, the *Cystoviridae* and *Birnaviridae* have

For a color version of all figures in this book, see www.iste.co.uk/saleh/virology.zip.

Virology,
coordinated by María-Carla SALEH and Félix AUGUSTO REY. © ISTE Ltd 2021

semi-conservative mechanisms, whereas the *Reoviridae* have a fully conservative replication mechanism. In packaging the genome, most dsRNA viruses surround the ribonucleoprotein complex with a non-enveloped capsid of one or more protein layers. In some instances, the capsid is further surrounded by an envelope (as with the *Cystoviridae*), and in other cases no true capsid has been described and the dsRNA genome/RdRp are surrounded by host-derived lipid vesicles (as with the *Hypoviridae*) (Mertens 2004).

Despite the noted variations among the dsRNA viruses, many have evolved to overcome similar hurdles in order to replicate successfully in the cytoplasm of their host. Because dsRNA molecules cannot function as mRNA templates for translation, nor can they serve as templates for the replication of their genome, these viruses must provide their own transcriptase and capping enzymes, which need to be contained in a viral particle to be able to initiate a productive infection. In fact, for a majority of the dsRNA viruses, the genome remains within the viral core throughout the replication cycle. Since naked dsRNA is recognized in the cells as a signal for the induction of antiviral defense mechanisms including modification of host cell translation apparatus, interferon (IFN) production in vertebrates, RNA silencing in plants, etc. (Sen and Sarkar 2007; Wu *et al.* 2010). Viruses belonging to this family hide their genomes within their capsids and also induce the formation of special structures called viral factories, viroplasms or viral inclusion bodies in the cytoplasm of infected cells. These cytoplasmic inclusions, which contain viral RNA, viral proteins and several cellular components, serve as the sites where RNA transcription, replication and viral assembly take place (Mertens 2004).

These characteristics, shared by many of the dsRNA viruses, result not only in similarities in the replication strategies of this virus family but also make it possible to identify cognate proteins with similar structures and functions, particularly in the proteins that form the inner layers of the viral particles. In comparison, the outer layer proteins and the non-structural proteins differ most among the various families, reflecting their specific interactions with different hosts.

Currently, the International Committee on Taxonomy of Viruses[1] recognizes 12 distinct families of dsRNA viruses that infect different species[2] (Box 2.1). By far the best characterized is the *Reoviridae* family which is also the largest family of dsRNA viruses.

---

1 https://talk.ictvonline.org/taxonomy/.

2 https://viralzone.expasy.org/293.

Family	Host
Birnaviridae *Picobirnaviridae* *Reoviridae*	Mammals, insects, plants, fish, reptiles, birds, arachnids, fungi, arthropods, and crustaceans
Chrysoviridae *Hypoviridae* *Totiviridae* *Megabirnaviridae* *Quadriviridae*	Fungi
Partitiviridae *Endornaviridae*	Fungi, plants
Amalgaviridae	Plants
Cystoviridae	Bacteria

Box 2.1. *dsRNA viruses*

In general, members of the *Reoviridae* are non-enveloped icosahedral particles with diameters that vary between 60 and 80 nm, and that are formed by one, two or three concentric layers of protein that surround the genome, composed of 10, 11 or 12 dsRNA segments. Within this family, there are 15 distinct genera that have been divided into two subfamilies, based on the morphology of the viral particle: The *Spinareovirinae* includes viruses that have spikes or turrets located at the 12 vertices of the icosahedral particle, and the *Sedoreovirinae* includes unturreted viruses that have a smooth, almost spherical, appearance. The members of each subfamily are shown in Box 2.2.

The most studied members of this family are the reoviruses, rotaviruses and orbiviruses, of which the last two cause important health problems in humans and in several animal species. Despite their common characteristics, these viruses also have their particularities and use different strategies for entry, replication, morphogenesis and egress from the host cells. In this chapter, we describe the general aspects of the replication cycle of these viruses and the unique ways in which they infect their hosts.

A cryo-electron microscopy reconstruction of the viral particles of rotavirus, reovirus and bluetongue virus is shown to compare the different structures of these three closely related viruses (Figure 2.1). As can be seen, these are all multi-layered particles in which the structures of the core and the intermediate layer are similar, but the structure of the outermost layer is distinctive for each virus.

Subfamily

Spinareovirinae		*Sedoreovirinae*	
Genus	Host	**Genus**	Host
Aquareovirus	Fish, mollusks	**Cardoreovirus**	Crustaceans
Coltivirus	Mammals, arthropods	**Mimoreovirus**	Marine protists
Cypovirus	Insects	**Orbivirus**	Mammals, birds, arthropods
Dinovernavirus	Insects	**Phytoreovirus**	Plants, insects
Fijivirus	Plants, insects	**Rotavirus**	Mammals, birds
Idnoreovirus	Insects	**Seadornavirus**	Humans, insects
Mycoreovirus	Fungi	-	-
Orthoreovirus	Mammals, birds, reptiles	-	-
Oryzavirus	Plants, insects	-	-

Box 2.2. *Family Reoviridae*

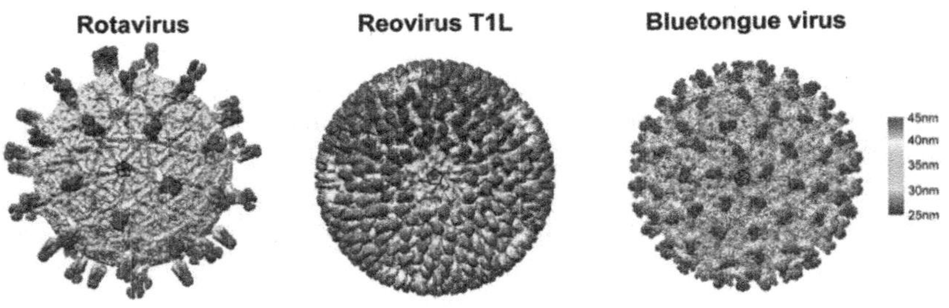

Figure 2.1. *Cryo-electron microscopy reconstruction of Rotavirus, Reovirus, and Bluetongue virus particles. The surface of each virus was viewed along a five-fold axis. The cutaway views of Rotavirus and Reovirus are shown with protein chains colored and labeled. The surface of the Bluetongue virus is radially colored using rainbow colors, from blue to red. Rotavirus (3.8Å, PDB 4V7Q) (Settembre* et al., *2011); Reovirus T1L (8.2Å, EMD-8916) (Snyder* et al., *2019); Bluetongue virus (7Å, EMD-5147) (Zhang* et al., *2010). Figure courtesy of Liya Hu, and B.V.V. Prasad, Baylor College of Medicine*

2.2. Rotaviruses

Rotaviruses represent one of the most important causes of life-threatening gastroenteritis in children under two years of age, and in the young of several animal species (Estes 2013). The pathophysiology of the diarrhea caused by rotavirus is multifactorial. These viruses primarily infect the mature enterocytes that line the tips of the villi within the small intestine. The infected cells are killed and detach from the villi, exposing immature cells that are not competent for absorption of water and solutes in the intestine, and this is one of the main causes of the pathology of rotaviruses (Lundgren and Svensson 2001). In addition, the non-structural protein, NSP4, functions as a secreted viral enterotoxin that triggers a signal transduction pathway to mobilize intracellular calcium and induce chloride secretion. Further, a secretory component of rotavirus, diarrhea appears to result from stimulation of the enteric nervous system (Ramig 2004).

In general, rotavirus infection causes diarrhea, vomiting, malaise and fever. The infection lasts between 3 and 7 days and is self-limiting, ending when most of the susceptible enterocytes have been shed. Although re-infections occur, the first infection is the most severe and is responsible for a high rate of hospitalizations in children between 6 and 24 months of age (Parashar *et al.* 2006). Since initial licensing in 2006, rotavirus vaccines have been introduced globally and their effectiveness has reduced the mortality rate of this infection by more that 50%. Unfortunately, rotaviruses are still responsible for about 200,000 deaths per year, mainly in poorly developed regions of the world where vaccines appear to be less effective (Tate *et al.* 2016).

2.2.1. *Virion structure*

As members of the *Reoviridae* family, rotaviruses are non-enveloped, icosahedral viral particles that contain a genome composed of 11 segments of dsRNA. The viral particle is formed by three concentric layers of protein that have different functions in the viral cycle. These structural proteins are involved in defining the tropism of the virus, its entry into the host cell and they also catalyze the transcription and replication of the viral genome (Estes 2013). The host-specific tropism of rotaviruses is defined in part by the proteins that form the outermost layer of the mature, triple-layered viral particle (TLP), the glycoprotein VP7 and the protease-sensitive protein VP4 (or its proteolytic products VP5 and VP8), which play important roles in cell attachment, receptor recognition and in determining the mechanism of cell entry and vesicle transport before the release of the double layered particles (DLPs) into the cytosolic compartment (López and Arias 2004). During virion entry, the TLP uncoats and the outer-surface proteins are removed.

Loss of the outer layer yields the DLP, consisting of the intermediate layer, formed by the most abundant viral protein VP6 surrounding the inner capsid, or core, formed by VP2. The innermost shell of the particle formed by VP2 contains within it VP1, the RNA-dependent RNA polymerase (RdRp), VP3, a multifunctional protein that has guanylyl–methyltransferase and phosphodiesterase activities, and the viral genome segments (Estes 2013).

The incoming DLP is transcriptionally active and synthesizes large amounts of viral transcripts that direct the synthesis of six structural viral proteins (VPs) and six non-structural proteins (NSPs), which have important roles in viral replication, morphogenesis and control of the antiviral responses of the cell. Viral transcripts also serve as templates for the synthesis of the complementary strand of the genomic RNA. Replication of the viral genome takes place in large cytoplasmic inclusions, termed viroplasms, where the non-structural NSP2 and NSP5 form a scaffold in which viral RNA, viral proteins and cellular components are concentrated to provide an environment that favors RNA replication and the assembly of progeny DLPs. Viroplasms, which resemble several of the RNA granules of the cell (like stress granules or P-bodies), require components of lipid droplets for their formation (Cheung *et al.* 2010).

The rotavirus outer-capsid assembly process is unique among members of the *Reoviridae*. Newly assembled DLPs are recruited to the ER via the transmembrane protein NSP4, which serves as an intracellular receptor for DLPs. As they bud through the ER membrane, DLPs temporarily acquire a transient lipid envelope, onto which the outer-layer proteins are assembled. Finally, the lipid envelope is eliminated by a poorly defined mechanism, and the mature TLPs are released from the cell, either by lysis or by a Golgi-independent non-classical vesicular transport mechanism, depending on the cell line (Arias *et al.* 2015; Trejo-Cerro *et al.* 2018).

2.2.2. *Genome*

The rotavirus genome is formed by 11 dsRNA genome segments that vary in size, ranging from about 660 base pairs (bp) (segment 11) to 3,300 bp, with a total genome length of approximately 18,560 bp. Each genome segment, except for the smallest one, encodes one protein. Genome segment 11 contains two open reading frames that code for NSP5 and NSP6 (Estes 2013).

The 5' end of each RNA genome segment begins with a guanidine followed by a conserved sequence that forms part of the 5' untranslated region (5' UTR); this region is followed by an open reading frame encoding the protein product, and ends with a stop codon and 3' UTR, which contains a consensus tetranucleotide sequence

(GACC 3') at the 3' end. There is no polyadenylation signal at the 3' end of the RNA genome segments. The 5' and 3' UTRs contain important sequences for transcription and genome replication. The size of the 3' and 5' UTR vary among different genome segments, but these sequences are highly conserved in homologous strains (Estes 2013).

2.2.3. *Virus entry*

The binding of rotaviruses to host cells is mediated by the VP8 domain of VP4, which interacts with molecules on the surface of the cells that include gangliosides (such as GM1 and GD1) and histo-blood group antigens (Arias *et al.* 2015). After the initial attachment, VP7 and the VP5 domain of VP4 interact with several other proteins that have been proposed as receptors for rotavirus cell entry, such as the integrins $\alpha2\beta1$, $\alpha V\beta3$ and $\alpha X\beta2$, the heat shock cognate protein hsc70 (López and Arias 2004), and some of the proteins that form part of the tight junction, such as junctional adhesion molecule A, occludin and the tight junction protein ZO-1 (Torres-Flores *et al.* 2015). Whether attachment is promiscuous or sequential, or if there is a specific order of interactions, has not been determined. Apparently, not all rotavirus strains bind to integrins, while all the strains tested depend on the interaction with hsc70 for efficient infection (Perez-Vargas *et al.* 2006). Ganglioside GM1, integrin subunits $\alpha2$ and $\beta3$ and hsc70 are localized in detergent-resistant membrane domains, and the integrity of these domains is fundamental for virus cell entry (Isa *et al.* 2004).

Depending on the rotavirus strain, the viral particle is internalized by either clathrin-dependent or clathrin-independent and caveolin-independent endocytic pathways (Arias *et al.* 2015). In any case, entry of all rotavirus strains depends on the presence of cholesterol in the cell membrane and on dynamin for their internalization (Gutierrez *et al.* 2010). Independently of the nature of the cell surface receptor employed, and of the endocytic pathway used for cell internalization, all rotavirus strains converge in the early endosomes during cell entry (Wolf *et al.* 2012; Silva-Ayala *et al.* 2013; Diaz-Salinas *et al.* 2014). In addition, all strains tested seem to require a functional ESCRT system during this process. Although the precise role of the ESCRT machinery during rotavirus infection is not defined, several of its components are known to participate in the intravesicular traffic of the entering viral particles (Silva-Ayala *et al.* 2013). Some rotavirus strains are released into the cell cytoplasm from maturing endosomes, while other strains are released from late endosomes. This characteristic allowed their classification into early-penetrating or late-penetrating rotaviruses (Arias *et al.* 2015). During this intravesicular trafficking, the pH decreases and calcium levels drop inside the endosomal vesicles causing the release of the outer-capsid-layer proteins. The membrane penetration step of rotaviruses depends on conformational rearrangements of the outer layer proteins. Within the endosome, the low levels of calcium dissociate the VP7 trimers and trigger the transition of the trypsin-primed VP5 to a fold-back conformation, exposing the hydrophobic loops that interact with the endosomal membrane and destabilize it (Estes 2013; Salgado *et al.* 2018).

The newly synthesized viral (+)ssRNAs direct the synthesis of all viral proteins, and once a critical mass of viral protein is made, viroplasms are formed. The structural organization of these replication centers is complex and the relative localization of some viral proteins has just been unraveled using super-resolution microscopy. The picture emerging from these studies suggests that the viral proteins analyzed are organized as concentric layers, in which NSP5 localizes at the center and is surrounded by a layer of NSP2 and NSP4. An intermediate region contains VP1, VP2 and VP6, and the outermost region is a ring of VP4 surrounded by a layer of VP7 (Garces Suarez *et al.* 2019).

2.2.4. *Transcription, replication and genome segment sorting*

Within viroplasms, the viral (+)ssRNA transcripts serve as templates for the synthesis of the negative-strand RNA during genome replication. VP1, the viral RdRp, catalyzes the synthesis of both the positive-strand RNA and the replication of the dsRNA segments. Unlike many other viral RdRps, VP1 is active only when associated with VP2, the main component of the core shell, and it has been proposed that VP2 may act as an activator or modulator of the polymerase (McDonald and Patton 2011; Long and McDonald 2017; Steger *et al.* 2019). Thus, during transcription, VP1 synthesizes positive-strand RNAs within the DLPs, while genome replication occurs within replicative intermediates in the early stages of virion particle assembly (Trask *et al.* 2012; Borodavka *et al.* 2018; Ding *et al.* 2019).

During viral replication, each rotavirus particle selectively encapsidates each of the 11 genome segments, and it has been proposed that genome segment assortment takes place concomitantly with replication. It is thought that multiple sequence-specific RNA-RNA interactions between (+)ssRNAs of genome segments are needed to assort a complete multi-segmented viral RNA genome within the VP2 shell. The molecular mechanism and the RNA sequences involved in this process are still unsolved but it has been proposed that NSP2, which has RNA-binding and helix destabilizing activities, might act as an RNA chaperone stabilizing RNA–RNA interactions between different segments, playing an important role in sorting the viral genome segments (Borodavka *et al.* 2017) (Figure 2.2).

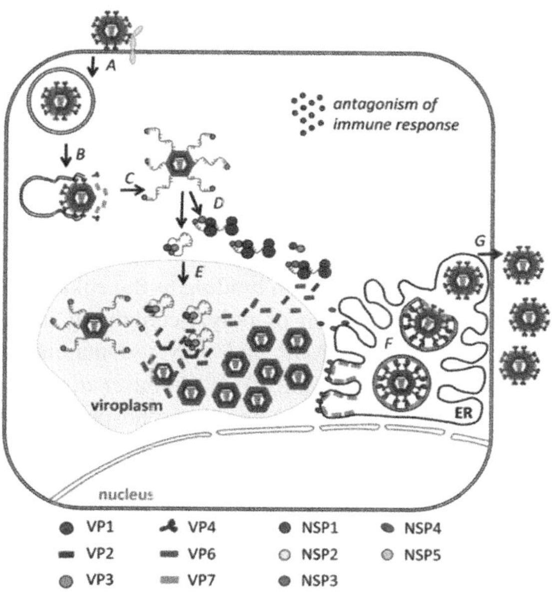

Figure 2.2. *Rotavirus replication cycle. (A) The rotavirus lifecycle initiates when the outer-capsid proteins, VP7 and VP4 (VP5 and VP8), interact with glycans and with several other proteins that have been proposed as receptors for rotavirus cell entry, such as the integrins $\alpha 2\beta 1$, $\alpha V\beta 3$ and $\alpha X\beta 2$, the heat shock cognate protein hsc70, and some of the proteins that form part of the tight junction. The viral particle enters the cell by endocytosis and is transported to a low-pH endosomal compartment where (B) the outer-capsid proteins disassemble and membrane penetration occurs, allowing the DLP to enter the cytosol. (C) The DLP is transcriptionally active, and the synthesis of capped but non-poly-adenylated mRNA begins. (D) Viral transcripts direct the synthesis of viral proteins. When a critical amount of viral protein is produced, viroplasms are formed. (E) Within the viroplasms, viral transcripts associate with VP1, VP3 and NSP2 and serve as templates for the synthesis of dsRNA. Assembly of the core takes place concomitantly with replication, and VP6 assembles on top of the core, forming nascent DLPs which are also transcriptionally active. (F) DLPs bud into the ER lumen through their interaction with NSP4, acquiring a transient lipid envelope. In the ER, the viral envelope is lost and VP4 and VP7 assemble on the DLP. (G) Mature viral particles exit the cell by lysis or by exocytosis depending on the cell line*

2.2.5. *Host cell interactions: protein synthesis*

During infection, rotaviruses establish intimate and complicated interactions with the host cell to take advantage of several cellular processes, and to counteract the antiviral responses elicited by the cell (Arnold *et al.* 2013b; López *et al.* 2016a, 2016b). Soon after entering the cell, rotaviruses take over the host translation machinery, causing a severe shutoff of cellular protein synthesis so that by the end of the replication cycle, most of the newly synthesized protein is of viral origin. Rotavirus (+)ssRNAs are capped but they are not polyadenylated. Instead, they have a consensus sequence at their 3' end that is conserved in all 11 viral genome segments. During infection, the translation of poly(A)-containing cellular mRNAs is prevented by at least three different mechanisms: (1) the non-structural protein NSP3 evicts polyA binding protein (PABP) from binding to the eukaryotic initiation factor 4G (eIF4G), preventing the translation of polyA-containing mRNAs (Piron *et al.* 1998; Montero *et al.* 2006); (2) PABP and polyA containing mRNAs become sequestered in the nucleus (Rubio *et al.* 2013; Gratia *et al.* 2015); and (3) the α subunit of translation factor eIF2 (eIF2α) becomes phosphorylated (Montero *et al.* 2008; Rojas *et al.* 2010). Under these conditions, the synthesis of most cellular proteins is arrested, while viral translation is unaffected. Despite these severe inhibitory conditions for cell protein synthesis, viral transcripts are efficiently translated. However, the precise mechanism by which viral protein synthesis takes place, and the translation factors involved, have not been identified.

2.2.6. *Innate immune evasion*

As in any other successful virus infection, rotaviruses have developed several strategies to counteract the innate immune responses of the cell (Arnold *et al.* 2013b; López *et al.* 2016b). At least three viral proteins are committed to this endeavor. NSP1 interacts with several proteins of the interferon (IFN) signaling cascade and induces their degradation in a proteasome-dependent or -independent manner. Degradation of these proteins prevents the activation of transcription factors that induce the synthesis of IFNs and cytokines, and inhibits the activity of ISGF3, thus blocking the activation of IFN-stimulated genes (ISGs) (Arnold *et al.* 2013a; Ding *et al.* 2016). The capping enzyme activity of VP3 protects viral mRNAs from being detected by RIG-I-like receptors (RLRs) early in the infection. VP3 also has a 2'-5'-phosphodiesterase activity that degrades 2'-5'-oligoadenylates to prevent the activation of RNase L (Zhang *et al.* 2013; Sánchez-Tacuba *et al.* 2015). VP3 may also target MAVS for proteasome degradation in a host-restricted manner (Ding *et al.* 2018). Finally, by preventing the interaction of cellular mRNAs with the initiation complex eIF4F, and by causing the accumulation of PABP and poly(A)-containing mRNAs in the cell nucleus, NSP3 prevents the translation of most cellular mRNAs, thus blocking the translation of many of the transcripts induced during the IFN response of the cell (López *et al.* 2016a).

2.3. Reoviruses

Mammalian orthoreovirus (reovirus) is the prototype member of the *Reoviridae* family of viruses. Reovirus has been used extensively as a model system for studies of viral replication, virus–host interactions and pathogenesis, and was crucial in the discovery of the methylated cap structure that modifies most eukaryotic mRNAs (Dermody *et al.* 2013). Reovirus is generally considered to be non-pathogenic in humans but has been shown to be a possible trigger of celiac disease (Bouziat *et al.* 2017). In addition, the oncolytic properties of reoviruses offer therapeutic potential, and several reovirus formulations are currently in clinical trials as anti-cancer therapeutics (Zhao *et al.* 2016).

2.3.1. *The use of reovirus as an anti-cancer agent*

Although reovirus naturally infects humans, it has been shown to preferentially infect transformed cell lines and causes cell lysis. The oncolytic properties of reovirus are possibly connected to activated Ras signaling, but there is a limited understanding of the mechanism by which Ras activation leads to enhanced replication and lysis (Strong *et al.* 1998). Wild-type reovirus has limited virulence in normal human cells, thus the preferential lysis of transformed cells has made it a potential oncolytic agent. Currently, the T3D strain of reovirus has been extensively studied as a treatment for several types of cancer, including gastric, pancreatic, ovarian and breast cancer (Fukuhara *et al.* 2016) yet there is no explanation for the susceptibility of certain cancers to reovirus oncolysis while other cancers are unaffected by infection.

An important problem to overcome in the use of reovirus as a cancer therapy is antibody-mediated clearance occurring prior to therapeutic benefit. Asymptomatic infections with reovirus are common in childhood, thus seropositivity can be >50% in children 5 years of age reaching up to 100% in adults (Minuk *et al.* 1985; Tai *et al.* 2005). Studies examining the use of reovirus as an anti-cancer therapy have detected an increase in serum anti-reovirus antibody titers, which coincided with regrowth of tumors (Hirasawa *et al.* 2003). Immune-mediated clearance of reovirus may be a barrier to effective treatment, thus use of reovirus may require screening for patient seropositivity or use of alternative delivery strategies to avoid detection by antibodies.

2.3.2. *Virion structure*

Reovirus contains 10 dsRNA genome segments surrounded by two concentric protein shells, the inner core and the outer capsid. The virion measures approximately 85 nm in diameter. The shell of the inner core is formed by the λ1 and σ2 viral proteins (Reinisch *et al.* 2000), which surround the viral polymerase λ3 and the minor core protein μ2. Turrets formed by the capping enzyme λ2 are located at the icosahedral vertices of the viral particle. Together, these five proteins form a core particle that contains the enzymatic activity necessary for transcription, methylation and capping of viral mRNAs. The outer shell of the virion is assembled onto the core and consists of three proteins, μ1, σ1 and σ3, which constitute the delivery system of the virus (Dryden *et al.* 1993).

2.3.3. *Genome*

The reovirus genome is composed of 10 segments of dsRNA that range in size from 1,196 nucleotides to 3,916 nucleotides with a total of approximately 23,500 base pairs. The genome segments are grouped and named according to their migration in polyacrylamide gels: large (L1–L3), medium (M1–M3) and small (S1–S4) (Dermody *et al.* 2013). The names of the proteins encoded by each genome segment are designated by Greek letters that corresponds to large (λ), medium (μ) and small (σ) sizes, which are encoded by genome segments of the L, M and S classes, respectively. Each segment encodes one protein with two exceptions: S1 codes for the structural σ1 protein and the non-structural σ1s protein, and M3 codes for the non-structural proteins μNS and μNSC (Dermody *et al.* 2013). The 5' end of each RNA genome segment harbors a short untranslated region that begins with a conserved GCUA sequence. The 3' end contains a somewhat longer untranslated region that ends with a conserved UCAUC sequence. The sequences required for transcription and genome replication may extend beyond the untranslated regions of the RNAs into the open reading frame of each genome segment. As with other members of this virus family, the 3' ends of the RNA genome and (+)ssRNAs are not polyadenylated.

2.3.4. *Virus entry*

Reovirus initiates infection by tethering to the cell surface via the attachment protein σ1, which engages sialylated glycans (Reiter *et al.* 2011) followed by higher affinity binding to protein receptors on the cell surface. Known protein receptors include JAM-A, which is expressed in tight junctions and on hematopoietic cells (Barton *et al.* 2001; Campbell *et al.* 2005), and the Nogo receptor NgR1, which is expressed on neurons (Konopka-Anstadt *et al.* 2014). Virions are internalized by clathrin-mediated endocytosis (Sturzenbecker *et al.* 1987; Ehrlich *et al.* 2004). Within the endosomes, acid-dependent cathepsin proteases cause degradation of the outer-capsid protein σ3, and the capsid protein μ1 is cleaved into μ1δ and Φ, which remain associated with the particle (Ebert *et al.* 2002). This disassembly intermediate is called an infectious subviral particle (ISVP) (Borsa *et al.* 1981; Sturzenbecker *et al.* 1987). Further disassembly of the ISVP occurs when μ1 trimers separate (Liemann *et al.* 2002; Zhang *et al.* 2006) and μ1δ is cleaved into μ1N and δ through autocatalytic activity (Odegard *et al.* 2004; Nibert *et al.* 2005). The attachment protein σ1 is lost at this stage and the resulting particle is called ISVP* (Chandran *et al.* 2002). Release of μ1N and Φ pore-forming peptides mediates membrane penetration, and the viral core is then delivered into the host cytoplasm.

2.3.5. *Transcription and protein synthesis*

Upon disassembly of the reovirus virion and release of the core particle into the cytoplasm, transcription of viral (+)ssRNAs is initiated. As with other members of the *Reoviridae* family, the dsRNA genome remains packaged within the core particle throughout the transcription cycle, which helps in preventing the dsRNAs from being recognized by the host innate immune response. Within the core particle, the viral polymerase λ3 is located at the base of the turrets, which are made up of the capping enzyme λ2 (Tao *et al.* 2002; Zhang *et al.* 2003). The strands of genomic dsRNA must be unwound to allow the negative strand to serve as a template for mRNA synthesis by the viral enzymes. Reovirus transcripts are capped as they are extruded through the turrets of the core particle into the cytoplasmic environment. Reovirus (+)ssRNAs contain a 5' cap, but lack 3' polyadenylated tails. Instead, short untranslated regions (UTRs) are found at the 5' and 3' ends (Kozak 1977; Kozak and Shatkin 1978).

As with all other viruses, reovirus mRNAs are translated by host ribosomes. The 10 (+)ssRNAs produced by the viral core are synthesized into 11 proteins (8 structural proteins and 3 non-structural proteins). The reovirus σ3 protein has been shown to stimulate translation of viral mRNAs but only those without a cap structure. This protein behaves like a translation initiation factor and associates with ribosomes (Lemay and Millward 1986; Lemieux *et al.* 1987). σ3 has also been implicated in inhibiting host cell protein synthesis, likely by binding to viral dsRNA, which could in turn inhibit the protein kinase PKR (Sharpe and Fields 1982; Yue and Shatkin 1997). The σ1s protein has also been shown to contribute to efficient synthesis of viral proteins in some cell types but the mechanism is unknown (Phillips *et al.* 2018). The enhancement of viral (+)ssRNA translation and inhibition of cellular mRNA translation in reovirus-infected cells is not entirely understood and warrants further study.

As viral proteins accumulate, non-structural proteins reorganize membranes from the endoplasmic reticulum to form the replication and particle assembly sites, called inclusions or viral factories (Fields *et al.* 1971; Fernández de Castro *et al.* 2014). Initially, viral factories form as punctate spots localized throughout the cytoplasm, but as transcription and translation of viral proteins progress during the course of infection, the factories grow in size and localize closer to the nucleus (Broering *et al.* 2002; Bussiere *et al.* 2017). The scaffolding of viral factories is formed by the non-structural protein μNS, which is also responsible for recruitment and coordination of other viral proteins (Broering *et al.* 2002, 2004; Miller *et al.* 2010). The σNS protein also localizes to factories and is thought to increase the half-life of viral (+)ssRNAs, thereby functioning as a stability factor necessary for genome replication (Zamora *et al.* 2018).

Viral factories are not bound by membranes but have been shown to recruit cellular membranes that are important for nascent particle assembly (Fernández de Castro *et al.* 2014). Host translational machinery appears to be recruited to viral factories, likely to aid in the localized synthesis of reovirus proteins in these sites of genome replication and particle assembly (Desmet *et al.* 2014). Because translation is energetically expensive, cells often transiently sequester RNAs by inducing stress granule formation. Stress granules are induced early in reovirus-infected cells but dissociate later during infection, possibly to allow viral protein synthesis (Smith *et al.* 2006; Qin *et al.* 2009, 2011). It is possible that viral (+)ssRNAs are transiently stored in stress granules until viral factories are formed, which would help the virus to avoid recognition by the host.

2.3.6. *RNA packaging and virion assembly*

Reovirus assembles nascent viral cores and packages (+)ssRNA in the protected environment of the viral factories. Within the factories, viral (+)ssRNAs are sorted and assembled into replicase, or precore, complexes that synthesize new dsRNA genomes (Morgan and Zweerink 1975; Murray and Nibert 2007). In addition to the polymerase, reovirus replicase complexes include a few minor core proteins with ATPase- and RNA-binding abilities. The genome segments of reovirus dsRNA are thought to form ordered complexes that nucleate the assembly of the surrounding core proteins. The RNA-dependent RNA polymerase λ3 assembles along with a genomic dsRNA segment inside the interior core face, near the 5-fold vertices. The viral enzyme responsible for capping (+)ssRNAs forms turrets that extend through the core protein layer at these 5-fold vertices (unlike rotaviruses, which package their capping enzymes inside the core particle).

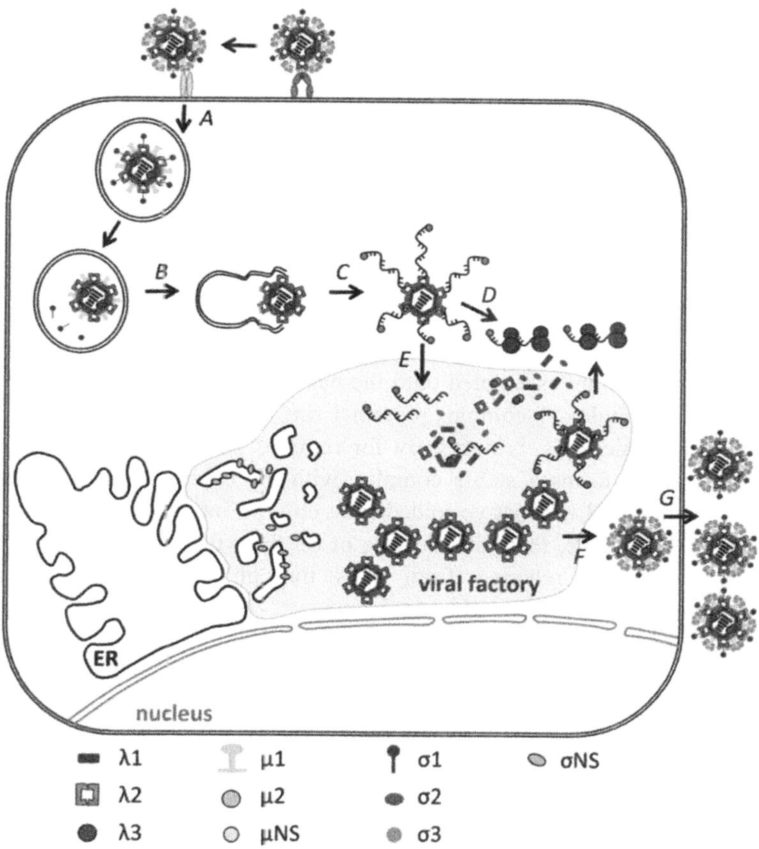

Figure 2.3. *Reovirus replication cycle. (A) The reovirus lifecycle initiates when the outer-capsid protein σ1 attaches to cell-surface carbohydrates and specific protein receptors (dependent on cell type). The viral particle enters the cell by receptor-mediated endocytosis, and the outer capsid undergoes acid-dependent proteolysis into ISVP and ISVP* forms. (B) μ1 cleavage fragments mediate membrane penetration, releasing the transcriptionally active core into the cytoplasm. (C) The core particle directs synthesis of capped but non-poly-adenylated mRNAs. (D) Viral transcripts direct the synthesis of viral proteins. When a critical amount of viral protein is produced, viral factories are formed. (E) ER remodeling takes place during the formation of viral factories, and membrane fragments within the ER may serve as assembly sites for new viral particles. Viral transcripts serve as templates for the synthesis of dsRNA, and core assembly takes place concomitantly with replication. (F) Core particles transit to the edge of viral factories where outer-capsid proteins are assembled. (G) Mature viral particles exit the cell by lysis or by exocytosis, depending on the cell line*

To ensure that one of each (of 10) genomic dsRNA segments is included within the viral core, packaging signals are located at the 5' and 3' ends of the (+)ssRNAs. The untranslated regions of reovirus RNAs are short, ranging from 12 to 32 nucleotides at the 5' end and 35–83 nucleotides at the 3' end. However, the packaging signals extend into the adjacent open reading frame (ORF) sequences at both ends (Roner *et al.* 2004; Roner and Steele 2007). Using reverse genetics systems, packaging signals have been specifically mapped to the 5'-terminal 125-200 nucleotides and 3'-terminal 180–285 nucleotides (Kobayashi *et al.* 2007).

Newly synthesized cores are transcriptionally active and produce secondary rounds of viral (+)ssRNAs within the factories. Transcription is terminated when outer-capsid proteins are assembled onto the nascent cores (Farsetta *et al.* 2000). A host chaperonin complex, T-complex protein-1 ring (TRiC), which promotes protein folding in eukaryotic cells, is necessary for reovirus particle assembly (Knowlton *et al.* 2018). TRiC forms a stable complex with the outer-capsid protein σ3 and appears to maintain σ3 in a native folded state until an interaction partner, possibly μ1 or another chaperone, facilitates release of σ3 from the TRiC complex. Mature reovirus particles are released from cells through an unknown mechanism (Figure 2.3).

2.3.7. *Innate immune evasion*

The nature of the pathogen-associated molecular pattern (PAMP) recognized by the host during reovirus infection is not well characterized. The segmented dsRNA genome of reovirus could conceivably be recognized by host dsRNA sensors such as RIG-I (Goubau *et al.* 2014); however, because the genome remains packaged within the core of incoming viral particles, it is not clear if the genome segments are exposed during the viral replication cycle. Instead, it is possible that the host recognizes extensively folded or incompletely capped viral (+)ssRNAs. Host recognition of viral infection via dsRNA sensors typically leads to activation of the type I interferon (IFN) response, which leads to the production of antiviral IFN-stimulated genes that inhibit viral replication.

Reovirus represses the type I IFN response in a strain-specific manner via the μ2 protein, which causes a nuclear accumulation of IFN regulatory factor 9 (IRF9) and thus prevents IFN signaling (Zurney *et al.* 2009). In infected cells, μ2 predominantly localizes to viral factories, but it has also been shown to localize to the nucleus (Mbisa *et al.* 2000; Parker *et al.* 2002; Kobayashi *et al.* 2009). The type I Lang (T1L) strain of reovirus, but not the type 3 Dearing strain, encodes a μ2 that complexes with the pre-mRNA splicing factor SRSF2 in nuclear speckles (Rivera-Serrano *et al.* 2017). The interaction of μ2 with SRSF2 results in alteration of transcript splicing for genes involved in RNA processing and maturation. Several lines of evidence indicate that the reovirus σ3 protein also blocks the IFN response during infection. In other viral systems, σ3 can substitute for PKR repressors or analogous dsRNA-binding proteins (Beattie *et al.* 1995; Gainey *et al.* 2008). However, activation of PKR does not appear to inhibit reovirus replication, and reovirus may benefit from the activation of PKR (Smith *et al.* 2005; Zhang and Samuel 2007).

2.4. Orbiviruses

Orbiviruses are arthropod-borne viruses that comprise the *Orbivirus* genus of the family *Reoviridae.* Orbiviruses can cause non-contagious infectious disease in vertebrates, arthropods and plants. The most important pathogenic orbiviruses are bluetongue virus (BTV) and African horsesickness virus (AHSV). Both bluetongue (BT) and African horsesickness (AHS) are OIE-listed[3] diseases that are subject to strict control with regard to international trade and movement of animals. Certain biting hematophagous midge vectors of the genus *Culicoides* (*Diptera: Ceratopogonidae*) are the biological vectors transmitting BTV and AHSV between their hosts (Du Toit 1944; Roy 2013). There are 27 serotypes of BTV (Schulz *et al.* 2016) and 9 serotypes of AHSV (Howell 1962). After recovery from natural infection, animals have a solid, life-long immunity to the homologous serotype but only partial or no protection against heterologous serotypes. Protective immunity has generally been associated with the presence of serotype-specific neutralizing antibodies.

Bluetongue (BT) is a disease of domestic and wild ruminants, first recognized in South Africa in the 19th Century when it manifested itself in imported European breeds of sheep, particularly fine-wool breeds such as the Merino (Verwoerd 2012). During the 1800s and 1900s, BT was mainly endemic in subtropical regions, especially in sub-Saharan Africa. However, the first outbreaks of clinical BT in northern Europe occurred in 2006 (Roy 2013). The clinical signs include fever, oedema of the face, lips, muzzle and ears, excessive salivation, hyperaemia of the oral mucosa, and profuse serous nasal discharge. Sometimes, cyanosis of the tongue and oral mucous membranes results in purple discoloration. Severely affected sheep develop focal haemorrhages and ulcers in the oral cavity which can be so painful that animals will not eat or drink. Lameness and stiffness caused by coronitis and myopathy can be severe and breaks in the wool are common (Verwoerd 2012).

3 OIE – World Organisation for Animal Health.

African horsesickness (AHS) is an infectious, non-contagious, arthropod-borne viral disease of equids, such as horses and donkeys. It is mostly confined to sub-Saharan Africa, but there are sporadic outbreaks in North Africa, Mediterranean countries and the Middle East. The mortality rate of AHS is over 90% in fully susceptible horses but zebra and African donkeys rarely show clinical signs (Lecatsas *et al.* 1973). In horses, AHSV infection may result in disease characterized by four, somewhat overlapping, clinical syndromes: a peracute form characterized by fulminant pulmonary oedema; a subacute form characterized by severe oedema of the head and neck; an acute form typified by characteristics of both the peracute form and the subacute form; and a mild form characterized by subtle clinical signs of which the transient increase in rectal temperature is the most readily detected (Erasmus 1972).

Vaccination is the most achievable and cost-effective way to minimize economic losses due to BT and AHS. Conventional live-attenuated vaccines that induce serotype-specific protection were developed in South Africa for both BT and AHS and are still widely used. These live-attenuated vaccines are effective and cheap to produce, but can cause disease in some situations due to under-attenuation, onward spread, reversion to virulence, and reassortment events. Inactivated BT and AHS vaccines were also developed and are effective and safe but relatively expensive and not in general use. Neither of these vaccines allow differentiation between infected and vaccinated animals (DIVA) and protection is limited to the respective serotype.

Several different experimental next generation recombinant vaccine candidates and platforms are being developed. The protein subunit, viral vectored candidate vaccines have DIVA potential but are less effective and likely more costly per protected animal than current vaccines. Several vaccine platforms, based on replicating viruses, are being investigated to accommodate multiple serotypes by exchanging serotype determining outer shell proteins through reverse genetics. These rationally designed transcapsidated viruses are attenuated or cannot replicate fully in vaccinated animals. This approach makes it possible to prevent disease and spread of vaccine virus, as well as reversion to virulence, and to induce humoral and T-cell mediated immune responses and can be manipulated to enable DIVA tests. Most of these replicating vaccines can be produced in a similar way to currently marketed vaccines. Several experimental BT and AHS vaccines show promising improvements and could compete with marketed vaccines in terms of their vaccine profile, but none of these next generation vaccines have yet been licensed (van Rijn 2019).

The focus of the remaining part of this chapter will be on BTV.

2.4.1. *Virion structure*

Orbiviruses are non-enveloped icosahedral particles that encapsulate a segmented dsRNA genome. Seven structural proteins, VP1–VP7, form the BTV virion. These proteins are arranged in three concentric layers that form the surface of the three distinct particles of BTV: the virion, the core (the equivalent to the rotavirus DLP and reovirus core particle) and the subcore (equivalent to the rotavirus SLP and reovirus inner-core particle). Two proteins form the outer-capsid surface of the virion, VP2 and VP5. Both are needed for virus entry in mammalian cells, but are removed from the virion after infection. The particle without VP2 and VP5 is known as the core particle that is transcriptionally active. VP7 forms the surface of the core particle. The VP7 trimers are arranged in typical ring structures visible by electron microscopy from which the name "orbi" (*orbi*, Latin for ring) is derived. Removal of VP7 releases the subcore. VP3 forms the surface of the subcore that encloses a viral genome of 10 double-stranded RNA (dsRNA) segments and the viral transcription complex of three proteins, VP1, VP4 and VP6. Electron microscopy revealed that subcores have a distinctive hexagonal pattern (Roy 2013).

2.4.2. *Genome*

The genome of BTV consists of 10 dsRNA genome segments that range in size from 822 to 3,954 base pairs, with a total of 19,200 base pairs. Each virion contains one complete equimolar set of genome segments. The dsRNA genome segments have conserved terminal sequences of six nucleotides that are mostly identical for all 10 dsRNA segments. The genome segments encode the seven structural proteins (VP1–VP7), which form the virus capsid, as well as four non-structural proteins (NS1, NS2, NS3/NS3A and NS4), which are synthesized in infected cells. The eight largest genome segments each encode one protein. Genome segment 9 encodes both VP6 and NS4 and genome segment 10 codes for NS3 and NS3A (Roy 2013)

2.4.3. *Replication cycle*

The basic features of the replication cycle of BTV are similar to those of reoviruses, rotaviruses and many other viruses, having the stages of entry into cells, replication, assembly and egress. However, unlike reoviruses or rotaviruses, BTV multiplies in both arthropod and vertebrate hosts; therefore, some aspects of BTV replication and morphogenesis are unique.

2.4.4. *Virus entry*

BTV appears to be one of the viruses that can exploit multiple endocytosis pathways for infectious entry.

Proteases in the saliva of vector *Culicoides spp*. modify the structure and infectivity of BTV particles by cleaving VP2 to form infectious subviral particles (ISVP) which have enhanced infectivity for adult *Culicoides* and for KC cells, a cell-line derived from *C. sonorensis* (Darpel *et al.* 2011). Furthermore, BTV cores, i.e. particles lacking both VP2 and VP5, are also infectious for invertebrate cells. The role of the arginine, glycine, aspartic acid (RGD) motif of VP7 in entry into *Culicoides* cells is not completely understood and needs to be investigated.

Bluetongue virus is inoculated into its mammalian host's bloodstream by the saliva of *Culicoides* during a blood meal. In mammals, both clathrin-mediated endocytosis and clathrin independent macropinocytosis-like entry mechanisms have been observed (Gold *et al.* 2010). Clathrin-mediated endocytosis infection occurs in two steps. First, outer-capsid protein VP2 binds to cell surface glycoproteins and facilitates clathrin-mediated endocytosis of the virion. VP2 senses the pH in the early endosome (6.5–6.0) and detaches. Then, the remaining particle continues to the late endosome, where VP5 senses a lower pH (~5.5), unfolds to disrupt the endosomal membrane, which releases the transcriptionally active viral core into the cytoplasm (Zhang *et al.* 2016).

2.4.5. *Transcription, (+)ssRNA selection and packaging, replication*

In the cytoplasm, the core particles repeatedly use the 10 dsRNA genome segments as templates for the transcription of 10 capped, non-polyadenylated (+)ssRNAs by the core-associated enzymes VP1 (RNA-dependent RNA polymerase), VP4 (capping enzyme) and VP6 (helicase/RNA packaging) (Roy 2013). These (+)ssRNAs are released from the core particle into the cytoplasm where they act as templates for translation of viral proteins (mRNAs) and for negative-strand viral RNA synthesis (replication) to generate genomic dsRNAs. Genome replication of the *Reoviridae* is fully conservative.

The four BTV non-structural proteins, NS1–NS4, which are synthesized in infected cells, play important roles in virus replication. NS1 forms tubules in the cytoplasm and promotes preferential translation of BTV (+)ssRNAs, enhancing viral protein synthesis. NS2 is the major component of viral inclusion bodies (VIBs) (Roy 2013), which recruit viral (+)ssRNAs and proteins required for (+)ssRNA packaging, replication and the assembly of cores (Kar *et al.* 2007; Patel and Roy 2014).

NS3/NS3A plays a critical role in virus intracellular trafficking and egress (Roy 2013). NS4 inhibits the innate immune response (Ratinier *et al.* 2016).

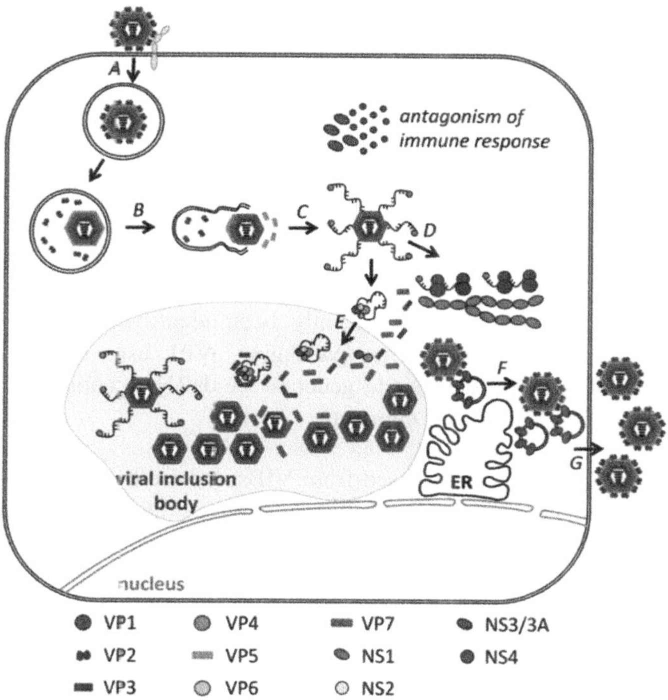

Figure 2.4. *Bluetongue virus replication cycle in mammalian cells. (A) The bluetongue virus lifecycle in mammalian cells initiates when the outer-capsid protein VP2 binds to cell surface glycoproteins and facilitates clathrin-mediated endocytosis of the virion. VP2 disassembles from the particle in the early endosome. (B) The remaining particle continues to the late endosome, where VP5 senses a lower pH, resulting in membrane penetration and the release of the transcriptionally active core into the cytoplasm. (C) The core particle directs synthesis of capped but non-poly-adenylated mRNAs. (D) Viral transcripts direct the synthesis of viral proteins. NS1 forms tubules in the cytoplasm and promotes preferential translation of BTV (+)ssRNAs. When a critical amount of viral protein is produced, viral inclusion bodies are formed. (E) Within the viral inclusion bodies, viral transcripts associate with VP1, VP4 and VP6 and serve as templates for the synthesis of dsRNA. Concomitantly with replication, the assembly of the subcore takes place. VP7 assembles on top of the core forming nascent cores, which are also transcriptionally active. (F) Cores are released from viral inclusion bodies and are trafficked on exocytotic vesicles by the interaction of NS3 with calpactin. VP5 and VP2 are assembled onto cores to generate complete viral particles. (G) Mature viral particles exit the cell by lysis or by exocytosis depending on the cell line*

All 10 BTV dsRNA genome segments have a common complementary short 5' and 3' untranslated regions (UTRs) of variable length that include highly conserved identical hexanucleotides at both ends. Selective packaging of the BTV (+)ssRNAs starts by recruiting them via these UTRs. Intersegment interactions leading to complex RNA networks are the driving force for sorting and packaging of a complete set of (+)ssRNAs for each of the genome segments. There is a specific order of (+)ssRNA recruitment, starting from the smallest for genome segment S10, which triggers the correct secondary structures required for the sequential RNA–RNA interactions of all 10 BTV (+)ssRNAs and their packaging into the capsid (Sung and Roy 2014; Fajardo *et al.* 2015; Boyce *et al.* 2016; Roy 2017). The packaging signals in each genome segment's (+)ssRNA are highly specific and dispersed largely in the UTRs, as well as in the coding regions. A viral RNA-packaging motif in VP6 has recently been identified to be essential for packaging (Sung *et al.* 2019). After packaging, VP1 uses the (+)ssRNAs as templates for minus-strand synthesis to generate the dsRNA genome, at which time progeny core particle formation is complete.

Once assembled, cores are released from VIBs and are trafficked on exocytotic vesicles by the interaction of NS3 with calpactin. During this process, the outer-capsid proteins VP5 and VP2 are acquired to generate a complete progeny particle. These new virions exit the cell via budding mediated by the interaction of Tsg101 (human tumor-susceptibility gene 101 protein) with NS3 or by lysis of the host cell (Roy 2013) (Figure 2.4).

2.4.6. *Innate immune evasion*

Replication of BTV in phagocytic cells (dendritic cells and macrophages) and endothelium cells leads to the generation of the innate and adaptive immune responses that mediate both initial virus clearance and subsequent resistance to infection with the homologous virus serotype.

The secretion of type I IFN, composed mainly of IFN-α and IFN-β, induces the expression of many gene products that are essential for the establishment of an antiviral state within virus-infected cells and neighboring uninfected cells. BTV is a strong inducer of IFN in non-hematopoietic cells (Vitour *et al.* 2014). BTV non-structural proteins NS3 and NS4 are antagonists of the type I IFN system and disrupt the cellular innate immune response to BTV infection (Chauveau *et al.* 2012; Ratinier *et al.* 2016).

The RLR/MAVS pathway controls both the sensing and the pro-inflammatory and antiviral responses to BTV in non-hematopoietic cells. The RNA helicases retinoic acid-inducible gene-I (RIG-I) and melanoma differentiation-associated gene 5 (MDA5) are involved in the expression of IFN-β and in the control of BTV infection in non-hematopoietic cells (Chauveau *et al.* 2012). In contrast, induction of IFN-α and IFN-β synthesis in sheep primary plasmacytoid dendritic cells (pDCs) requires the MyD88 adaptor independently of the TLR7, as well as the kinases dsRNA-activated protein kinase (PKR) and stress-activated protein kinase (SAPK)/Jun N-terminal protein kinase (JNK) (Chauveau *et al.* 2012; Vitour *et al.* 2014).

2.5. Concluding remarks and future challenges to understand dsRNA virus biology

Viruses in the *Reoviridae* family have a number of similar features in their virion structure, genomic content and overall viral lifecycle. Common to all of these viruses is the need for an efficient RNA-dependent polymerase that is capable of transcribing and replicating the dsRNA genome segments (Table 2.1). The recently resolved structures by high-resolution cryo-EM of the RdRp, and their associated RNAs, of several dsRNA viruses have revealed the emergence of common themes for the regulation of the RdRp during transcription and replication (Ogden 2019). Each of these viruses must also hide their dsRNA genome using aggregates of RNA and protein (viroplasms, viral factories or viral inclusion bodies) to prevent the activation of the innate immune sensors.

However, it is clear that the replicative cycle of each of these viruses, and the cellular mechanisms that they hijack to successfully conquer their host, have many variations. Studying and comparing the details of each of these viruses will provide important insights into conserved and unique strategies that are essential for invasion and replication. A number of broad questions about the lifecycle of rotavirus, reovirus and bluetongue virus still require investigation, including:

What is the mechanism used by members of the *Reoviridae*, which are non-enveloped particles, to penetrate the host membrane during entry?

Is the recruitment of translational machinery to the sites of viral replication common among the *Reoviridae*? If so, what is the mechanism of recruitment?

What is the detailed mechanism for genome segment assortment during dsRNA virus assembly, and is there a common solution to this problem?

What is the mechanism of virus egress from polarized epithelial cells? How do viral particles exit the cell before lysis? How do proteins with no transmembrane domains associate with host membranes?

Recent methodological advances in cryo-EM, high-resolution microscopy, organoid culture and reverse genetics systems can be applied to the study of the viral cycles of the *Reoviridae* and will advance our knowledge of these viruses.

	Rotavirus		Reovirus		Orbivirus	
Function of viral protein	*Protein*	*Genome Segment*	*Protein*	*Genome Segment*	*Protei n*	*Genome Segment*
RNA polymerase	VP1	1	λ3	L1	VP1	1
Polymerase scaffold (inner-capsid structure)	VP2	2	λ1/σ2	L3/S2	VP3	3
Guanylyl- and methyltransferase	VP3	3	λ2	L2	VP4	4
Particle intermediate layer	VP6	6	μ1	M2	VP7	7
Receptor attachment, membrane penetration, neut. antigens	VP4/VP7	4/7	σ3/μ1/σ 1	S4/M2/S1	VP2/ VP5	2/5
IFN modulation	NSP1	5	μ2/σ3	M1/S4	NS3/ NS4	9/10
Helicase/NTPase	NSP2	8	λ1/μ2	L3/M1	VP6	9
Viroplasms/viral factories/inclusion bodies	NSP2/ NSP5	8/11	μNS/ σNS	M3/S3	NS2	8
Translational control	NSP3	7	σ3	S4	NS1	6
Membrane interacting proteins	NSP4	10	?	?	NS3/ NS3A	10

* NS denotes non-structural proteins.

Table 2.1. *Equivalent functions between dsRNA viruses*

2.6. References

Arias, C.F., Silva-Ayala, D., and López, S. (2015). Rotavirus entry: A deep journey into the cell with several exits. *J Virol*, 89, 890–893.

Arnold, M.M., Barro, M., and Patton, J.T. (2013a). Rotavirus NSP1 mediates degradation of interferon regulatory factors through targeting of the dimerization domain. *J Virol*, 87, 9813–9821.

Arnold, M.M., Sen, A., Greenberg, H.B., and Patton, J.T. (2013b). The battle between rotavirus and its host for control of the interferon signaling pathway. *PLoS Pathog*, 9, e1003064.

Barton, E.S., Forrest, J.C., Connolly, J.L., Chappell, J.D., Liu, Y., Schnell, F.J., Nusrat, A., Parkos, C.A., and Dermody, T.S. (2001). Junction adhesion molecule is a receptor for reovirus. *Cell*, 104, 441–451.

Beattie, E., Denzler, K.L., Tartaglia, J., Perkus, M.E., Paoletti, E., and Jacobs, B.L. (1995). Reversal of the interferon-sensitive phenotype of a vaccinia virus lacking E3L by expression of the reovirus S4 gene. *J Virol*, 69, 499–505.

Borodavka, A., Dykeman, E.C., Schrimpf, W., and Lamb, D.C. (2017). Protein-mediated RNA folding governs sequence-specific interactions between rotavirus genome segments. *Elife*, 6, e27453.

Borodavka, A., Desselberger, U., and Patton, J.T. (2018). Genome packaging in multi-segmented dsRNA viruses: Distinct mechanisms with similar outcomes. *Curr Opin Virol*, 33, 106–112.

Borsa, J., Sargent, M.D., Lievaart, P.A., and Copps, T.P. (1981). Reovirus: Evidence for a second step in the intracellular uncoating and transcriptase activation process. *Virology*, 111, 191–200.

Bouziat, R., Hinterleitner, R., Brown, J.J., Stencel-Baerenwald, J.E., Ikizler, M., Mayassi, T., Meisel, M., Kim, S.M., Discepolo, V., Pruijssers, A.J., Ernest, J.D., Iskarpatyoti, J.A., Costes, L.M., Lawrence, I., Palanski, B.A., Varma, M., Zurenski, M.A., Khomandiak, S., Mcallister, N., Aravamudhan, P., Boehme, K.W., Hu, F., Samsom, J.N., Reinecker, H.C., Kupfer, S.S., Guandalini, S., Semrad, C.E., Abadie, V., Khosla, C., Barreiro, L.B., Xavier, R.J., Ng, A., Dermody, T.S., and Jabri, B. (2017). Reovirus infection triggers inflammatory responses to dietary antigens and development of celiac disease. *Science*, 356, 44–50.

Boyce, M., McCrae, M.A., Boyce, P., and Kim, J.T. (2016). Inter-segment complementarity in orbiviruses: A driver for co-ordinated genome packaging in the *Reoviridae*? *J Gen Virol*, 97, 1145–57.

Broering, T.J., Parker, J.S., Joyce, P.L., Kim, J., and Nibert, M.L. (2002). Mammalian reovirus nonstructural protein microNS forms large inclusions and colocalizes with reovirus microtubule-associated protein micro2 in transfected cells. *J Virol*, 76, 8285–8297.

Broering, T.J., Kim, J., Miller, C.L., Piggott, C.D., Dinoso, J.B., Nibert, M.L., and Parker, J.S. (2004). Reovirus nonstructural protein mu NS recruits viral core surface proteins and entering core particles to factory-like inclusions. *J Virol*, 78, 1882–1892.

Bussiere, L.D., Choudhury, P., Bellaire, B., and Miller, C.L. (2017). Characterization of a replicating mammalian orthoreovirus with tetracysteine-tagged muNS for live-cell visualization of viral factories. *J Virol*, 91, e01371-17.

Campbell, J.A., Schelling, P., Wetzel, J.D., Johnson, E.M., Forrest, J.C., Wilson, G.A., Aurrand-Lions, M., Imhof, B.A., Stehle, T., and Dermody, T.S. (2005). Junctional adhesion molecule a serves as a receptor for prototype and field-isolate strains of mammalian reovirus. *J Virol*, 79, 7967–7978.

Chandran, K., Farsetta, D.L., and Nibert, M.L. (2002). Strategy for nonenveloped virus entry: A hydrophobic conformer of the reovirus membrane penetration protein micro 1 mediates membrane disruption. *J Virol*, 76, 9920–9933.

Chauveau, E., Doceul, V., Lara, E., Adam, M., Breard, E., Sailleau, C., Viarouge, C., Desprat, A., Meyer, G., Schwartz-Cornil, I., Ruscanu, S., Charley, B., Zientara, S., and Vitour, D. (2012). Sensing and control of bluetongue virus infection in epithelial cells via RIG-I and MDA5 helicases. *J Virol*, 86, 11789–11799.

Cheung, W., Gill, M., Esposito, A., Kaminski, C.F., Courousse, N., Chwetzoff, S., Trugnan, G., Keshavan, N., Lever, A., and Desselberger, U. (2010). Rotaviruses associate with cellular lipid droplet components to replicate in viroplasms, and compounds disrupting or blocking lipid droplets inhibit viroplasm formation and viral replication. *J Virol*, 84, 6782–6798.

Darpel, K.E., Langner, K.F., Nimtz, M., Anthony, S.J., Brownlie, J., Takamatsu, H.H., Mellor, P.S., and Mertens, P.P. (2011). Saliva proteins of vector Culicoides modify structure and infectivity of bluetongue virus particles. *PLoS One*, 6, e17545.

Dermody, T.S., Parker, J.S.L., and Sherry, B. (2013). Orthoreoviruses. In *Fields Virology*, Knipe, D.M. and Howley, P.M. (eds), 6th edition. Lippincott Williams & Wilkins, Philadelphia.

Desmet, E.A., Anguish, L.J., and Parker, J.S. (2014). Virus-mediated compartmentalization of the host translational machinery. *MBio*, 5, e01463-14.

Diaz-Salinas, M.A., Silva-Ayala, D., López, S., and Arias, C.F. (2014). Rotaviruses reach late endosomes and require the cation-dependent mannose-6-phosphate receptor and the activity of cathepsin proteases to enter the cell. *J Virol*, 88, 4389–4402.

Ding, S., Mooney, N., Li, B., Kelly, M.R., Feng, N., Loktev, A.V., Sen, A., Patton, J.T., Jackson, P.K., and Greenberg, H.B. (2016). Comparative proteomics reveals strain-specific beta-TrCP degradation via rotavirus NSP1 hijacking a host Cullin-3-Rbx1 complex. *PLoS Pathog*, 12, e1005929.

Ding, S., Zhu, S., Ren, L., Feng, N., Song, Y., Ge, X., Li, B., Flavell, R.A., and Greenberg, H.B. (2018). Rotavirus VP3 targets MAVS for degradation to inhibit type III interferon expression in intestinal epithelial cells. *Elife*, 7, e39494.

Ding, K., Celma, C.C., Zhang, X., Chang, T., Shen, W., Atanasov, I., Roy, P., and Zhou, Z.H. (2019). *In situ* structures of rotavirus polymerase in action and mechanism of mRNA transcription and release. *Nat Commun*, 10, 2216.

Dryden, K.A., Wang, G., Yeager, M., Nibert, M.L., Coombs, K.M., Furlong, D.B., Fields, B.N., and Baker, T.S. (1993). Early steps in reovirus infection are associated with dramatic changes in supramolecular structure and protein conformation: Analysis of virions and subviral particles by cryoelectron microscopy and image reconstruction. *J Cell Biol*, 122, 1023–1041.

Ebert, D.H., Deussing, J., Peters, C., and Dermody, T.S. (2002). Cathepsin L and cathepsin B mediate reovirus disassembly in murine fibroblast cells. *J Biol Chem*, 277, 24609-17.

Ehrlich, M., Boll, W., Van Oijen, A., Hariharan, R., Chandran, K., Nibert, M.L., and Kirchhausen, T. (2004). Endocytosis by random initiation and stabilization of clathrin-coated pits. *Cell*, 118, 591–605.

Erasmus, B.J. (1972). The pathogenesis of African horsesickness. *Proceedings of the 3rd International Conference of Equine Infectious Diseases*, Karger, Basel, Switzerland.

Estes, M.K. and Greenberg, H.B. (2013). Rotaviruses. In *Fields Virology*, Knipe, D.M. and Howley, P.M. (eds), 6th edition. Lippincott Williams & Wilkins, Philadelphia.

Fajardo Jr., T., Sung, P.Y., and Roy, P. (2015). Disruption of specific RNA-RNA interactions in a double-stranded RNA virus inhibits genome packaging and virus infectivity. *PLoS Pathog*, 11, e1005321.

Farsetta, D.L., Chandran, K., and Nibert, M.L. (2000). Transcriptional activities of reovirus RNA polymerase in recoated cores. Initiation and elongation are regulated by separate mechanisms. *J Biol Chem*, 275, 39693–39701.

Fernández De Castro, I., Zamora, P.F., Ooms, L., Fernández, J.J., Lai, C.M., Mainou, B.A., Dermody, T.S., and Risco, C. (2014). Reovirus forms neo-organelles for progeny particle assembly within reorganized cell membranes. *MBio*, 5. e00931-13.

Fields, B.N., Raine, C.S., and Baum, S.G. (1971). Temperature-sensitive mutants of reovirus type 3: Defects in viral maturation as studied by immunofluorescence and electron microscopy. *Virology*, 43, 569–578.

Fukuhara, H., Ino, Y., and Todo, T. (2016). Oncolytic virus therapy: A new era of cancer treatment at dawn. *Cancer Sci*, 107, 1373–1379.

Gainey, M.D., Dillon, P.J., Clark, K.M., Manuse, M.J., and Parks, G.D. (2008). Paramyxovirus-induced shutoff of host and viral protein synthesis: Role of the P and V proteins in limiting PKR activation. *J Virol*, 82, 828–839.

Garces Suarez, Y., Martinez, J.L., Torres Hernandez, D., Hernandez, H.O., Perez-Delgado, A., Mendez, M., Wood, C.D., Rendon-Mancha, J.M., Silva-Ayala, D., López, S., Guerrero, A., and Arias, C.F. (2019). Nanoscale organization of rotavirus replication machineries. *Elife*, 8, e42906.

Gold, S., Monaghan, P., Mertens, P., and Jackson, T. (2010). A clathrin independent macropinocytosis-like entry mechanism used by bluetongue virus-1 during infection of BHK cells. *PLoS One*, 5, e11360.

Goubau, D., Schlee, M., Deddouche, S., Pruijssers, A.J., Zillinger, T., Goldeck, M., Schuberth, C., Van der Veen, A.G., Fujimura, T., Rehwinkel, J., Iskarpatyoti, J.A., Barchet, W., Ludwig, J., Dermody, T.S., Hartmann, G., and Reis e Sousa, C. (2014). Antiviral immunity via RIG-I-mediated recognition of RNA bearing 5'-diphosphates. *Nature*, 514, 372–375.

Gratia, M., Sarot, E., Vende, P., Charpilienne, A., Baron, C.H., Duarte, M., Pyronnet, S., and Poncet, D. (2015). Rotavirus NSP3 is a translational surrogate of the poly(A) binding protein-poly(A) complex. *J Virol*, 89, 8773–8782.

Gutierrez, M., Isa, P., Sánchez-San Martin, C., Perez-Vargas, J., Espinosa, R., Arias, C.F., and López, S. (2010). Different rotavirus strains enter MA104 cells through different endocytic pathways: The role of clathrin-mediated endocytosis. *J Virol*, 84, 9161–9169.

Hirasawa, K., Nishikawa, S.G., Norman, K.L., Coffey, M.C., Thompson, B.G., Yoon, C.S., Waisman, D.M., and Lee, P.W. (2003). Systemic reovirus therapy of metastatic cancer in immune-competent mice. *Cancer Res*, 63, 348–53.

Howell, P.G. (1962). The isolation and identification of further antigenic types of African horsesickness virus. *Onderstepoort J Vet Res*, 29, 139–149.

Isa, P., Realpe, M., Romero, P., López, S., and Arias, C.F. (2004). Rotavirus RRV associates with lipid membrane microdomains during cell entry. *Virology*, 322, 370–381.

Kar, A.K., Bhattacharya, B., and Roy, P. (2007). Bluetongue virus RNA binding protein NS2 is a modulator of viral replication and assembly. *BMC Mol Biol*, 8, 4.

Knowlton, J.J., Fernández de Castro, I., Ashbrook, A.W., Gestaut, D.R., Zamora, P.F., Bauer, J.A., Forrest, J.C., Frydman, J., Risco, C., and Dermody, T.S. (2018). The TRiC chaperonin controls reovirus replication through outer-capsid folding. *Nat Microbiol*, 3, 481–493

Kobayashi, T., Antar, A.A., Boehme, K.W., Danthi, P., Eby, E.A., Guglielmi, K.M., Holm, G.H., Johnson, E.M., Maginnis, M.S., Naik, S., Skelton, W.B., Wetzel, J.D., Wilson, G.J., Chappell, J.D., and Dermody, T.S. (2007). A plasmid-based reverse genetics system for animal double-stranded RNA viruses. *Cell Host Microbe*, 1, 147–157.

Kobayashi, T., Ooms, L.S., Chappell, J.D., and Dermody, T.S. (2009). Identification of functional domains in reovirus replication proteins muNS and mu2. *J Virol*, 83, 2892–2906.

Konopka-Anstadt, J.L., Mainou, B.A., Sutherland, D.M., Sekine, Y., Strittmatter, S.M., and Dermody, T.S. (2014). The Nogo receptor NgR1 mediates infection by mammalian reovirus. *Cell Host Microbe*, 15, 681–691.

Kozak, M. (1977). Nucleotide sequences of 5'-terminal ribosome-protected initiation regions from two reovirus messages. *Nature*, 269, 391–394.

Kozak, M. and Shatkin, A.J. (1978). Identification of features in 5' terminal fragments from reovirus mRNA which are important for ribosome binding. *Cell*, 13, 201–212.

Lecatsas, G., Erasmus, B.J., and Els, H.J. (1973). Electron microscopic studies on equine encephalosis virus. *Onderstepoort J Vet Res*, 40, 53–57.

Lemay, G. and Millward, S. (1986). Expression of the cloned S4 gene of reovirus serotype 3 in transformed eucaryotic cells: Enrichment of the viral protein in the crude initiation factor fraction. *Virus Res*, 6, 133–140.

Lemieux, R., Lemay, G., and Millward, S. (1987). The viral protein σ3 participates in translation of late viral mRNA in reovirus-infected L cells. *J Virol*, 61, 2472–2479.

Liemann, S., Chandran, K., Baker, T.S., Nibert, M.L., and Harrison, S.C. (2002). Structure of the reovirus membrane-penetration protein, Mu1, in a complex with is protector protein σ3. *Cell*, 108, 283–295.

Long, C.P. and McDonald, S.M. (2017). Rotavirus genome replication: Some assembly required. *PLoS Pathog*, 13, e1006242.

López, S. and Arias, C.F. (2004). Multistep entry of rotavirus into cells: A Versaillesque dance. *Trends Microbiol*, 12, 271–278.

López, S., Oceguera, A., and Sandoval-Jaime, C. (2016a). Stress response and translation control in rotavirus infection. *Viruses*, 8, 162.

López, S., Sánchez-Tacuba, L., Moreno, J., and Arias, C.F. (2016b). Rotavirus strategies against the innate antiviral system. *Annu Rev Virol*, 3, 591–609.

Lundgren, O. and Svensson, L. (2001). Pathogenesis of rotavirus diarrhea. *Microbes Infect*, 3, 1145–1156.

Mbisa, J.L., Becker, M.M., Zou, S., Dermody, T.S., and Brown, E.G. (2000). Reovirus mu2 protein determines strain-specific differences in the rate of viral inclusion formation in L929 cells. *Virology*, 272, 16–26.

McDonald, S.M. and Patton, J.T. (2011). Rotavirus VP2 core shell regions critical for viral polymerase activation. *J Virol*, 85, 3095–3105.

Mertens, P. (2004). The dsRNA viruses. *Virus Res*, 101, 3–13.

Miller, C.L., Arnold, M.M., Broering, T.J., Hastings, C.E., and Nibert, M.L. (2010). Localization of mammalian orthoreovirus proteins to cytoplasmic factory-like structures via nonoverlapping regions of microNS. *J Virol*, 84, 867–882.

Minuk, G.Y., Paul, R.W., and Lee, P.W. (1985). The prevalence of antibodies to reovirus type 3 in adults with idiopathic cholestatic liver disease. *J Med Virol*, 16, 55-60.

Montero, H., Arias, C.F., and López, S. (2006). Rotavirus nonstructural protein NSP3 is not required for viral protein synthesis. *J Virol*, 80, 9031–9038.

Montero, H., Rojas, M., Arias, C.F., and López, S. (2008). Rotavirus infection induces the phosphorylation of eIF2α but prevents the formation of stress granules. *J Virol*, 82, 1496–1504.

Morgan, E.M. and Zweerink, H.J. (1975). Characterization of transcriptase and replicase particles isolated from reovirus-infected cells. *Virology*, 68, 455–466.

Murray, K.E. and Nibert, M.L. (2007). Guanidine hydrochloride inhibits mammalian orthoreovirus growth by reversibly blocking the synthesis of double-stranded RNA. *J Virol*, 81, 4572–4584.

Nibert, M.L., Odegard, A.L., Agosto, M.A., Chandran, K., and Schiff, L.A. (2005). Putative autocleavage of reovirus mu1 protein in concert with outer-capsid disassembly and activation for membrane permeabilization. *J Mol Biol*, 345, 461–474.

Odegard, A.L., Chandran, K., Zhang, X., Parker, J.S., Baker, T.S., and Nibert, M.L. (2004). Putative autocleavage of outer capsid protein micro1, allowing release of myristoylated peptide micro1N during particle uncoating, is critical for cell entry by reovirus. *J Virol*, 78, 8732–8745.

Parashar, U.D., Gibson, C.J., Bresee, J.S., and Glass, R.I. (2006). Rotavirus and severe childhood diarrhea. *Emerg Infect Dis*, 12, 304–306.

Parker, J.S., Broering, T.J., Kim, J., Higgins, D.E., and Nibert, M.L. (2002). Reovirus core protein mu2 determines the filamentous morphology of viral inclusion bodies by interacting with and stabilizing microtubules. *J Virol*, 76, 4483–4496.

Patel, A. and Roy, P. (2014). The molecular biology of Bluetongue virus replication. *Virus Res*, 182, 5–20.

Perez-Vargas, J., Romero, P., López, S., and Arias, C.F. (2006). The peptide-binding and ATPase domains of recombinant hsc70 are required to interact with rotavirus and reduce its infectivity. *J Virol*, 80, 3322–3331.

Phillips, M.B., Stuart, J.D., Simon, E.J., and Boehme, K.W. (2018). Nonstructural protein σ1s is required for optimal reovirus protein expression. *J Virol*, 92. e02259-17.

Piron, M., Vende, P., Cohen, J., and Poncet, D. (1998). Rotavirus RNA-binding protein NSP3 interacts with eIF4GI and evicts the poly(A) binding protein from eIF4F. *EMBO J*, 17, 5811–5821.

Qin, Q., Hastings, C., and Miller, C.L. (2009). Mammalian orthoreovirus particles induce and are recruited into stress granules at early times postinfection. *J Virol*, 83, 11090–11101.

Qin, Q., Carroll, K., Hastings, C., and Miller, C.L. (2011). Mammalian orthoreovirus escape from host translational shutoff correlates with stress granule disruption and is independent of eIF2α phosphorylation and PKR. *J Virol*, 85, 8798–8810.

Ramig, R.F. (2004). Pathogenesis of intestinal and systemic rotavirus infection. *J Virol*, 78, 10213–10220.

Ratinier, M., Shaw, A.E., Barry, G., Gu, Q., Di Gialleonardo, L., Janowicz, A., Varela, M., Randall, R.E., Caporale, M., and Palmarini, M. (2016). Bluetongue virus NS4 protein is an interferon antagonist and a determinant of virus virulence. *J Virol*, 90, 5427–5439.

Reinisch, K.M., Nibert, M.L., and Harrison, S.C. (2000). Structure of the reovirus core at 3.6 A resolution. *Nature*, 404, 960967.

Reiter, D.M., Frierson, J.M., Halvorson, E.E., Kobayashi, T., Dermody, T.S., and Stehle, T. (2011). Crystal structure of reovirus attachment protein σ1 in complex with sialylated oligosaccharides. *PLoS Pathog*, 7, e1002166.

van Rijn, P. A. (2019). Prospects of next-generation vaccines for bluetongue. *Front Vet Sci*, 6, 407.

Rivera-Serrano, E.E., Fritch, E.J., Scholl, E.H., and Sherry, B. (2017). A cytoplasmic RNA virus alters the function of the cell splicing protein SRSF2. *J Virol*, 91. e02488-16.

Rojas, M., Arias, C.F., and López, S. (2010). Protein kinase R is responsible for the phosphorylation of eIF2α in rotavirus infection. *J Virol*, 84, 10457–10466.

Roner, M.R. and Steele, B.G. (2007). Localizing the reovirus packaging signals using an engineered m1 and s2 ssRNA. *Virology*, 358, 89–97.

Roner, M.R., Bassett, K., and Roehr, J. (2004). Identification of the 5' sequences required for incorporation of an engineered ssRNA into the Reovirus genome. *Virology*, 329, 348–360.

Roy, P. (2013). Orbiviruses. In *Fields Virology*, Knipe, D.M. and Howley, P.M. (eds), 6th edition. Lippincott Williams & Wilkins, Philadelphia.

Roy, P. (2017). Bluetongue virus structure and assembly. *Curr Opin Virol*, 24, 115–123.

Rubio, R.M., Mora, S.I., Romero, P., Arias, C.F., and López, S. (2013). Rotavirus prevents the expression of host responses by blocking the nucleocytoplasmic transport of polyadenylated mRNAs. *J Virol*, 87, 6336–6345.

Salgado, E.N., Garcia Rodriguez, B., Narayanaswamy, N., Krishnan, Y., and Harrison, S.C. (2018). Visualization of calcium ion loss from rotavirus during cell entry. *J Virol*, 92, e01327-18.

Sánchez-Tacuba, L., Rojas, M., Arias, C.F., and López, S. (2015). Rotavirus controls activation of the 2'-5'-oligoadenylate synthetase/RNase L pathway using at least two distinct mechanisms. *J Virol*, 89, 12145–12153.

Schulz, C., Breard, E., Sailleau, C., Jenckel, M., Viarouge, C., Vitour, D., Palmarini, M., Gallois, M., Hoper, D., Hoffmann, B., Beer, M. and Zientara, S. (2016). Bluetongue virus serotype 27: Detection and characterization of two novel variants in Corsica, France. *J Gen Virol*, 97, 2073–2083.

Sharpe, A.H. and Fields, B.N. (1982). Reovirus inhibition of cellular RNA and protein synthesis: Role of the S4 gene. *Virology*, 122, 381–391.

Silva-Ayala, D., López, T., Gutierrez, M., Perrimon, N., López, S., and Arias, C.F. (2013). Genome-wide RNAi screen reveals a role for the ESCRT complex in rotavirus cell entry. *Proc Natl Acad Sci USA*, 110, 10270–10275.

Smith, J.A., Schmechel, S.C., Williams, B.R., Silverman, R.H., and Schiff, L.A. (2005). Involvement of the interferon-regulated antiviral proteins PKR and RNase L in reovirus-induced shutoff of cellular translation. *J Virol*, 79, 2240–2250.

Smith, J.A., Schmechel, S.C., Raghavan, A., Abelson, M., Reilly, C., Katze, M.G., Kaufman, R.J., Bohjanen, P.R., and Schiff, L.A. (2006). Reovirus induces and benefits from an integrated cellular stress response. *J Virol*, 80, 2019–2033.

Steger, C.L., Boudreaux, C.E., Laconte, L.E., Pease, J.B., and McDonald, S.M. (2019). Group A rotavirus VP1 polymerase and VP2 core shell proteins: Intergenotypic sequence variation and *in vitro* functional compatibility. *J Virol*, 93. e01642-18.

Strong, J.E., Coffey, M.C., Tang, D., Sabinin, P., and Lee, P.W. (1998). The molecular basis of viral oncolysis: Usurpation of the Ras signaling pathway by reovirus. *EMBO J*, 17, 3351–3362.

Sturzenbecker, L.J., Nibert, M., Furlong, D., and Fields, B.N. (1987). Intracellular digestion of reovirus particles requires a low pH and is an essential step in the viral infectious cycle. *J Virol*, 61, 2351–2361.

Sung, P.Y. and Roy, P. (2014). Sequential packaging of RNA genomic segments during the assembly of bluetongue virus. *Nucleic Acids Res*, 42, 13824–13838.

Sung, P.Y., Vaughan, R., Rahman, S.K., Yi, G., Kerviel, A., Kao, C.C., and Roy, P. (2019). The interaction of bluetongue virus VP6 and genomic RNA is essential for genome packaging. *J Virol*, 93. e02023-18.

Tai, J.H., Williams, J.V., Edwards, K.M., Wright, P.F., Crowe Jr., J.E., and Dermody, T.S. (2005). Prevalence of reovirus-specific antibodies in young children in Nashville, Tennessee. *J Infect Dis*, 191, 1221-4.

Tao, Y., Farsetta, D.L., Nibert, M.L., and Harrison, S.C. (2002). RNA synthesis in a cage – Structural studies of reovirus polymerase lambda3. *Cell*, 111, 733–745.

Tate, J.E., Burton, A.H., Boschi-Pinto, C., Parashar, U.D., and World Health Organization – Coordinated Global Rotavirus Surveillance Network (2016). Global, regional, and national estimates of rotavirus mortality in children <5 years of age, 2000–2013. *Clin Infect Dis*, 62(Suppl 2), S96–S105.

Torres-Flores, J.M., Silva-Ayala, D., Espinoza, M.A., López, S., and Arias, C.F. (2015). The tight junction protein JAM-A functions as coreceptor for rotavirus entry into MA104 cells. *Virology*, 475, 172–178.

Trask, S.D., McDonald, S.M., and Patton, J.T. (2012). Structural insights into the coupling of virion assembly and rotavirus replication. *Nat Rev Microbiol*, 10, 165–77.

Trejo-Cerro, O., Eichwald, C., Schraner, E.M., Silva-Ayala, D., López, S., and Arias, C.F. (2018). Actin-dependent nonlytic rotavirus exit and infectious virus morphogenetic pathway in nonpolarized cells. *J Virol*, 92. e02076-17.

Verwoerd, D.W. (2012). History of orbivirus research in South Africa. *J S Afr Vet Assoc*, 83, 532.

Vitour, D., Doceul, V., Ruscanu, S., Chauveau, E., Schwartz-Cornil, I., and Zientara, S. (2014). Induction and control of the type I interferon pathway by bluetongue virus. *Virus Res*, 182, 59–70.

Whelan, S. (2013). Viral replication strategies. In *Fields Virology*, Knipe, D.M. and Howley, P.M. (eds), 6th edition, Lippincott Williams & Wilkins, Philadelphia.

Wolf, M., Deal, E.M., and Greenberg, H.B. (2012). Rhesus rotavirus trafficking during entry into MA104 cells is restricted to the early endosome compartment. *J Virol*, 86, 4009–4013.

Yue, Z. and Shatkin, A.J. (1997). Double-stranded RNA-dependent protein kinase (PKR) is regulated by reovirus structural proteins. *Virology*, 234, 364–371.

Zamora, P.F., Hu, L., Knowlton, J.J., Lahr, R.M., Moreno, R.A., Berman, A.J., Prasad, B.V.V., and Dermody, T.S. (2018). Reovirus nonstructural protein σNS acts as an RNA stability factor promoting viral genome replication. *J Virol*, 92. e00563-18.

Zhang, P. and Samuel, C.E. (2007). Protein kinase PKR plays a stimulus- and virus-dependent role in apoptotic death and virus multiplication in human cells. *J Virol*, 81, 8192–8200.

Zhang, X., Walker, S.B., Chipman, P.R., Nibert, M.L., and Baker, T.S. (2003). Reovirus polymerase lambda 3 localized by cryo-electron microscopy of virions at a resolution of 7.6 A. *Nat Struct Biol*, 10, 1011–1018.

Zhang, L., Chandran, K., Nibert, M.L., and Harrison, S.C. (2006). Reovirus mu1 structural rearrangements that mediate membrane penetration. *J Virol*, 80, 12367–12376.

Zhang, X., Boyce, M., Bhattacharya, B., Zhang, X., Schein, S., Roy, P., and Zhou, Z.H. (2010). Bluetongue virus coat protein VP2 contains sialic acid-binding domains, and VP5 resembles enveloped virus fusion proteins. *Proc Natl Acad Sci USA*, 107, 6292–6297.

Zhang, Z.Q., Kim, T., Bao, M.S., Facchinetti, V., Jung, S.Y., Ghaffari, A.A., Qin, J., Cheng, G.H., and Liu, Y.J. (2011). DDX1, DDX21, and DHX36 helicases form a complex with the adaptor molecule TRIF to sense dsRNA in dendritic cells. *Immunity*, 34, 866–878.

Zhang, R., Jha, B.K., Ogden, K.M., Dong, B., Zhao, L., Elliott, R., Patton, J.T., Silverman, R.H., and Weiss, S.R. (2013). Homologous 2',5'-phosphodiesterases from disparate RNA viruses antagonize antiviral innate immunity. *Proc Natl Acad Sci USA*, 110, 13114–13119.

Zhang, X., Patel, A., Celma, C.C., Yu, X., Roy, P., and Zhou, Z.H. (2016). Atomic model of a nonenveloped virus reveals pH sensors for a coordinated process of cell entry. *Nat Struct Mol Biol*, 23, 74–80.

Zhao, X., Chester, C., Rajasekaran, N., He, Z., and Kohrt, H.E. (2016). Strategic combinations: The future of oncolytic virotherapy with reovirus. *Mol Cancer Ther*, 15, 767–773.

Zurney, J., Kobayashi, T., Holm, G.H., Dermody, T.S., and Sherry, B. (2009). Reovirus mu2 protein inhibits interferon signaling through a novel mechanism involving nuclear accumulation of interferon regulatory factor 9. *J Virol*, 83, 2178–2187.

3

Negative-strand RNA Viruses

Rachel FEARNS
Boston University School of Medicine, USA

3.1. Introduction

The negative-strand RNA viruses are a large and diverse group of viruses. They are united by the fact that their RNA genomes must be transcribed into mRNA for viral protein expression to occur. This field has undergone two major advances in recent years. First, the breadth of viruses in this group has dramatically expanded, with the advent of low-cost, high-throughput sequencing enabling virus discovery at an unprecedented rate In step with this expansion, the taxonomic classifications to categorize negative-strand RNA viruses have undergone rapid changes and will likely continue to do so as more viruses are sequenced, and their molecular biology is better characterized[1]. The second major advance lies in recent breakthroughs in determining atomic-resolution structures of the polymerases of some of these viruses. This has transformed the field and provided dazzling new insight into the molecular mechanisms that these viruses employ to transcribe and replicate their genomes. It is this machinery and the tasks it undertakes in the replication cycles of different negative-strand RNA viruses that will take center stage in this chapter. In the first part of this chapter, well-characterized groups of negative-strand RNA viruses will be introduced, and their replication cycles will be briefly described. The second part of this chapter will provide a more in-depth description of the molecular

---

For a color version of all figures in this book, see www.iste.co.uk/saleh/virology.zip.

1. See https://talk.ictvonline.org/ for up to date taxonomic information.

Virology,
coordinated by María-Carla SALEH and Félix AUGUSTO REY. © ISTE Ltd 2021

mechanisms underlying viral transcription and genome replication with particular focus on the viral polymerases that undertake these activities.

3.2. Replication cycles of negative-strand RNA viruses

There are two major groups, or sub-phyla, of negative-strand RNA viruses: the non-segmented and the segmented negative-strand RNA viruses. In the non-segmented negative-strand RNA viruses, we will focus on the order *Mononegavirales,* which includes a large number of human pathogens, such as respiratory syncytial virus (RSV), human metapneumovirus (HMPV), measles, mumps, rabies and ebolaviruses. In the segmented negative-strand RNA viruses, we will focus on two groups: the order *Bunyavirales*, which encompasses a large number of families and includes deadly human pathogens hantavirus, Lassa fever virus and Crimean Congo hemorrhagic virus, and the *Orthomyxoviridae* family, in the order *Articulavirales*, which includes the influenza viruses. Despite significant differences between these viruses, they all encode their own RNA-dependent RNA polymerase (RdRp) and have an RNA genome that is encapsidated along its length with nucleoprotein (N or NP; for simplicity, N will be used throughout this chapter). In addition to these viruses, hepatitis delta virus (HDV), in the genus *Deltavirus*, will also be described. HDV uses an entirely different mechanism to express and replicate its genome, and is presented as a contrast to the other well-known negative-strand RNA viruses.

3.2.1. *The order* Mononegavirales

The order *Mononegavirales* encompasses multiple virus families, including the *Rhabdoviridae*, *Paramyxoviridae*, *Pneumoviridae* and *Filoviridae*. The mononegavirus genomes are a single strand of negative-sense RNA. Throughout the viral replication cycle, the genome RNA is intimately associated along its length with a linear, homopolymeric assembly of N protein to form a long filament (Finch and Gibbs 1970; Longhi 2009; Ruigrok *et al.* 2011). The N protein protects the RNA from nucleases and enables it to be recognized as a template by the viral polymerase. The polymerase is a complex of at least two proteins, the large polymerase subunit L and phosphoprotein, P (or VP35 in the case of the filoviruses). The N-RNA and polymerase associate to form a ribonucleoprotein complex (RNP). In the virus particle, the RNP is surrounded by an oligomeric matrix protein and an envelope containing one or more glycoprotein spikes. The viral replication cycle begins when the glycoproteins enable attachment to a cellular receptor and fusion of the viral envelope with cellular membranes, resulting in delivery of the RNP into the cytoplasm of a host cell. For most mononegaviruses, the viral genome is transcribed and replicated in the cytoplasm by the viral polymerase, independently of cellular RNA synthesis processes (Whelan *et al.* 2004; Fearns and Plemper 2017; Hume and Muhlberger 2019). However, there are some viruses in this order, such as Borna disease virus, which can enter the nucleus and take advantage of nuclear RNA modifying enzymes (Honda and Tomonaga 2013). In addition to the N, P/VP35, L, matrix and glycoprotein(s), many viruses in the order express additional, virus-specific proteins with functions as innate immune antagonists, mediators or regulators of transcription and/or replication and/or additional packaging proteins.

The viral genes lie sequentially on the viral genome (Figure 3.1). Each gene is flanked with short *cis*-acting elements approximately 10–13 nucleotides in length; at the beginning of each gene is a gene start signal, and at the end is a gene end signal. Typically, the genes are separated by a short non-coding intergenic region, but some viruses have one or more overlapping gene junctions, in which the gene start signal for the downstream gene overlaps with, or lies upstream of, the gene end signal for the upstream gene (Fearns and Collins 1999; Whelan *et al.* 2004; Brauburger *et al.* 2014; Noton and Fearns 2015). At the 3' and 5' ends of the genome are extragenic regions, known as the leader and trailer regions, respectively. The leader region is approximately 50 nucleotides long and contains signals to direct transcription and genome replication initiation, while the complement of the trailer, which lies at the 3' end of the genome, contains signals for replication initiation (Whelan *et al.* 2004; Noton and Fearns 2015; Hume and Muhlberger 2019). Transcription and replication both begin at the 3' end of the genome. The polymerase associates with the N-RNA template by virtue of interactions between P/VP35 and N and makes contact with specific RNA bases in the leader region (Calain and Roux 1995; Li and Pattnaik 1999; Vulliemoz and Roux 2001; Fearns *et al.* 2002; Leung *et al.* 2015; Longhi *et al.* 2017). The polymerase then moves along the template, transiently displacing N protein as the RNA feeds into the polymerase (Severin *et al.* 2016). The exact mechanism of transcription initiation differs between viruses, but in all cases, the polymerase gains access to the first gene start signal. Having initiated RNA synthesis at the gene start, the polymerase co-transcriptionally adds a cap to the 5' end of the RNA (Tekes *et al.* 2011). This occurs via a distinctive enzymatic activity known as GDP polyribonucleotidyltransferase and involves an interaction between the complement of the gene start signal at the 5' end of the mRNA and the capping active site on the polymerase (Ogino and Banerjee 2007; Wang *et al.* 2007; Ogino and Green 2019). The cap is then methylated, either by the viral polymerase, in the case of cytoplasmic viruses, or by cellular methyltransferases, in the case of nuclear viruses (Ogino and Green 2019). Capping allows the polymerase to become processive and to elongate the RNA to the end of the gene (Stillman and Whitt 1999; Braun *et al.* 2017). The gene end signal contains a polyuridine tract and prompts the polymerase to stutter, creating a polyadenylate tail on the mRNA

(Barr *et al.* 1997; Barr and Wertz 2001). The polymerase releases the RNA and scans back and forth along the template to locate the next gene start signal, where it reinitiates RNA synthesis (Fearns and Collins 1999). By continuing in this fashion, the polymerase is able to generate the individual viral mRNAs, which are then translated by cellular machinery. Because the genome contains a single promoter at its 3' end, expression of each of the genes is dependent on the polymerase initiating at the promoter and transcribing the preceding genes. Because the polymerase sometimes fails to reinitiate at the gene junctions, there is a gradient of transcription with genes at the 3' end of the genome being transcribed at a higher level than those at the 5' end (Whelan *et al.* 2004). This allows a passive mechanism of gene regulation, in which genes that need to be highly expressed, such as the *N* gene, are positioned at or near the 3' end of the genome, whereas genes that only require low-level expression, such as the *L* gene for the catalytic polymerase, are at the 5' end.

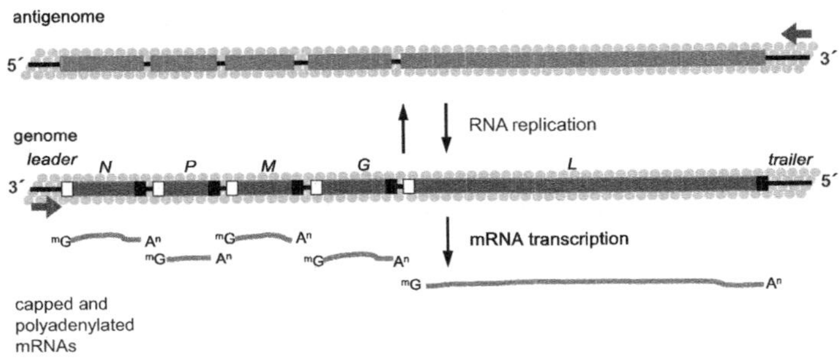

Figure 3.1. *Schematic diagram illustrating the genome organization of VSV, a member of the* Mononegavirales, *and the encoded transcription and replication products. Each gene is shown as a long rectangle, and is flanked with gene start and gene end signals, shown as white and black rectangles, respectively. The leader region at the 3' end of the genome contains promoter sequences for transcription and genome replication. Transcription yields subgenomic mRNAs corresponding to each of the genes. The mRNAs contain a methylated cap (^{m}G) and a poly A tail (A^{n}). Replication yields an antigenome RNA intermediate, which contains a promoter at its 3' end, to signal further rounds of genome synthesis. The antigenome and genome RNAs are encapsidated along their length with N (gray circles). Promoters are indicated with block arrows*

To replicate the genome, the polymerase initiates at the 3' end of the leader region. In this case, the polymerase does not cap the RNA, and it disregards all the gene junction signals, elongating the nascent RNA to the end of the genome to generate a positive-sense RNA, known as an antigenome. The antigenome becomes encapsidated with N as it is synthesized and it is thought that concurrent encapsidation increases polymerase processivity, enabling the polymerase to ignore the gene end signals (Gubbay *et al.* 2001; McGivern *et al.* 2005). The 3' end of the antigenome, the complement of the trailer region, contains a promoter that signals initiation of genome synthesis. Like the antigenome, the genome RNA is concurrently encapsidated, and never exists as completely naked RNA. Because N is required for the synthesis of antigenome and genome RNA, replication can only occur after sufficient N has accumulated in the infected cell. In some mononegaviruses, N concentration plays a role in regulating the polymerase between transcription and replication. However, this is not the case for other viruses, likely due to differences in mechanisms of transcription initiation (Noton and Fearns 2015). Newly synthesized genome N-RNA complexes bind the newly synthesized viral polymerase proteins, then associate with the matrix protein, which in turn associates with the cytoplasmic tails of the transmembrane glycoproteins. New particles are released by budding.

3.2.2. *The order* Bunyavirales

Like most mononegaviruses, bunyaviruses replicate in the cell cytoplasm, but in contrast to the mononegaviruses, bunyavirus genomes are segmented, such that different genes are on different strands of RNA. Infectious virus particles contain multiple segments of the genome, known as viral RNAs (vRNAs). Each vRNA is encapsidated with N and associated with a single copy of the viral polymerase to form an RNP. In contrast with the mononegaviruses, the bunyavirus polymerase can bind to both ends of each vRNA segment, such that the RNP forms a panhandle structure. The particles are enveloped with a lipid bilayer, studded with glycoprotein spikes (Elliott 2014; Ferron *et al.* 2017). Following binding to a receptor on the cell surface, the virus particles are internalized by endocytosis. Fusion of the virus envelope with the endosomal membrane releases the viral RNPs into the cytoplasm, where the polymerase transcribes and replicates the genome segments. The best studied bunyaviruses have two or three genome segments, although some have more. In some cases, one or more segments are ambisense, with functional open reading frames on both the vRNA and the replicative intermediate, complementary RNAs (cRNAs). In the case of bunyaviruses that have two genome segments, the segments are termed small (S) and large (L) and both are ambisense, with a total of four open reading frames. The S segment codes for the N protein and glycoproteins, and the L segment codes for the viral polymerase and Z protein, a protein involved in polymerase regulation and virus assembly (Ferron *et al.* 2017). In the case of bunyaviruses with three genome segments, they are termed small (S), medium (M) and large (L). The S segment codes for the N protein, and in some viruses also codes for a non-structural S (NSs) protein, which plays a role in subverting innate immune responses. The M segment codes for the viral glycoprotein, which is processed to form two polypeptides, Gn and Gc. Some bunyaviruses code for a non-structural M (NSm) protein, which also plays a role in virus assembly. The L segment codes for the viral polymerase, which is responsible for viral transcription and genome replication (Elliott 2014; Ferron *et al.* 2017).

The bunyavirus vRNA segments have non-translated regions at their ends, flanking the open reading frame(s) (Figure 3.2). The extreme terminal sequences (approximately 11 nt) are conserved between the different segments and are essential for RNA synthesis (Barr and Wertz 2004; Kohl *et al.* 2006; Ferron *et al.* 2017). The ends of the bunyavirus genome segments have a high degree of complementarity, with some conserved mismatches (Barr and Wertz 2004, 2005; Kohl *et al.* 2004; Ferron *et al.* 2017). To some extent, this reflects the conserved roles of 3' ends of the genomic and antigenomic RNAs. However, it is also due to a requirement for the 3' and 5' termini to cooperate together in RNA synthesis (Perez and de la Torre 2003; Barr and Wertz 2004; Hass *et al.* 2006; Pyle and Whelan 2019). There is a transcription termination signal after the end of the open reading frame, but before the end of the template. This termination signal is typically a short (e.g. pentanucleotide) sequence (Barr *et al.* 2006; Albarino *et al.* 2007; Ikegami *et al.* 2007; Lara *et al.* 2011; Blakqori *et al.* 2012). In the case of the ambisense viruses, the replicative intermediate cRNA contains sequences that enable transcription initiation in addition to replication. In these viruses, the two genes residing on the complementary template strands are separated with a structured intergenic sequence, which plays an important role in transcription termination (Auperin *et al.* 1984a, 1984b, 1986; Wilson and Clegg 1991; López and Franze-Fernández 2007).

Bunyavirus transcription results in the synthesis of a capped mRNA (Raju *et al.* 1990; Jin and Elliott 1993; Meyer and Southern 1993; Polyak *et al.* 1995; van Poelwijk *et al.* 1996; Cheng and Mir 2012). The bunyavirus polymerase does not have the enzymatic activities required to generate a cap, and instead "steals" it from host mRNAs (Figure 3.2). Studies with some bunyaviruses have shown that they recruit cap donor mRNAs from the cellular decapping machinery, which is normally involved in mRNA turnover (Mir *et al.* 2008; Hopkins *et al.* 2013). In these cases, viral N protein binds with high affinity to the mRNA caps, protecting them from the decapping enzymes (Mir *et al.* 2008). The polymerase has an endonuclease activity that can cleave the 5' end of the cellular mRNA. Following engagement of the transcription promoter in the double-stranded RNA of the template panhandle, it can begin RNA synthesis using the cleaved capped RNA as a primer (Barr and Wertz

2005; Kranzusch *et al.* 2010; Vogel *et al.* 2019) (Figure 3.2). The polymerase can then elongate the RNA until it reaches the termination signal, where the RNA is released. Because bunyavirus transcription occurs in the cytoplasm, ribosomes can be recruited to the nascent mRNA as it is being synthesized, and in some bunyaviruses, transcription and translation are coupled. In these cases, the translocating ribosomes play an important role in transcription by preventing the polymerase from terminating at sequences resembling termination sites in the open reading frame, allowing mRNA elongation until the authentic termination site (Bellocq and Kolakofsky 1987; Barr 2007; Lara *et al.* 2011). Unlike the mRNAs of other negative-strand RNA viruses, the bunyavirus termination signal does not contain a polyuridine tract and the mRNAs are not polyadenylated (Elliott 2014; Ferron *et al.* 2017). The termination signals of many bunyaviruses contain a sequence that forms a hairpin structure. It seems likely that this facilitates termination by forming a structure in the nascent mRNA, pulling it out of the polymerase active site and causing chain release (Barr 2007). Bunyavirus mRNAs are translated in a cap-dependent manner, but because their mRNAs lack poly A tails, their translation does not involve cellular poly A binding protein and instead employs a non-canonical cap-dependent mechanism (Blakqori *et al.* 2009; Foscaldi *et al.* 2017). This provides an opportunity for the virus to promote translation of its own mRNAs at the expense of that of host mRNAs. For example, in the case of bunyamwera virus, viral N binds and sequesters poly A binding protein, resulting in inhibition of translation of polyadenylated cellular mRNAs, but not viral mRNAs (Blakqori *et al.* 2009).

Bunyavirus replication occurs via cRNA intermediates, with N protein encapsidating the replicative RNA as it is synthesized, as in the case of the *Mononegavirales*. Replication initiation occurs by a distinct mechanism from transcription initiation and does not involve a primer derived from host RNAs. Instead, bunyavirus genome ends have repetitive sequences. This allows the polymerase to initiate replication by a prime-realign mechanism, in which it initiates internally, synthesizes a short primer, realigns to position the primer at the end or overhanging the end of the template mechanism and then continues to elongate the RNA (Garcin and Kolakofsky 1992; Garcin *et al.* 1995; Polyak *et al.* 1995). Concurrent encapsidation presumably prevents the nascent RNA from forming secondary structures that would precipitate termination, and allows elongation to the end of the template. Like the mononegaviruses, N protein accumulation is required for bunyavirus replication (Pinschewer *et al.* 2003). Newly synthesized RNPs are targeted to regions of Golgi membrane containing the inserted Gn and Gc proteins, and the particles bud into Golgi-derived vesicles. The vesicles are then transported to the cell surface where vesicle membrane fusion with the plasma membrane results in virus particle release (Elliott 2014; Ferron *et al.* 2017; Hallam *et al.* 2018).

Because bunyavirus genes are separated onto different segments, they can be differentially expressed. Segment-specific sequences combined with variations in segment length allow each RNA segment to accumulate to different levels (Barr *et al.* 2003; Gauliard *et al.* 2006). In addition, arrangement of genes on an ambisense genome affords those viruses a mechanism for temporal regulation of gene expression. For example, in the two-segment arenaviruses, N and L mRNAs are transcribed from vRNA strands, whereas GP and Z mRNAs are transcribed from the replicative intermediate cRNAs, meaning that N and L proteins are expressed earlier in infection than GP and Z protein (Auperin *et al.* 1984b; Fuller-Pace and Southern 1988). The arenavirus Z protein also plays a key role in regulating RNA synthesis, inhibiting RNA synthesis by locking the polymerase into an inactive state on its promoter (Cornu and de la Torre 2001; Kranzusch and Whelan 2011). This probably allows the polymerase–template complexes to be silenced in preparation for packaging.

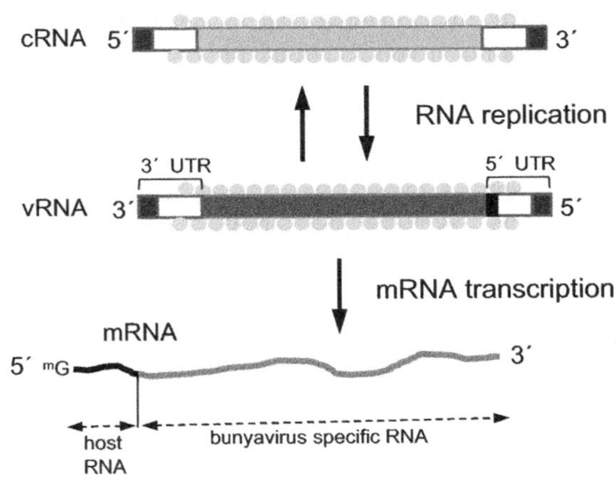

Figure 3.2. *Schematic diagram illustrating the organization of a single bunyavirus vRNA segment and its transcription and replication products. The gene on the vRNA is shown as a long rectangle and the 3' and 5' untranslated regions (UTR) are indicated. The terminal sequences, shown as green rectangles, are complementary and both the 3' and 5' ends bind to the polymerase and confer promoter activity. Transcription of mRNA depends on a primer derived from a capped host cell mRNA that has been cleaved by the viral endonuclease. The mRNA is released at a termination signal (black rectangle). Replication occurs via a cRNA intermediate. The vRNA and cRNA are encapsidated with N protein (gray circles)*

3.2.3. *The order* Articulavirales

The order *Articulavirales* includes two families, the *Amnoonviridae* and the *Orthomyxoviridae*. Of these, the *Orthomyxoviridae*, which includes influenza viruses A, B and C, are the best characterized. The orthomyxovirus genome is segmented, similarly to that of the bunyaviruses, with the influenza A virus genome comprising eight segments. Also like the bunyaviruses, the 3' and 5' ends of the genome are associated with the viral polymerase to form a panhandle structure, while the remainder is encapsidated in N protein (Zheng and Tao 2013). The influenza virus replication cycle begins with attachment of the enveloped viral particle to receptors on the cell surface by virtue of the viral hemagglutinin (HA) protein. The virus enters the cell by receptor-mediated endocytosis, whereby the reduction in pH triggers a conformational change in HA that results in fusion pore formation, and entry of protons through the M2 ion channel in the virus envelope elicits dissociation of the matrix protein from the RNPs. The RNPs are trafficked to the nucleus, and it is here that the viral genome is transcribed and replicated.

The influenza virus promoters are composed of the 3' 12 nucleotides and 5' 13 nucleotides of each vRNA (Fodor 2013; Te Velthuis and Fodor 2016). Like the bunyavirus polymerase, the influenza virus polymerase binds both the 3' and 5' termini during transcription initiation and uses a cap-snatching mechanism to generate a capped mRNA (Fodor *et al.* 1994). Although the bunyaviruses take advantage of cytoplasmic decapping sites to generate capped primers, the influenza virus polymerase takes advantage of transcribing molecules of cellular RNA polymerase II (RNAP II). The influenza virus polymerase binds to the carboxyl terminal domain of RNAP II during the initial stages of cellular transcription before the RNA becomes associated with cellular proteins involved in pre-mRNA processing and transport (Engelhardt *et al.* 2005). It cleaves the 5' end of the capped cellular RNA with a polymerase-associated endonuclease activity and then uses the RNA as a primer to initiate RNA synthesis (Dias *et al.* 2009). The 5' end of the vRNA remains bound to the polymerase during transcription. It is thought that continued association of the polymerase to the 5' terminus of the vRNA during mRNA transcription sterically hinders polymerase translocation as it nears the 5' end of the template, prompting it to reiteratively stutter on a polyuridine tract approximately 17–22 nt before the end of the template, causing polyadenylation and then mRNA release (Poon *et al.* 1998, 1999; Reich *et al.* 2014). Because influenza virus replicates in the nucleus, it is able to expand its genome coding capacity by taking advantage of cellular splicing machinery.

Influenza virus genomes are replicated through synthesis of a cRNA intermediate. Similar to bunyavirus genome replication, this process differs from transcription in that it does not require the provision of an exogenous primer for initiation, and the polymerase terminates RNA synthesis at the end of the template, rather than at the U-tract. Synthesis of full-length replication products requires soluble N to encapsidate the newly synthesized RNA. Replication also depends on the promoter sequences at the 3' and 5' ends of the vRNA and cRNA, but intriguingly, the mechanism of replication initiation differs for the two templates (Kim *et al.* 1997; Crow *et al.* 2004; Deng *et al.* 2006). The differences in the initiation mechanisms for mRNA, cRNA and vRNA synthesis probably play a key role in allowing temporal regulation of these different synthetic events over the course of an infection. The influenza virus polymerase depends on cellular RNAP II for transcription of its mRNAs, but induces RNAP II degradation (Rodriguez *et al.* 2007). Limitations in RNAP II availability as the infection proceeds could allow the viral polymerase to adopt a different conformation appropriate for replication. In addition, while a monomeric form of the polymerase is capable of performing mRNA and cRNA synthesis, a dimeric form is required for it to engage in vRNA synthesis (Fan *et al.* 2019). This limits vRNA synthesis to occurring at later times in an infection, when the polymerase is more abundant. Newly synthesized vRNPs are then exported from the nucleus and associate with other viral proteins to generate progeny virus particles (Zheng and Tao 2013).

3.2.4. *The genus* Deltavirus

The *Deltavirus* genus is very different from the viruses described above. Currently, it does not belong in a family or order, and HDV is its sole member (Magnius *et al.* 2018). HDV is a defective virus that relies on hepatitis B virus (HBV) to provide the envelope proteins that allow for the packaging of the HDV virus particle, and subsequent entry into new cells. However, HBV plays no role in synthesizing HDV RNAs. Instead, the components required to transcribe and replicate the HDV genome are provided by HDV and the host cell.

Transcription and replication of the HDV genome occur in the nucleus of a host cell. The HDV genome is a circular RNA ~1,700 nt in length. It is high in G/C content and is capable of extensive intramolecular base-pairing, allowing it to form a rod-like structure (Kos *et al.* 1986; Wang *et al.* 1986). Transcription yields an 800 nt capped and polyadenylated mRNA (Gudima *et al.* 2000). Remarkably, this mRNA is synthesized by cellular RNAP II, using the circularized HDV genome RNA as a template (Filipovska and Konarska 2000; Modahl *et al.* 2000; Moraleda and Taylor 2001). Initiation of mRNA synthesis is governed by a promoter that lies in the stem-loop at the end of the genome rod-like structure, which recruits a preinitiation complex consisting of RNAP II and cellular transcription factors

(Macnaughton *et al.* 1993; Beard *et al.* 1996; Modahl and Lai 1998; Filipovska and Konarska 2000; Gudima *et al.* 2000; Abrahem and Pelchat 2008; Haussecker *et al.* 2008). The nascent RNA becomes capped and methylated by cellular enzymes that associate with RNAP II during transcription. Following elongation, the mRNA is terminated and polyadenylated in response to an AAUAAA and downstream CA consensus sequence that mimics the polyadenylation signals in cellular genes (Hsieh and Taylor 1991).

The mechanisms underlying HDV genome replication are less well understood. Evidence indicates that replication occurs by a rolling circle mechanism in which the polymerase initiates RNA synthesis and then continuously synthesizes multiple copies of the RNA (Macnaughton *et al.* 2002; Nie *et al.* 2004). Both the genome and antigenome RNAs contain self-cleaving ribozyme sequences, which cleave the products into monomer lengths (Kuo *et al.* 1988; Sharmeen *et al.* 1988). These RNA products are then circularized by ligation of the termini, possibly by a host ligase (Reid and Lazinski 2000). Because the replicative intermediate antigenome and genome RNA products are circularized, it is not possible to know with certainty where they are initiated from. There is also debate in the field regarding the identity of the polymerase responsible for their synthesis (Taylor 2015). One model is that RNAP II is responsible for all HDV RNA synthesis. According to this model, the decision between mRNA versus antigenome synthesis depends on whether the RNA is processed by polyadenylation and termination (to yield mRNA) or by ribozyme cleavage (to yield antigenome) (Brown *et al.* 2008). A second model is that while the mRNA and genome RNA are synthesized by RNAP II, the antigenome is synthesized by a different polymerase (RNAP I or III) (Modahl *et al.* 2000; Macnaughton *et al.* 2002). According to this model, the mRNA would be polyadenylated and terminated because it is being transcribed by RNAP II, whereas the antigenome, being transcribed by another polymerase, would not be, and so by default would be cleaved by the ribozyme instead.

The mRNA produced from the HDV genome codes for the delta antigen protein. However, two forms of protein are produced, a small and large delta antigen protein (HDAgS and HDAgL, respectively), with HDAgL containing an additional 19 or 20 amino acids at the carboxyl terminus. The HDAg-S and HDAg-L proteins are functionally distinct. HDAg-S is required for RNA replication (Kuo *et al.* 1989). It interacts with RNAP II and stimulates HDV elongation of genome RNA from the antigenomic template (Yamaguchi *et al.* 2001). In contrast, HDAg-L is inhibitory to genome RNA synthesis and instead plays a key role in viral packaging (Chao *et al* 1990; Chang *et al.* 1991; Glenn and White 1991). The ability of the virus to generate two forms of protein arises from post-transcriptional editing of the antigenome by a cellular enzyme, double-stranded RNA-specific adenosine deaminase 1, or ADAR 1

(Luo *et al.* 1990; Casey and Gerin 1995; Polson *et al.* 1996). This manifests as an A-to-G transition in the UAG stop codon of the HDAg mRNA, substituting the stop codon in the mRNA with tryptophan and extending the HDAg open reading frame. Both unedited and edited genome RNAs can become packaged into virus particles, but only particles containing unedited genomes (and able to express HDAg-S) are infectious (Sheu 2002). At early times, HDAg-S is produced, stimulating mRNA transcription and genome replication, resulting in an accumulation of HDAg-S, and antigenome and genome RNAs. ADAR-mediated editing of the antigenome results in edited genome RNA, and the proportion of edited genomes increases over time, as existing edited genomes are replicated and unedited antigenomes become edited. This results in an accumulation of HDAg-L, causing a suppression of genome replication and an initiation of genome packaging (Casey 2012).

3.2.5. *Summary of viral replication cycles*

It can be seen from the above descriptions that negative-strand RNA viruses have evolved a variety of mechanisms to transcribe and replicate their genomes. They range from one extreme, in which the viral polymerase is responsible for all enzymatic activities involved in transcription and replication, including RNA synthesis, capping, cap methylation and polyadenylation, as seen in the mononegaviruses, to the other extreme involving subversion of cellular RNA polymerases and other enzymes to perform all enzymatic activities involved in mRNA and genome RNA synthesis, as in the case of *Deltavirus*. The bunyaviruses and orthomyxoviruses lie in between on this spectrum, capable of synthesizing and polyadenylating RNA, but relying on cellular mRNAs to provide the mRNA methylguanosine caps. Clearly, the *Articulavirales*, *Bunyavirales* and *Mononegavirales* share properties in common, including their ability to encode an RdRp, the encapsidated nature of their genomes and the role of N protein in enabling replication elongation. This commonality is reflected in the structures of their polymerases, as described in the following section.

3.3. The transcription and replication machinery of the negative-strand RNA viruses

3.3.1. *Overview of the different negative-strand RNA virus polymerases*

As noted above, the polymerase of the *Mononegavirales* consists of an L–P (or L-VP35) protein complex. In the case of the mononegaviruses that replicate in the cytoplasm, the L protein is approximately 250 kDa in size and contains three enzymatic domains. These include the RdRp domain, the capping domain responsible for GDP polyribonucleotidyltransferase activity and the methyltransferase domain (Fearns and Plemper 2017). The bunyavirus polymerase is a single large, multifunctional L protein, ranging in size from 200 kDa up to 450 kDa in molecular weight. In addition to the RdRp domain, it contains domains for cap binding and endonuclease activity, which are required for cleaving capped cellular mRNA to generate a primer (Reguera *et al.* 2016a). Whereas the bunyavirus polymerase is a single protein, the orthomyxoviruses have a heterotrimeric polymerase complex, with separate proteins for the different activities. The polymerase basic protein 1 (PB1) contains the RdRp domain. Polymerase basic protein 2 (PB2) contains the cap-binding domain, and polymerase acidic protein (PA) harbors the endonuclease domain, necessary for cleaving the capped RNA to generate a primer. The total molecular weight of the polymerase complex is approximately 250 kDa, similar to that of the other negative-strand RNA viruses, described above (Fodor 2013). There are three-dimensional polymerase structures available for representatives of each of these groups of viruses, allowing comparisons to be made.

3.3.2. *Orthomyxovirus polymerases and their transcription and replication mechanisms*

The greatest wealth of structural data exists for the orthomyxovirus polymerases, with structures available for influenza viruses A, B, C and D in association with different ligands, and consequently, with different conformations (Pflug *et al.* 2014; Reich *et al.* 2014; Hengrung *et al.* 2015; Thierry *et al.* 2016; Lukarska *et al.* 2017; Serna Martin *et al.* 2018; Fan *et al.* 2019; Kouba *et al.* 2019; Peng *et al.* 2019). A crystal structure of the PB1, PB2, PA heterotrimer of a bat influenza A virus revealed that the three individual proteins are intimately associated with one another, with the PB1 protein acting as a central scaffold forming extensive contacts with the PB2 and PA proteins (Pflug *et al.* 2014). Overall, the trimeric complex adopts a U-shaped structure, with the RdRp domain at the base, and with the cap binding and endonuclease domains forming protruding arms. The RdRp domain has a structure shared by many other polymerases, resembling a cupped right hand with fingers and thumb subdomains coming into proximity with each other. The RdRp domain also includes motifs conserved with other polymerases. Phosphodiester bond formation uses a two-metal mechanism of catalysis in which two aspartic acid residues in the palm subdomain are responsible for anchoring the divalent metal cations that enable catalysis (van Dijk *et al.* 2004), and these residues could be identified in the influenza virus RdRp. Other motifs involved in stabilizing the incoming NTP base-pair with the template strand and the 3' end of the transcript, the substrate for catalysis, could also be identified. The structure also revealed a narrow tunnel lined with positively charged amino acids, which likely helps to attract NTPs into the active site.

As described above, the influenza virus promoter sequences are located in both the 3' and 5' termini of the template. The first resolved structures of influenza virus polymerase showed it in complex with the vRNA 3' and 5' ends (Pflug *et al.* 2014; Reich *et al.* 2014). The 3' and 5' RNA termini formed base-pairing interactions between nucleotides 10–13 from the 3' end (UCUC) and nucleotides 11–14 (AGAG) from the 5' end. Nucleotides 1–9 of the 3' end of the RNA were either unresolved (indicating that they were flexible), or were single-stranded and bound on the surface of the polymerase, rather than in the template entry channel. This exterior binding site involved multiple base contacts, indicating that it is functionally important, even though RNA bound to this site is not positioned to act as a template. This form of the polymerase is referred to as the pre-initiation form. There are several possible explanations for the polymerase adopting this conformation. For example, it might be important for maintaining the polymerase-RNA template in close association, but only in an inactive state, for packaging into viral particles. Alternatively, it might help the polymerase retain contact with the 3' end of newly synthesized vRNA as it emerges from the transcript channel during genome replication. This could allow the newly synthesized RNA to remain in close proximity to the polymerase for initiation of the next round of RNA synthesis. In contrast with the 3' terminal sequence, the 5' terminal 10 nucleotides of the RNA folded into a stem-loop or hook, which bound another pocket close to the 3' terminal sequence. In the absence of the promoter RNA, or in the presence of the 5' end of the cRNA promoter, the polymerase adopted an alternative structure, with significant differences in the structures of the endonuclease and cap-binding domains, and occlusion of the cap binding pocket, indicating that the vRNA promoter sequence functions to stabilize the polymerase in a form capable of transcription (Hengrung *et al.* 2015; Thierry *et al.* 2016).

To perform transcription, the polymerase needs to bind to capped pre-mRNA and cleave it to generate a primer. This primer then needs to enter the polymerase so that it is positioned with its 3' terminal nucleotide, the substrate for RdRp-mediated polymerization, in the active site, and the 3' end of the vRNA needs to be in the template channel. As noted above, the influenza virus polymerase steals caps from pre-mRNAs as they are being synthesized. RNAP II acquires different post-translational modifications at different stages of RNA transcription, and influenza virus polymerase binds specifically to RNAP II containing serine 5 phosphorylated residues in its C-terminal domain, allowing it to become associated with capped cellular RNAs before they become associated with other pre-mRNA binding proteins (Lukarska *et al.* 2017). Binding involves slightly different sites in influenza

A, B and C viruses, but in each case, the RNAP II binding site lies at the base of the U structure of the influenza virus polymerase, distant from the RdRp active site and promoter binding site (Lukarska *et al.* 2017; Serna Martin *et al.* 2018). This binding interaction helps to stabilize the viral polymerase in a preinitiation conformation and promotes cap cleavage and transcription initiation (Serna Martin *et al.* 2018; Walker and Fodor 2019). The cap binding and endonuclease domains lie at the ends of the two "arms" of the U-structure. As noted above, binding of the vRNA promoter enables the polymerase to adopt a conformation in which it is primed for cap-binding and endonuclease activity (Hengrung *et al.* 2015; Thierry *et al.* 2016). Comparison of different influenza virus polymerase structures revealed that the cap-binding domain in PB2 can adopt different conformations (Pflug *et al.* 2014; Reich *et al.* 2014). In one conformation, the capped pre-mRNA is positioned appropriately to allow endonuclease cleavage 10–15 nt from the 5' end. The cap-binding domain has the capacity to then rotate, shielding the capped primer from further cleavage and positioning the 3' end of the RNA into the active site of the RdRp (Reich *et al.* 2014). In the active site, a priming loop, an element required specifically for replication, is positioned to provide space for the transcription initiation complex. Initiation of mRNA elongation results in full displacement of the priming loop to accommodate the RNA duplex that is formed and allows template exit (Kouba *et al.* 2019). The nascent and template strands of the duplex pass at either side of a helical lid domain, which forces strand separation, enabling processive elongation (Kouba *et al.* 2019).

During genome replication, the influenza virus polymerase initiates RNA synthesis using different mechanisms on the cRNA and vRNA templates (Figure 3.3). Initiation of cRNA synthesis begins opposite positions 1 and 2 of the vRNA template using a *de novo* (primer-independent) initiation mechanism. In contrast, initiation of vRNA synthesis involves a prime-realign mechanism, in which the polymerase initiates opposite positions 4 and 5 of the cRNA 3' end and generates a dinucleotide pppApG primer, which then becomes repositioned opposite the 3' UC residues at positions 1 and 2 and elongated (Deng *et al.* 2006). These two initiation mechanisms require the polymerase to have different interactions with the template. Initiation of cRNA synthesis at the 3' end of the vRNA template depends on the priming loop structure, which helps to stabilize the initiation complex. The priming loop is a beta-hairpin, which protrudes into the RdRp active site. It contains a proline priming residue, which can form base-stacking interactions with the initiating NTPs. This priming residue is not required for transcription initiation, or for vRNA synthesis from the cRNA promoter; it is also not required for elongation of a pApG primer. This indicates that the priming residue is specifically required for the formation of the first dinucleotide at the 3' end of the template, where the complex is thermodynamically less stable than at the internal vRNA initiation site on the cRNA promoter (Te Velthuis *et al.* 2016). The 3' end of the vRNA promoter can bind the polymerase in two different conformations, one of which is bound on the exterior of the polymerase, and an initiation state, in which the 3' end of the template has entered the template channel and abuts the priming loop (Reich *et al.* 2014). There is a dynamic relationship between the two states with the polymerase able to fluctuate back and forth between them (Robb *et al.* 2019). Formation of the pppApG, the rate-limiting step in initiation, stabilizes the polymerase in the initiation conformation (Robb *et al.* 2019). It can then continue to elongate the RNA, undergoing conformational changes to allow the priming loop to retract and allow RNA egress (Pflug *et al.* 2014).

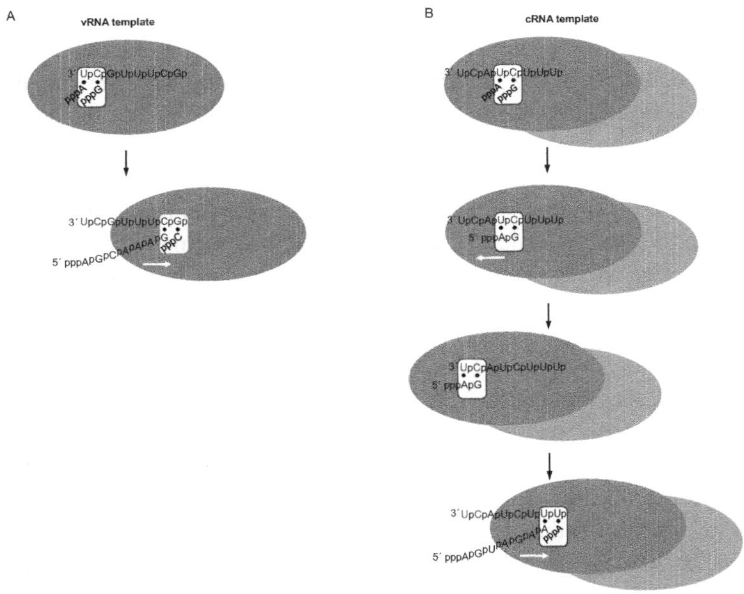

Figure 3.3. *Schematic diagram illustrating events that occur during influenza virus replication initiation on the vRNA and cRNA templates. (A) The monomeric form of polymerase initiates cRNA synthesis on the vRNA template. The polymerase recruits ATP and GTP, which form Watson–Crick base-pairs opposite positions 1U and 2C of the template, as indicated with black circles. Phosphodiester bond formation results in the synthesis of a pppApG dinucleotide, which is then elongated. (B) The dimeric form of the polymerase initiates vRNA synthesis on the cRNA template. In this case, the polymerase recruits ATP and GTP opposite positions 4U and 5C of the template and generates a pppApG dinucleotide. The pppApG dinucleotide, in conjunction with the RdRp, becomes repositioned opposite positions 1U and 2C of the template. It can then be elongated. The white arrows show direction of polymerase movement relative to the template. For simplicity, only the 3' end of the template is shown*

In contrast to the vRNA promoter, the cRNA does not have a dynamic relationship with the polymerase and directs internal initiation at a lower rate (Reich *et al.* 2017; Robb *et al.* 2019). A recent study showed that the trimeric influenza virus polymerase can adopt a dimeric structure through interaction of the PA and PB2 domains, resulting in rearrangements in the active site (Fan *et al.* 2019). Interestingly, dimerization was necessary for the initiation of vRNA synthesis from the cRNA promoter, but not for mRNA or cRNA synthesis (Fan *et al.* 2019). More detailed analysis indicated that dimerization was not necessary for the formation of the dinucleotide primer but instead for accurate realignment of the primer with the 3' end of the template. This indicates that dimerization stabilizes the initiation complex to enable the correct positioning of the 3' end of the template for the binding and elongation of the dinucleotide primer. Comparison of monomeric and dimeric forms of the polymerase showed that in the monomeric form, the 3' end of the template was resolved and in the binding site in which it is not in the template channel, whereas in the dimeric form, it was in the template entry channel, but not resolved, indicating that it was highly dynamic. Dimerization of the polymerase created rearrangements in the polymerase active site, with the priming loop retracted. These changes presumably provide the polymerase and RNA with the flexibility that is required for the back-tracking action that would occur during realignment following primer synthesis on the cRNA promoter (Figure 3.3).

3.3.3. *The bunyavirus polymerase*

Three different structures of the La Crosse bunyavirus polymerase have been resolved (Gerlach *et al.* 2015). These include the apo-enzyme, the enzyme in complex with nucleotides 1–16 of the 3' end of the vRNA, with base-paired RNA from nucleotides 9–16, and the latter complex with nucleotides 1–10 from the 5' terminus of the vRNA soaked in. The overall structure of the polymerase is similar in the absence or presence of the RNA ligands and resembles the influenza virus polymerase in complex with the 3' and 5' termini of its vRNA. The bunyavirus polymerase contains a globular core, containing the RdRp domain, from which protrudes a flexible endonuclease domain that allows cleavage of host capped mRNAs (Morin *et al.* 2010; Reguera *et al.* 2010, 2016b; Wallat *et al.* 2014; Fernández-Garcia *et al.* 2016; Rosenthal *et al.* 2017). The C-terminus of the protein was not resolved in these structures, but has been resolved in isolation. It is structurally similar to that of influenza virus PB2 protein, and is thought to be the cap-binding domain (Rosenthal *et al.* 2017; Gogrefe *et al.* 2019). The RdRp domain shares similar structural motifs as other RNA polymerases, including the finger, palm and thumb domains. Like the influenza virus polymerase, the thumb domain contains the putative priming loop. In the resolved structure, the 3' end of the template is not in the active site, but is tightly and specifically bound by a clamp close to, but not in, the template entry channel, somewhat similar to the pre-initiation form of the influenza virus polymerase. The 5' end of the RNA is also bound specifically to a pocket, but despite partial complementarity with the 3' end of the RNA, the 3' and 5' termini are not base-paired, and instead bind distinct sites (Gerlach *et al.* 2015). The 5' terminal RNA adopts a hook formation, with base-pairing of some of the residues in this 5' terminal sequence, as is the case with influenza virus polymerase. Binding of the 5' terminal RNA causes structural rearrangement of the fingers domain and the RdRp active site (Gerlach *et al.* 2015), likely activating polymerase activity (Pyle and Whelan 2019).

The La Crosse bunyavirus polymerase structure also revealed for the first time the channels that allow the RNA template to enter and exit the polymerization active site (Gerlach *et al.* 2015). The template threads into the central cavity of the RdRp and then makes a sharp U-turn to exit, with the templating base being positioned to face the catalytic residues. These are both on the same face of the polymerase, allowing N protein to be displaced from the template as the RNA feeds into the polymerization domain and to rapidly reassociate with the template RNA as it emerges. Because N molecules in the RNP form an interlinked chain independently of the RNA, this positioning of the template entry and exit channels could allow the N chain to be maintained and to rapidly fold back onto the template RNA as it emerges. The transcript product channel is on the other side of the polymerase, allowing the nascent RNA to emerge without the chance of it rehybridizing to the template strand. Ribosomes, required for transcription, and soluble N protein, required for encapsidation, could be positioned to associate with this RNA as it emerges.

3.3.4. *The mononegavirus polymerases and their transcription and replication mechanisms*

To date, there are polymerase structures resolved for the rhabdovirus, VSV, and for the pneumoviruses, RSV and HMPV (Liang *et al.* 2015; Gilman *et al.* 2019; Pan *et al.* 2020). Intriguingly, while each of these polymerases are in the apo-form, the rhabdovirus and pneumovirus polymerases were captured in different conformations, which might provide insight into the initiation versus elongation states of the polymerase. The VSV polymerase structure is of the L protein in association with a fragment of P (which is not visible) (Liang *et al.* 2015). The RdRp domain has a similar conformation to that of other RdRps, with thumb, palm and fingers domains, and the same conserved motifs. Template and NTP entry channels could also be observed. The capping and methyltransferase domains form two separate lobes, separated from each other by an additional connector domain, of unknown function. There is also a C-terminal domain proximal to the methyltransferase. The capping domain forms a lobe that folds over the RdRp domain, enclosing the RdRp active site. While a loop in the thumb domains of the influenza and bunyavirus RdRps provides the priming residue to enable the initiation of RNA synthesis from the 3' end of the template, in VSV L, the priming residue is located on a loop contiguous with the capping domain, which protrudes into the RdRp catalytic cavity (Ogino *et al.* 2019). The capping domain and the priming loop are positioned such that they would block the exit channels for the template and nascent transcript, indicating that following initiation of RNA synthesis opposite position 1 of the template, the polymerase undergoes conformational changes that allow the priming loop to retract out of the way.

The RSV and HMPV polymerase structures are consistent with this model (Gilman *et al.* 2019; Pan *et al.* 2020). In these two polymerase structures, the putative priming loop is fully retracted out of the RdRp active site and is more intimately associated with the capping domain. This creates more space for the RNA duplex in the central cavity of the RdRp, and modeling revealed exit channels that separate the template and transcript as they emerge (Pan *et al.* 2019). The essential residues of the capping domain are more tightly organized than in the VSV L structure, suggesting that this conformation might represent the active form of the capping domain, and while the methyltransferase, connector and C-terminal domains were present in the protein that was analyzed, they were not detectable in the structure, indicating that they are flexible relative to the RdRp and capping domains (Gilman *et al.* 2019; Pan *et al.* 2020). The RSV and HMPV polymerase structures are of the L protein in association with its cofactor, P. P forms a tetramer, with a central oligomerization domain (Castagne *et al.* 2004; Llorente *et al.* 2008; Leyrat *et al.* 2013; Pereira *et al.* 2017; Renner *et al.* 2017). This central domain showed extensive contacts with the RdRp domain of L, while the N- and C-terminal domains of P were largely unstructured. The C-terminal domains of P have been shown to bind to the RNP and might help displace N from the RNA as it feeds into the template entry channel, while the N-terminal arms of P bind to soluble N and a pneumovirus-specific transcription elongation factor, M2-1 (Asenjo *et al.* 2006; Renner *et al.* 2016, 2017; Esneau *et al.* 2019). Thus, these arms could allow recruitment of these proteins during replication and transcription, respectively.

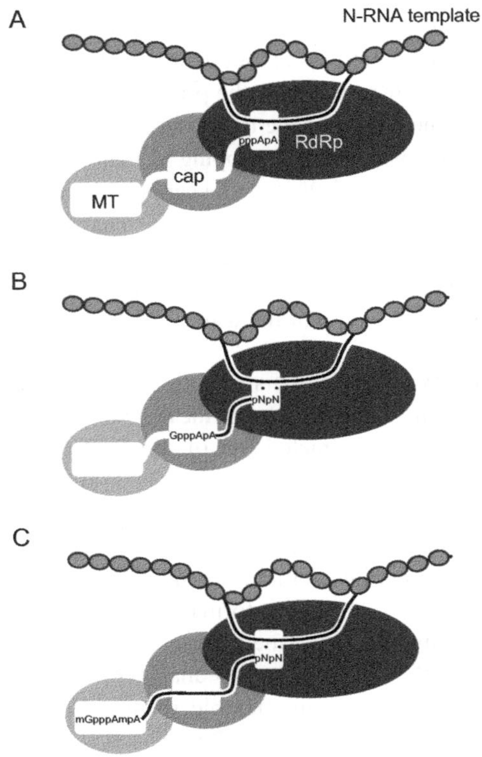

Figure 3.4. *Schematic diagrams illustrating the products of the different enzymatic activities of a mononegavirus polymerase during mRNA transcription. (A) The RdRp initiates RNA synthesis and elongates the RNA; this example shows the initiating nucleotides of VSV mRNA. (B) When the elongating mRNA transcript reaches the capping domain (cap), its 5' end is modified with a guanosine cap. (C) Following cap addition, the 5' end of the mRNA is methylated by the methyltransferase (MT) to introduce the 2'O and N7 methyl groups, as indicated with "m". The template and transcript RNA are illustrated with black lines, except for the details shown in the active sites. The polymeric N protein that sheathes the template RNA is illustrated by a chain of gray ovals. It is shown being displaced as the RNA enters the template channel of the polymerase, as first suggested for the bunyavirus polymerase. It should be noted that the polymerase is represented schematically and the figure does not represent the actual relative positions of the different domains or the possible channels between them*

The two different polymerase conformations could also reflect different initiation mechanisms. Transcription initiation differs for different families within the *Mononegavirales*. The paramyxovirus polymerase initiates opposite the first nucleotide of the genome and synthesizes a short uncapped and unencapsidated leader transcript (Noton and Fearns 2015). It releases this transcript at the end of the leader and then reinitiates RNA synthesis at the first gene start signal. In the case of VSV, the polymerase appears to be able to initiate transcription either opposite the first nucleotide of the template, like the paramyxoviruses, or directly at the first gene start signal, depending on the environment (Whelan and Wertz 2002; Qanungo *et al.* 2004). The *Pneumoviridae* have evolved yet another mechanism in which the polymerase can initiate at two sites in the leader promoter. Initiation at position 1 begins the process of replication, whereas initiation at position 3 begins transcription (Tremaglio *et al.* 2013). It has been shown that the RSV polymerase can bind initiating NTPs independently of the template sequence and that initiation site selection is governed by which NTP is bound (Noton *et al.* 2010; Cressey *et al.* 2018; Noton *et al.* 2019). Thus, it is possible that the more open conformation of the pneumovirus polymerases represents a pre-initiation state and that the binding of NTPs alters its conformation, pulling the priming loop into the appropriate position to enable initiation at either position 1 or position 3 of the template. Having initiated RNA synthesis, the capping domain could reorganize, retracting the priming loop and creating an exit path. Consistent with this, amino acid substitutions in the RSV capping domain and priming loop cause defects in RNA elongation in the leader promoter region (Braun *et al.* 2017).

Regardless of the mechanism by which the polymerase initiates transcription, it is likely that having reached the gene start signal, the mechanisms underlying transcription are similar for different mononegaviruses. At this point, the template would already be threaded through the template entry and exit channels of the RdRp. Studies of the rabies virus polymerase have indicated that the priming residue is not required for initiation at an internal gene start-like sequence (Ogino *et al.* 2019). Therefore, it is likely that during RNA synthesis initiation at these signals, the priming loop is either fully or, at least, partially retracted, similar to the conformation observed in the pneumovirus polymerase structures. Following initiation opposite the first nucleotide of the gene start signal, the RNA is presumably transferred to the capping domain of the polymerase (Figure 3.4). As noted above, VSV cap addition occurs via a GDP polyribonucleotidyltransferase reaction, and this mechanism is likely conserved throughout the *Mononegavirales* (Ogino and Green 2019). For capping to occur, the nascent RNA transcript loses two phosphates from the 5' terminal triphosphate group and becomes covalently linked to a histidine residue in the capping domain. This monophosphorylated RNA is then transferred onto a GDP acceptor, resulting in the GpppN guanosine cap structure (Ogino and Banerjee 2007; Ogino *et al.* 2010; Ogino and Green 2019). Many of the amino acid residues essential for this reaction lie close to the transcript exit channel of the polymerase, providing ready access for the 5' end of the RNA to the capping active site (Neubauer *et al.* 2016; Gilman *et al.* 2019; Pan *et al.* 2020). Following addition of the cap, the polymerase then sequentially methylates the RNA, first at the 2'O position on the first encoded nucleotide of the mRNA and then at the N7 position on the guanosine cap, to generate a cap 1 structure 5' mGpppNm (Rahmeh *et al.* 2009; Paesen *et al.* 2015). Both methyltransferase reactions are performed by the same active site in the polymerase (Li *et al.* 2006; Rahmeh *et al.* 2009; Liang *et al.* 2015; Paesen *et al.* 2015). In the VSV polymerase structure, there is no obvious channel for the RNA to follow from the capping to methyltransferase domain. It is possible that the flexibility of the methyltransferase domain observed for RSV and HMPV polymerases allows it to come into proximity with the capped RNA, or that structural rearrangements that occur as the polymerase enters an elongation phase create a more unified capping/methylation domain that channel the RNA transcript appropriately. The enzyme activities that occur at the 5' end of the mRNA somehow affect the ability of the polymerase to respond appropriately to the gene end signal. A gene end signal artificially positioned too close to the gene start signal cannot be recognized, and factors that inhibit methylation can result in mRNA hyperpolyadenylation, indicating that they prevent the polymerase from releasing the mRNA in a timely manner (Hunt *et al.* 1984; Whelan *et al.* 2000; Galloway and Wertz 2008; Galloway *et al.* 2008). It is possible that the polymerase structural rearrangements that occur during initiation and early elongation of the mRNA transcript expose residues that enable gene end signal recognition. In most of the mononegaviruses, RNA replication is initiated opposite the 3' nucleotide of the genome or antigenome template and ends at the 5' end. In this case, the polymerase probably adopts the conformation observed for the VSV polymerase, with the priming loop extended into the RdRp active site and a priming residue is positioned to provide base-stacking interactions to stabilize the initiating NTPs opposite the first two nucleotides of the template (Liang *et al.* 2015; Ogino *et al.* 2019).

3.3.5. *Concluding remarks*

The structures described above reveal the negative-strand RNA virus polymerases to be amazingly intricate machines, capable of adopting multiple conformations to perform the variety of tasks involved in generating capped and polyadenylated mRNAs and replication products. While a structural understanding is valuable for helping us understand the mechanistic details underlying transcription and replication, it is also highly impactful for translational research. These polymerases are essential for their viruses to replicate, and this property, combined with their unique enzymatic activities, makes them highly attractive targets for drug development. It is anticipated that the next few years will yield further insight into different stages of transcription and replication, accompanied by a development of novel antiviral inhibitors that would have the potential to significantly improve public health.

3.4. References

Abrahem, A. and Pelchat, M. (2008). Formation of an RNA polymerase II preinitiation complex on an RNA promoter derived from the hepatitis delta virus RNA genome. *Nucleic Acids Res*, 36, 5201–5211.

Albarino, C.G., Bird, B.H., and Nichol, S.T. (2007). A shared transcription termination signal on negative and ambisense RNA genome segments of Rift Valley fever, sandfly fever Sicilian, and Toscana viruses. *J Virol*, 81, 5246–5256.

Asenjo, A., Calvo, E., and Villanueva, N. (2006). Phosphorylation of human respiratory syncytial virus P protein at threonine 108 controls its interaction with the M2-1 protein in the viral RNA polymerase complex. *J Gen Virol*, 87, 3637–3642.

Auperin, D.D., Galinski, M., and Bishop, D.H. (1984a). The sequences of the N protein gene and intergenic region of the S RNA of pichinde arenavirus. *Virology*, 134, 208–219.

Auperin, D.D., Romanowski, V., Galinski, M., and Bishop, D.H. (1984b). Sequencing studies of pichinde arenavirus S RNA indicate a novel coding strategy, an ambisense viral S RNA. *J Virol*, 52, 897–904.

Auperin, D.D., Sasso, D.R., and McCormick, J.B. (1986). Nucleotide sequence of the glycoprotein gene and intergenic region of the Lassa virus S genome RNA. *Virology*, 154, 155–167.

Barr, J.N. (2007). Bunyavirus mRNA synthesis is coupled to translation to prevent premature transcription termination. *RNA*, 13, 731–736.

Barr, J.N., Whelan, S.P., and Wertz, G.W. (1997). cis-Acting signals involved in termination of vesicular stomatitis virus mRNA synthesis include the conserved AUAC and the U7 signal for polyadenylation. *J Virol*, 71, 8718–8725.

Barr, J.N. and Wertz, G.W. (2001). Polymerase slippage at vesicular stomatitis virus gene junctions to generate poly(A) is regulated by the upstream 3'-AUAC-5' tetranucleotide: Implications for the mechanism of transcription termination. *J Virol*, 75, 6901–6913.

Barr, J.N. and Wertz, G.W. (2004). Bunyamwera bunyavirus RNA synthesis requires cooperation of 3'- and 5'-terminal sequences. *J Virol*, 78, 1129–1138.

Barr, J.N. and Wertz, G.W. (2005). Role of the conserved nucleotide mismatch within 3'- and 5'-terminal regions of Bunyamwera virus in signaling transcription. *J Virol*, 79, 3586–3594.

Barr, J.N., Elliott, R.M., Dunn, E.F., and Wertz, G.W. (2003). Segment-specific terminal sequences of Bunyamwera bunyavirus regulate genome replication. *Virology*, 311, 326–338.

Barr, J.N., Rodgers, J.W., and Wertz, G.W. (2006). Identification of the Bunyamwera bunyavirus transcription termination signal. *J Gen Virol*, 87, 189–198.

Beard, M.R., Macnaughton, T.B., and Gowans, E.J. (1996). Identification and characterization of a hepatitis delta virus RNA transcriptional promoter. *J Virol*, 70, 49864995.

Bellocq, C. and Kolakofsky, D. (1987). Translational requirement for La Crosse virus S-mRNA synthesis: A possible mechanism. *J Virol*, 61, 3960–3967.

Blakqori, G., Van Knippenberg, I., and Elliott, R.M. (2009). Bunyamwera orthobunyavirus S-segment untranslated regions mediate poly(A) tail-independent translation. *J Virol*, 83, 3637–3746.

Blakqori, G., Lowen, A.C., and Elliott, R.M. (2012). The small genome segment of Bunyamwera orthobunyavirus harbours a single transcription-termination signal. *J Gen Virol*, 93, 1449–1455.

Brauburger, K., Boehmann, Y., Tsuda, Y., Hoenen, T., Olejnik, J., Schumann, M., Ebihara, H., and Muhlberger, E. (2014). Analysis of the highly diverse gene borders in Ebola virus reveals a distinct mechanism of transcriptional regulation. *J Virol*, 88, 12558–12571.

Braun, M.R., Deflube, L.R., Noton, S.L., Mawhorter, M.E., Tremaglio, C.Z., and Fearns, R. (2017). RNA elongation by respiratory syncytial virus polymerase is calibrated by conserved region V. *PLoS Pathog*, 13, e1006803.

Brown, A.L., Perrotta, A.T., Wadkins, T.S., and Been, M.D. (2008). The poly(A) site sequence in HDV RNA alters both extent and rate of self-cleavage of the antigenomic ribozyme. *Nucleic Acids Res*, 36, 2990–3000.

Calain, P. and Roux, L. (1995). Functional characterisation of the genomic and antigenomic promoters of Sendai virus. *Virology*, 212, 163–173.

Casey, J.L. (2012). Control of ADAR1 editing of hepatitis delta virus RNAs. *Curr Top Microbiol Immunol*, 353, 123–143.

Casey, J.L. and Gerin, J.L. (1995). Hepatitis D virus RNA editing: Specific modification of adenosine in the antigenomic RNA. *J Virol*, 69,7593–7600.

Castagne, N., Barbier, A., Bernard, J., Rezaei, H., Huet, J.C., Henry, C., Da Costa, B., and Eleouet, J.F. (2004). Biochemical characterization of the respiratory syncytial virus P-P and P-N protein complexes and localization of the P protein oligomerization domain. *J Gen Virol*, 85, 1643–1653.

Chang, F.L., Chen, P.J , Tu, S.J , Wang, C.J., and Chen, D.S. (1991). The large form of hepatitis delta antigen is crucial for assembly of hepatitis delta virus. *Proc Natl Acad Sci USA*, 88, 8490–8494.

Chao, M., Hsieh, S.Y., and Taylor, J. (1990). Role of two forms of hepatitis delta virus antigen: Evidence for a mechanism of self-limiting genome replication. *J Virol*, 64, 5066–5069.

Cheng, E. and Mir, M.A. (2012). Signatures of host mRNA 5' terminus for efficient hantavirus cap snatching. *J Virol*, 86, 10173–10185.

Cornu, T.I. and De La Torre, J.C. (2001). RING finger Z protein of lymphocytic choriomeningitis virus (LCMV) inhibits transcription and RNA replication of an LCMV S-segment minigenome. *J Virol*, 75, 9415–9426.

Cressey, T.N., Noton, S.L., Nagendra, K., Braun, M.R., and Fearns, R. (2018). Mechanism for de novo initiation at two sites in the respiratory syncytial virus promoter. *Nucleic Acids Res*, 46, 6785–6796.

Crow, M., Deng, T., Addley, M., and Brownlee, G.G. (2004). Mutational analysis of the influenza virus cRNA promoter and identification of nucleotides critical for replication. *J Virol*, 78, 6263–6270.

Deng, T., Vreede, F.T., and Brownlee, G.G. (2006). Different de novo initiation strategies are used by influenza virus RNA polymerase on its cRNA and viral RNA promoters during viral RNA replication. *J Virol*, 80, 2337–2348.

Dias, A., Bouvier, D., Crepin, T., Mccarthy, A.A., Hart, D.J., Baudin, F., Cusack, S., and Ruigrok, R.W. (2009). The cap-snatching endonuclease of influenza virus polymerase resides in the PA subunit. *Nature*, 458, 914–918.

van Dijk, A.A., Makeyev, E.V., and Bamford, D.H. (2004). Initiation of viral RNA-dependent RNA polymerization. *J Gen Virol*, 85, 1077–1093.

Elliott, R.M. (2014). Orthobunyaviruses: Recent genetic and structural insights. *Nat Rev Microbiol*, 12, 673–685.

Engelhardt, O.G., Smith, M., and Fodor, E. (2005). Association of the influenza A virus RNA-dependent RNA polymerase with cellular RNA polymerase II. *J Virol*, 79, 5812–5818.

Esneau, C., Raynal, B., Roblin, P., Brule, S., Richard, C.A., Fix, J., Eleouet, J.F., and Galloux, M. (2019). Biochemical characterization of the respiratory syncytial virus N(0)-P complex in solution. *J Biol Chem*, 294, 3647–3660.

Fan, H., Walker, A.P., Carrique, L., Keown, J.R., Serna Martin, I., Karia, D., Sharps, J., Hengrung, N., Pardon, E., Steyaert, J., Grimes, J.M., and Fodor, E. (2019). Structures of influenza A virus RNA polymerase offer insight into viral genome replication. *Nature*, 573, 287–290.

Fearns, R. and Collins, P.L. (1999). Model for polymerase access to the overlapped L gene of respiratory syncytial virus. *J Virol*, 73, 388–397.

Fearns, R. and Plemper, R.K. (2017). Polymerases of paramyxoviruses and pneumoviruses. *Virus Res*, 234, 87–102.

Fearns, R., Peeples, M.E., and Collins, P.L. (2002). Mapping the transcription and replication promoters of respiratory syncytial virus. *J Virol*, 76, 1663–1672.

Fernàndez-Garcia, Y., Reguera, J., Busch, C., Witte, G., Sanchez-Ramos, O., Betzel, C., Cusack, S., Gunther, S., and Reindl, S. (2016). Atomic structure and biochemical characterization of an RNA endonuclease in the N terminus of Andes Virus L protein. *PLoS Pathog*, 12, e1005635.

Ferron, F., Weber, F., De La Torre, J.C., and Reguera, J. (2017). Transcription and replication mechanisms of Bunyaviridae and Arenaviridae L proteins. *Virus Res*, 234, 118–134.

Filipovska, J. and Konarska, M.M. (2000). Specific HDV RNA-templated transcription by pol II *in vitro*. *RNA*, 6, 41–54.

Finch, J.T. and Gibbs, A.J. (1970). Observations on the structure of the nucleocapsids of some paramyxoviruses. *J Gen Virol*, 6, 141–150.

Fodor, E. (2013). The RNA polymerase of influenza a virus: Mechanisms of viral transcription and replication. *Acta Virol*, 57, 113–122.

Fodor, E., Pritlove, D.C., and Brownlee, G.G. (1994). The influenza virus panhandle is involved in the initiation of transcription. *J Virol*, 68, 4092–4096.

Foscaldi, S., D'antuono, A., Noval, M.G., De Prat Gay, G., Scolaro, L., and Lopez, N. (2017). Regulation of tacaribe mammarenavirus translation: Positive 5' and negative 3' elements and role of key cellular factors. *J Virol*, 91(14), e00084-17.

Fuller-Pace, F.V. and Southern, P.J. (1988). Temporal analysis of transcription and replication during acute infection with lymphocytic choriomeningitis virus. *Virology*, 162, 260–263.

Galloway, S.E. and Wertz, G.W. (2008). S-adenosyl homocysteine-induced hyperpolyadenylation of vesicular stomatitis virus mRNA requires the methyltransferase activity of L protein. *J Virol*, 82, 12280–12290.

Galloway, S.E., Richardson, P.E., and Wertz, G.W. (2008). Analysis of a structural homology model of the 2'-O-ribose methyltransferase domain within the vesicular stomatitis virus L protein. *Virology*, 382, 69–82.

Garcin, D. and Kolakofsky, D. (1992). Tacaribe arenavirus RNA synthesis *in vitro* is primer dependent and suggests an unusual model for the initiation of genome replication. *J Virol*, 66, 1370–1376.

Garcin, D., Lezzi, M., Dobbs, M., Elliott, R.M., Schmaljohn, C., Kang, C.Y., and Kolakofsky, D. (1995). The 5' ends of Hantaan virus (Bunyaviridae) RNAs suggest a prime-and-realign mechanism for the initiation of RNA synthesis. *J Virol*, 69, 5754–5762.

Gauliard, N., Billecocq, A., Flick, R., and Bouloy, M. (2006). Rift Valley fever virus noncoding regions of L, M and S segments regulate RNA synthesis. *Virology*, 351, 170–179.

Gerlach, P., Malet, H., Cusack, S., and Reguera, J. (2015). Structural insights into bunyavirus replication and its regulation by the vRNA promoter. *Cell*, 161, 1267–1279.

Gilman, M.S.A., Liu, C., Fung, A., Behera, I., Jordan, P., Rigaux, P., Ysebaert, N., Tcherniuk, S., Sourimant, J., Eleouet, J.F., Sutto-Ortiz, P., Decroly, E., Roymans, D., Jin, Z., and Mclellan, J.S. (2019). Structure of the respiratory syncytial virus polymerase complex. *Cell*, 179, 193-204 e14.

Glenn, J.S. and White, J.M. (1991). Trans-dominant inhibition of human hepatitis delta virus genome replication. *J Virol*, 65, 2357–2361.

Gogrefe, N., Reindl, S., Gunther, S., and Rosenthal, M. (2019). Structure of a functional cap-binding domain in Rift Valley fever virus L protein. *PLoS Pathog*, 15, e1007829.

Gubbay, O., Curran, J., and Kolakofsky, D. (2001). Sendai virus genome synthesis and assembly are coupled: A possible mechanism to promote viral RNA polymerase processivity. *J Gen Virol*, 82, 2895–2903.

Gudima, S., Wu, S.Y., Chiang, C.M., Moraleda, G., and Taylor, J. (2000). Origin of hepatitis delta virus mRNA. *J Virol*, 74, 7204–7210.

Hallam, S.J., Koma, T., Maruyama, J., and Paessler, S. (2018). Review of Mammarenavirus biology and replication. *Front Microbiol*, 9, 1751.

Hass, M., Westerkofsky, M., Muller, S., Becker-Ziaja, B., Busch, C., and Gunther, S. (2006). Mutational analysis of the lassa virus promoter. *J Virol*, 80, 12414–12419.

Haussecker, D., Cao, D., Huang, Y., Parameswaran, P., Fire, A.Z., and Kay, M.A. (2008). Capped small RNAs and MOV10 in human hepatitis delta virus replication. *Nat Struct Mol Biol*, 15, 714–721.

Hengrung, N., El Omari, K., Serna Martin, I., Vreede, F.T., Cusack, S., Rambo, R.P., Vonrhein, C., Bricogne, G., Stuart, D.I., Grimes, J.M., and Fodor, E. (2015). Crystal structure of the RNA-dependent RNA polymerase from influenza C virus. *Nature*, 527, 114–117.

Honda, T. and Tomonaga, K. (2013). Nucleocytoplasmic shuttling of viral proteins in borna disease virus infection. *Viruses*, 5, 1978–1990.

Hopkins, K.C., McLane, L.M., Maqbool, T., Panda, D., Gordesky-Gold, B., and Cherry, S. (2013). A genome-wide RNAi screen reveals that mRNA decapping restricts bunyaviral replication by limiting the pools of Dcp2-accessible targets for cap-snatching. *Genes Dev*, 27, 1511–1525.

Hsieh, S.Y. and Taylor, J. (1991). Regulation of polyadenylation of hepatitis delta virus antigenomic RNA. *J Virol*, 65, 6438–6446.

Hume, A.J. and Muhlberger, E. (2019). Distinct genome replication and transcription strategies within the growing filovirus family. *J Mol Biol*, 431, 4290–4320.

Hunt, D.M., Smith, E.F., and Buckley, D.W. (1984). Aberrant polyadenylation by a vesicular stomatitis virus mutant is due to an altered L protein. *J Virol*, 52, 515–521.

Ikegami, T., Won, S., Peters, C.J., and Makino, S. (2007). Characterization of Rift Valley fever virus transcriptional terminations. *J Virol*, 81, 8421–8438.

Jin, H. and Elliott, R.M. (1993). Non-viral sequences at the 5’ ends of Dugbe nairovirus S mRNAs. *J Gen Virol*, 74(Pt 10), 2293–2297.

Kim, H.J., Fodor, E., Brownlee, G.G., and Seong, B.L. (1997). Mutational analysis of the RNA-fork model of the influenza A virus vRNA promoter *in vivo*. *J Gen Virol*, 78(Pt 2), 353–357.

Kohl, A., Dunn, E.F., Lowen, A.C., and Elliott, R.M. (2004). Complementarity, sequence and structural elements within the 3' and 5' non-coding regions of the Bunyamwera orthobunyavirus S segment determine promoter strength. *J Gen Virol*, 85, 3269–3278.

Kohl, A., Lowen, A.C., Leonard, V.H., and Elliott, R.M. (2006). Genetic elements regulating packaging of the Bunyamwera orthobunyavirus genome. *J Gen Virol*, 87, 177–187.

Kos, A., Dijkema, R., Arnberg, A.C., Van Der Meide, P.H., and Schellekens, H. (1986). The hepatitis delta (delta) virus possesses a circular RNA. *Nature*, 323, 558–560.

Kouba, T., Drncová, P., and Cusack, S. (2019). Structural snapshots of actively transcribing influenza polymerase. *Nat Struct Mol Biol*, 26, 460–470.

Kranzusch, P.J. and Whelan, S.P. (2011). Arenavirus Z protein controls viral RNA synthesis by locking a polymerase-promoter complex. *Proc Natl Acad Sci USA*, 108, 19743–19748.

Kranzusch, P.J., Schenk, A.D., Rahmeh, A.A., Radoshitzky, S.R., Bavari, S., Walz, T., and Whelan, S.P. (2010). Assembly of a functional Machupo virus polymerase complex. *Proc Natl Acad Sci USA*, 107, 20069–20074.

Kuo, M.Y., Sharmeen, L., Dinter-Gottlieb, G., and Taylor, J. (1988). Characterization of self-cleaving RNA sequences on the genome and antigenome of human hepatitis delta virus. *J Virol*, 62, 4439–4444.

Kuo, M.Y., Chao, M., and Taylor, J. (1989). Initiation of replication of the human hepatitis delta virus genome from cloned DNA: role of delta antigen. *J Virol*, 63, 1945–1950.

Lara, E., Billecocq, A., Leger, P., and Bouloy, M. (2011). Characterization of wild-type and alternate transcription termination signals in the Rift Valley fever virus genome. *J Virol*, 85, 12134–12145.

Leung, D.W., Borek, D., Luthra, P., Binning, J.M., Anantpadma, M., Liu, G., Harvey, I.B., Su, Z., Endlich-Frazier, A., Pan, J., Shabman, R.S., Chiu, W., Davey, R.A., Otwinowski, Z., Basler, C.F., and Amarasinghe, G.K. (2015). An intrinsically disordered peptide from Ebola Virus VP35 controls viral RNA synthesis by modulating nucleoprotein-RNA interactions. *Cell Rep*, 11, 376–389.

Leyrat, C., Renner, M., Harlos, K., and Grimes, J.M. (2013). Solution and crystallographic structures of the central region of the phosphoprotein from human metapneumovirus. *PLoS One*, 8, e80371.

Li, T. and Pattnaik, A.K. (1999). Overlapping signals for transcription and replication at the 3' terminus of the vesicular stomatitis virus genome. *J Virol*, 73, 444–452.

Li, J., Wang, J.T., and Whelan, S.P. (2006). A unique strategy for mRNA cap methylation used by vesicular stomatitis virus. *Proc Natl Acad Sci USA*, 103, 8493–8498.

Liang, B., Li, Z., Jenni, S., Rahmeh, A.A., Morin, B.M., Grant, T., Grigorieff, N., Harrison, S.C., and Whelan, S.P.J. (2015). Structure of the L protein of vesicular stomatitis virus from electron cryomicroscopy. *Cell*, 162, 314–327.

Llorente, M.T., Taylor, I.A., López-Viñas, E., Gomez-Puertas, P., Calder, L.J., García-Barreno, B., and Melero, J.A. (2008). Structural properties of the human respiratory syncytial virus P protein: Evidence for an elongated homotetrameric molecule that is the smallest orthologue within the family of paramyxovirus polymerase cofactors. *Proteins*, 72, 946–958.

Longhi, S. (2009). Nucleocapsid structure and function. *Curr Top Microbiol Immunol*, 329, 103–128.

Longhi, S., Bloyet, L.M., Gianni, S., and Gerlier, D. (2017). How order and disorder within paramyxoviral nucleoproteins and phosphoproteins orchestrate the molecular interplay of transcription and replication. *Cell Mol Life Sci*, 74, 3091–3118.

López, N. and Franze-Fernández, M.T. (2007). A single stem-loop structure in Tacaribe arenavirus intergenic region is essential for transcription termination but is not required for a correct initiation of transcription and replication. *Virus Res*, 124, 237–244.

Lukarska, M., Fournier, G., Pflug, A., Resa-Infante, P., Reich, S., Naffakh, N., and Cusack, S. (2017). Structural basis of an essential interaction between influenza polymerase and Pol II CTD. *Nature*, 541, 117–121.

Luo, G.X., Chao, M., Hsieh, S.Y., Sureau, C., Nishikura, K., and Taylor, J. (1990). A specific base transition occurs on replicating hepatitis delta virus RNA. *J Virol*, 64, 1021–1027.

Macnaughton, T.B., Beard, M.R., Chao, M., Gowans, E.J., and Lai, M.M. (1993). Endogenous promoters can direct the transcription of hepatitis delta virus RNA from a recircularized cDNA template. *Virology*, 196, 629–636.

Macnaughton, T.B., Shi, S.T., Modahl, L.E., and Lai, M.M. (2002). Rolling circle replication of hepatitis delta virus RNA is carried out by two different cellular RNA polymerases. *J Virol*, 76, 3920–3927.

Magnius, L., Taylor, J., Mason, W.S., Sureau, C., Deny, P., Norder, H., and ICTV Report Consortium (2018). ICTV Virus Taxonomy Profile: Deltavirus. *J Gen Virol*, 99, 1565–1566.

McGivern, D.R., Collins, P.L., and Fearns, R. (2005). Identification of internal sequences in the 3' leader region of human respiratory syncytial virus that enhance transcription and confer replication processivity. *J Virol*, 79, 2449–2460.

Meyer, B.J. and Southern, P.J. (1993). Concurrent sequence analysis of 5' and 3' RNA termini by intramolecular circularization reveals 5' nontemplated bases and 3' terminal heterogeneity for lymphocytic choriomeningitis virus mRNAs. *J Virol*, 67, 2621–2627.

Mir, M.A., Duran, W.A., Hjelle, B.L., Ye, C., and Panganiban, A.T. (2008). Storage of cellular 5' mRNA caps in P bodies for viral cap-snatching. *Proc Natl Acad Sci USA*, 105, 19294–19299.

Modahl, L.E. and Lai, M.M. (1998). Transcription of hepatitis delta antigen mRNA continues throughout hepatitis delta virus (HDV) replication: A new model of HDV RNA transcription and replication. *J Virol*, 72, 5449–5456.

Modahl, L.E., Macnaughton, T.B., Zhu, N., Johnson, D.L., and Lai, M.M. (2000). RNA-dependent replication and transcription of hepatitis delta virus RNA involve distinct cellular RNA polymerases. *Mol Cell Biol*, 20, 6030–6039.

Moraleda, G. and Taylor, J. (2001). Host RNA polymerase requirements for transcription of the human hepatitis delta virus genome. *J Virol*, 75, 10161–10169.

Morin, B., Coutard, B., Lelke, M., Ferron, F., Kerber, R., Jamal, S., Frangeul, A., Baronti, C., Charrel, R., De Lamballerie, X., Vonrhein, C., Lescar, J., Bricogne, G., Gunther, S., and Canard, B. (2010). The N-terminal domain of the arenavirus L protein is an RNA endonuclease essential in mRNA transcription. *PLoS Pathog*, 6, e1001038.

Neubauer, J., Ogino, M., Green, T.J., and Ogino, T. (2016). Signature motifs of GDP polyribonucleotidyltransferase, a non-segmented negative strand RNA viral mRNA capping enzyme, domain in the L protein are required for covalent enzyme-pRNA intermediate formation. *Nucleic Acids Res*, 44, 330–341.

Nie, X., Chang, J., and Taylor, J.M. (2004). Alternative processing of hepatitis delta virus antigenomic RNA transcripts. *J Virol*, 78, 4517–4524.

Noton, S.L. and Fearns, R. (2015). Initiation and regulation of paramyxovirus transcription and replication. *Virology*, 479–480, 545–554.

Noton, S.L., Cowton, V.M., Zack, C.R., McGivern, D.R., and Fearns, R. (2010). Evidence that the polymerase of respiratory syncytial virus initiates RNA replication in a nontemplated fashion. *Proc Natl Acad Sci USA*, 107, 10226–10231.

Noton, S.L., Tremaglio, C.Z., and Fearns, R. (2019). Killing two birds with one stone: How the respiratory syncytial virus polymerase initiates transcription and replication. *PLoS Pathog*, 15, e1007548.

Ogino, T. and Banerjee, A.K. (2007). Unconventional mechanism of mRNA capping by the RNA-dependent RNA polymerase of vesicular stomatitis virus. *Mol Cell*, 25, 85–97.

Ogino, T. and Green, T.J. (2019). RNA synthesis and capping by non-segmented negative strand RNA viral polymerases: Lessons from a prototypic virus. *Front Microbiol*, 10, 1490.

Ogino, T., Yadav, S.P., and Banerjee, A.K. (2010). Histidine-mediated RNA transfer to GDP for unique mRNA capping by vesicular stomatitis virus RNA polymerase. *Proc Natl Acad Sci USA*, 107, 3463–3468.

Ogino, M., Gupta, N., Green, T.J., and Ogino, T. (2019). A dual-functional priming-capping loop of rhabdoviral RNA polymerases directs terminal de novo initiation and capping intermediate formation. *Nucleic Acids Res*, 47, 299–309.

Paesen, G.C., Collet, A., Sallamand, C., Debart, F., Vasseur, J.J., Canard, B., Decroly, E., and Grimes, J.M. (2015). X-ray structure and activities of an essential Mononegavirales L-protein domain. *Nat Commun*, 6, 8749.

Pan, J., Qian, X., Lattmann, S., El Sahili, A., Yeo, T.H., Jia, H., Cressey, T., Ludeke, B., Noton, S., Kalocsay, M., Fearns, R., and Lescar, J. (2020). Structure of the human metapneumovirus polymerase phosphoprotein complex. *Nature*, 577, 275–279.

Peng, Q., Liu, Y., Peng, R., Wang, M., Yang, W., Song, H., Chen, Y., Liu, S., Han, M., Zhang, X., Wang, P., Yan, J., Zhang, B., Qi, J., Deng, T., Gao, G.F., and Shi, Y. (2019). Structural insight into RNA synthesis by influenza D polymerase. *Nat Microbiol*, 4, 1750–1759.

Pereira, N., Cardone, C., Lassoued, S., Galloux, M., Fix, J., Assrir, N., Lescop, E., Bontems, F., Eleouet, J.F., and Sizun, C. (2017). New insights into structural disorder in human respiratory syncytial virus phosphoprotein and implications for binding of protein partners. *J Biol Chem*, 292, 2120–2131.

Perez, M. and De La Torre, J.C. (2003). Characterization of the genomic promoter of the prototypic arenavirus lymphocytic choriomeningitis virus. *J Virol*, 77, 1184–1194.

Pflug, A., Guilligay, D., Reich, S., and Cusack, S. (2014). Structure of influenza A polymerase bound to the viral RNA promoter. *Nature*, 516, 355–360.

Pinschewer, D.D., Perez, M., and De La Torre, J.C. (2003). Role of the virus nucleoprotein in the regulation of lymphocytic choriomeningitis virus transcription and RNA replication. *J Virol*, 77, 3882–38877.

Polson, A.G., Bass, B.L., and Casey, J.L. (1996). RNA editing of hepatitis delta virus antigenome by dsRNA-adenosine deaminase. *Nature*, 380, 454–456.

Polyak, S.J., Zheng, S., and Harnish, D.G. (1995). 5' termini of Pichinde arenavirus S RNAs and mRNAs contain nontemplated nucleotides. *J Virol*, 69, 3211–3215.

Poon, L.L., Pritlove, D.C., Sharps, J., and Brownlee, G.G. (1998). The RNA polymerase of influenza virus, bound to the 5' end of virion RNA, acts in cis to polyadenylate mRNA. *J Virol*, 72, 8214–8219.

Poon, L.L., Pritlove, D.C., Fodor, E., and Brownlee, G.G. (1999). Direct evidence that the poly(A) tail of influenza A virus mRNA is synthesized by reiterative copying of a U track in the virion RNA template. *J Virol*, 73, 3473–3476.

Pyle, J.D. and Whelan, S.P.J. (2019). RNA ligands activate the Machupo virus polymerase and guide promoter usage. *Proc Natl Acad Sci USA*, 116, 10518–10524.

Qanungo, K.R., Shaji, D., Mathur, M., and Banerjee, A.K. (2004). Two RNA polymerase complexes from vesicular stomatitis virus-infected cells that carry out transcription and replication of genome RNA. *Proc Natl Acad Sci USA*, 101, 5952–5957.

Rahmeh, A.A., Li, J., Kranzusch, P.J., and Whelan, S.P. (2009). Ribose 2'-O methylation of the vesicular stomatitis virus mRNA cap precedes and facilitates subsequent guanine-N-7 methylation by the large polymerase protein. *J Virol*, 83, 11043–11050.

Raju, R., Raju, L., Hacker, D., Garcin, D., Compans, R., and Kolakofsky, D. (1990). Nontemplated bases at the 5' ends of Tacaribe virus mRNAs. *Virology*, 174, 53–59.

Reguera, J., Weber, F., and Cusack, S. (2010). Bunyaviridae RNA polymerases (L-protein) have an N-terminal, influenza-like endonuclease domain, essential for viral cap-dependent transcription. *PLoS Pathog*, 6, e1001101.

Reguera, J., Gerlach, P., and Cusack, S. (2016a). Towards a structural understanding of RNA synthesis by negative strand RNA viral polymerases. *Curr Opin Struct Biol*, 36, 75–84.

Reguera, J., Gerlach, P., Rosenthal, M., Gaudon, S., Coscia, F., Gunther, S., and Cusack, S. (2016b). Comparative structural and functional analysis of bunyavirus and arenavirus cap-snatching endonucleases. *PLoS Pathog*, 12, e1005636.

Reich, S., Guilligay, D., Pflug, A., Malet, H., Berger, I., Crepin, T., Hart, D., Lunardi, T., Nanao, M., Ruigrok, R.W., and Cusack, S. (2014). Structural insight into cap-snatching and RNA synthesis by influenza polymerase. *Nature*, 516, 361–366.

Reich, S., Guilligay, D., and Cusack, S. (2017). An *in vitro* fluorescence based study of initiation of RNA synthesis by influenza B polymerase. *Nucleic Acids Res*, 45, 3353–3368.

Reid, C.E. and Lazinski, D.W. (2000). A host-specific function is required for ligation of a wide variety of ribozyme-processed RNAs. *Proc Natl Acad Sci USA*, 97, 424–429.

Renner, M., Bertinelli, M., Leyrat, C., Paesen, G.C., Saraiva De Oliveira, L.F., Huiskonen, J.T., and Grimes, J.M. (2016). Nucleocapsid assembly in pneumoviruses is regulated by conformational switching of the N protein. *Elife*, 5, e12627.

Renner, M., Paesen, G.C., Grison, C.M., Granier, S., Grimes, J.M., and Leyrat, C. (2017). Structural dissection of human metapneumovirus phosphoprotein using small angle x-ray scattering. *Sci Rep*, 7, 14865.

Robb, N.C., Te Velthuis, A.J.W., Fodor, E., and Kapanidis, A.N. (2019). Real-time analysis of single influenza virus replication complexes reveals large promoter-dependent differences in initiation dynamics. *Nucleic Acids Res*, 47, 6466–6477.

Rodriguez, A., Perez-Gonzalez, A., and Nieto, A. (2007). Influenza virus infection causes specific degradation of the largest subunit of cellular RNA polymerase II. *J Virol*, 81, 5315–5324.

Rosenthal, M., Gogrefe, N., Vogel, D., Reguera, J., Rauschenberger, B., Cusack, S., Gunther, S., and Reindl, S. (2017). Structural insights into reptarenavirus cap-snatching machinery. *PLoS Pathog*, 13, e1006400.

Ruigrok, R.W., Crepin, T., and Kolakofsky, D. (2011). Nucleoproteins and nucleocapsids of negative-strand RNA viruses. *Curr Opin Microbiol*, 14, 504–51010.

Serna Martin, I., Hengrung, N., Renner, M., Sharps, J., Martinez-Alonso, M., Masiulis, S., Grimes, J.M., and Fodor, E. (2018). A mechanism for the activation of the influenza virus transcriptase. *Mol Cell*, 72, 396.

Severin, C., Terrell, J.R., Zengel, J.R., Cox, R., Plemper, R.K., He, B., and Luo, M. (2016). Releasing the genomic RNA sequestered in the mumps virus nucleocapsid. *J Virol*, 90, 10113–10119.

Sharmeen, L., Kuo, M.Y., Dinter-Gottlieb, G., and Taylor, J. (1988). Antigenomic RNA of human hepatitis delta virus can undergo self-cleavage. *J Virol*, 62, 2674–2679.

Sheu, G.T. (2002). Initiation of hepatitis delta virus (HDV) replication: HDV RNA encoding the large delta antigen cannot replicate. *J Gen Virol*, 83, 2507–2513.

Stillman, E.A. and Whitt, M.A. (1999). Transcript initiation and 5'-end modifications are separable events during vesicular stomatitis virus transcription. *J Virol*, 73, 7199–7209.

Taylor, J.M. (2015). Hepatitis D virus replication. *Cold Spring Harb Perspect Med*, 5.

Tekes, G., Rahmeh, A.A., and Whelan, S.P. (2011). A freeze frame view of vesicular stomatitis virus transcription defines a minimal length of RNA for 5' processing. *PLoS Pathog,* 7, e1002073.

Te Velthuis, A.J. and Fodor, E. (2016). Influenza virus RNA polymerase: Insights into the mechanisms of viral RNA synthesis. *Nat Rev Microbiol*, 14, 479–493.

Te Velthuis, A.J., Robb, N.C., Kapanidis, A.N., and Fodor, E. (2016). The role of the priming loop in influenza A virus RNA synthesis. *Nat Microbiol*, 1, 16029.

Thierry, E., Guilligay, D., Kosinski, J., Bock, T., Gaudon, S., Round, A., Pflug, A., Hengrung, N., El Omari, K., Baudin, F., Hart, D.J., Beck, M., and Cusack, S. (2016). Influenza polymerase can adopt an alternative configuration involving a radical repacking of PB2 domains. *Mol Cell*, 61, 125–137.

Tremaglio, C.Z., Noton, S.L., Deflube, L.R., and Fearns, R. (2013). Respiratory syncytial virus polymerase can initiate transcription from position 3 of the leader promoter. *J Virol*, 87, 3196–3207.

Van Poelwijk, F., Kolkman, J., and Goldbach, R. (1996). Sequence analysis of the 5' ends of tomato spotted wilt virus N mRNAs. *Arch Virol*, 141, 177–184.

Vogel, D., Rosenthal, M., Gogrefe, N., Reindl, S., and Gunther, S. (2019). Biochemical characterization of the Lassa virus L protein. *J Biol Chem*, 294, 8088–8100.

Vulliemoz, D. and Roux, L. (2001). "Rule of six": How does the Sendai virus RNA polymerase keep count? *J Virol*, 75, 4506–4518.

Walker, A.P. and Fodor, E. (2019). Interplay between influenza virus and the host RNA polymerase II transcriptional machinery. *Trends Microbiol*, 27, 398–407.

Wallat, G.D., Huang, Q., Wang, W., Dong, H., Ly, H., Liang, Y., and Dong, C. (2014). High-resolution structure of the N-terminal endonuclease domain of the Lassa virus L polymerase in complex with magnesium ions. *PLoS One*, 9, e87577.

Wang, K.S., Choo, Q.L., Weiner, A.J., Ou, J.H., Najarian, R.C., Thayer, R.M., Mullenbach, G.T., Denniston, K.J., Gerin, J.L., and Houghton, M. (1986). Structure, sequence and expression of the hepatitis delta (delta) viral genome. *Nature*, 323, 508–514.

Wang, J.T., McElvain, L.E., and Whelan, S.P. (2007). Vesicular stomatitis virus mRNA capping machinery requires specific cis-acting signals in the RNA. *J Virol*, 81, 11499-506.

Whelan, S.P. and Wertz, G.W. (2002). Transcription and replication initiate at separate sites on the vesicular stomatitis virus genome. *Proc Natl Acad Sci USA*, 99, 9178–9183.

Whelan, S.P., Barr, J.N., and Wertz, G.W. (2000). Identification of a minimal size requirement for termination of vesicular stomatitis virus mRNA: Implications for the mechanism of transcription. *J Virol*, 74, 8268–8276.

Whelan, S.P., Barr, J.N., and Wertz, G.W. (2004). Transcription and replication of nonsegmented negative-strand RNA viruses. *Curr Top Microbiol Immunol*, 283, 61–119.

Wilson, S.M. and Clegg, J.C. (1991). Sequence analysis of the S RNA of the African arenavirus Mopeia: An unusual secondary structure feature in the intergenic region. *Virology*, 180, 543–552.

Yamaguchi, Y., Filipovska, J., Yano, K., Furuya, A., Inukai, N., Narita, T., Wada, T., Sugimoto, S., Konarska, M.M., and Handa, H. (2001). Stimulation of RNA polymerase II elongation by hepatitis delta antigen. *Science*, 293, 124–127.

Zheng, W. and Tao, Y.J. (2013). Structure and assembly of the influenza A virus ribonucleoprotein complex. *FEBS Lett*, 587, 1206–1214.

4

Viral Epitranscriptomics

Rachel NETZBAND and Cara T. PAGER
University at Albany – SUNY, Albany, NY, USA

4.1. Introduction

Viral epitranscriptomics is a growing field examining the changes of chemical marks on ribonucleic acid (RNA) during viral infection. To date, more than 160 chemical marks, or modifications (Boccaletto *et al.* 2018), have been identified either on the nucleobase or ribose of RNA. These modifications occur after an RNA is synthesized, and, as such, they are considered to be post-transcriptional modifications (PTMs). In this chapter, we introduce RNA modifications, how these are identified, review the role of RNA modifications during viral infection and present a perspective for the field.

4.1.1. *What are epitranscriptomic marks?*

RNA modifications are chemical moieties deposited on RNAs to impact the structure and function of RNA. These modifications are chemically diverse and can be added, removed and read by both cellular and viral enzymes, thus creating an arena for synergy and/or competition between the host and the virus. RNA modifications impact numerous RNA processes in the cell, including localization of RNAs, translation of mRNA transcripts, RNA processing capabilities and regulation, immune response and others (Song and Yi 2017). The availability of high-throughput detection methods, breadth of chemical moieties and potential

For a color version of all figures in this book, see www.iste.co.uk/saleh/virology.zip.

Virology,
coordinated by María-Carla SALEH and Félix AUGUSTO REY. © ISTE Ltd 2021

impact on biological functions has sparked vast interest in this emerging field of epitranscriptomics.

The presence of RNA modifications on viral RNAs was first reported in the 1970s. However, recent emergence of select detection reagents, advances in next generation sequencing technologies and identification of enzymes and proteins involved in RNA modification function has reinvigorated the field of viral epitranscriptomics.

4.1.2. *How are epitranscriptomic marks installed?*

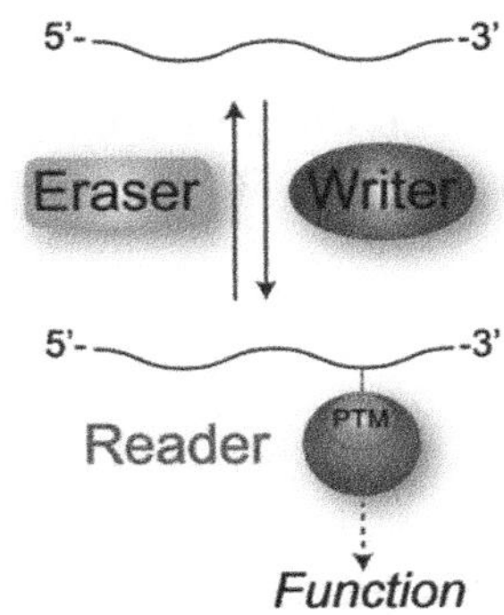

Figure 4.1. *Enzymes involved in the installation, removal and function of post-transcriptional RNA modifications (PTMs)*

The biogenesis and function of RNA modifications is directed by three enzymes: the writers, erasers and readers. Writer enzymes are those which deposit the chemical moiety onto RNA, while eraser enzymes catalyze the removal of the modification (Figure 4.1). The fact that these post-transcriptional modifications are transient and dynamic also presents a significant role for these chemical moieties in cellular signaling pathways. For example, a modification might be upregulated during viral infection to initiate an immune response, but levels can return to normal once the stress has been eliminated (Edupuganti *et al.* 2017). Reader enzymes are critical interpreters for the function of RNA modifications such as immune sensing or changes in RNA localization; however, very few of these enzymes have been identified.

4.2. The tools of RNA modification discovery

Different approaches have been employed to detect RNA modifications. Below, we describe methods that agnostically identify all RNA modifications as well as the recent next generation sequencing approaches that positionally resolve single RNA modifications on RNA.

4.2.1. *Chromatography and mass spectrometry*

RNA modifications were first detected through a combination of radiolabeling and paper chromatography approaches (Wei *et al.* 1975; Dimock and Stoltzfus 1977; Kellner, Burhenne *et al.* 2010). Specifically, cells were first labeled with tritiated (^{3}H) methionine which facilitated the incorporation of radiolabeled methyl-containing RNA modifications through the methyl donor S-adenosyl methionine (SAM). SAM is a pivotal metabolite that is incorporated into macromolecules through different methyltransferases to function in various cellular processes (Fontecave *et al.* 2004). Radiolabeled methylated-RNAs were then digested into mononucleotides and resolved by thin-layer chromatography which separated the mononucleotides based on size and shape (Figure 4.2A). This chromatographic approach together with known standards is able to differentiate different methylated nucleotides such as 2'O-methyladenosine (Am) from N6-methyladenosine (m^{6}A) and N6,N6-dimethylated adenosine (m$^6{}_6$A), etc. (Kruse *et al.* 2011). Moreover, quantification of the radiolabeled nucleotides permitted an extrapolation of the extent of methylation in RNA. While this method was present for the birth of the viral epitranscriptomics field, it has fallen out of favor in recent decades as new methods have become more available.

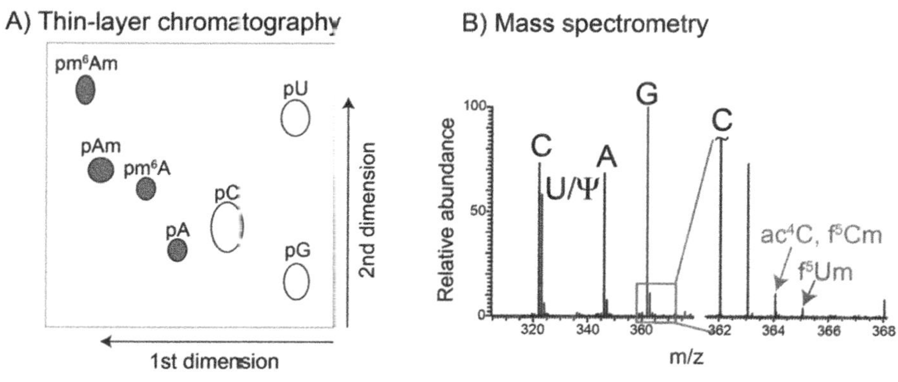

Figure 4.2. *Schematic of two methods to identify multiple PTMs in RNA. (A) Thin-layer chromatography can be used to analyze multiple PTMs within a radiolabeled RNA. Schematic adapted from Kruse* et al. *2011. (B) Mass spectrometry identifies and determines the relative abundance of the canonical nucleotides (C, U/Ψ, A and G) as well as that of lower abundant modified nucleotides*

Mass spectrometry, similar to chromatography, permits the detection of numerous RNA modifications, albeit at increased sensitivity and resolution (Figure 4.2B). There are two main types of mass spectrometric analysis. The first involves chromatography beforehand to separate similar modifications and the second type uses fragmentation to separate related species. Similar to the sample preparation for thin-layer chromatography, mass spectrometric analysis of RNA modifications requires a pool of mono-ribonucleotides. Specifically, RNA is digested using the exonuclease nuclease P1 which severs the nucleotide linkages, followed by phosphodiesterase treatment which removes the phosphate groups from all of the mononucleotides (Rose *et al.* 2016; McIntyre *et al.* 2018). To date, both liquid chromatography mass spectrometry (LC-MS) and mass spectrometry ionization techniques have been employed to analyze RNA modifications. LC-MS combines the benefit of a separation technique such as chromatography with the high-resolution and sensitive detection of mass spectrometry. Here, the pool of mononucleotides is first separated based on charge, size or a combination thereof using a liquid matrix and then directed for MS analysis. While LC-MS enables high-resolution detection and quantification of modifications, this method requires chemical standards that are either run alongside a sample or spiked into the sample for comparison. To date, limited standards are commercially available and, if available, may be prohibitively expensive. To circumvent the limitation of LC-MS, mass spectrometry ionization techniques such as matrix-assisted laser desorption/ionization (MALDI), electrospray ionization (ESI), nanospray, etc. as well as various ion detection techniques (time of flight (TOF), ion mobility (IMS), Orbitrap, Quadrupole, etc.) have been used to detect RNA modifications (Rose *et al.* 2016; McIntyre *et al.* 2018). Because mass spectrometry detects mass-to-charge ratios (m/z), depending on the instrument, molecules in the attomole range can be identified. Thus, mass spectrometry ionization is an unbiased and highly sensitive analytical approach for detecting RNA modifications.

Because mass spectrometry ionization detects mass-to-charge ratios, RNA modifications with a single mass (e.g. m^6A, m^8A, m^1A, m^2A, Am; Figure 4.3A and B) in the same single signal may be identified. These isomeric species may be further dissected by additional analyses (Figure 4.3B and C). In particular, a selected ion can be further fragmented such that other components or separations can be applied towards identification. Such methods of separation include chromatography separation by charge (Chan *et al.* 2010), ion mobility mass spectrometry (Clemmer and Jarrold 1997; Verbeck *et al.* 2002; Rose *et al.* 2015) and others (McLuckey *et al.* 1992; McIntyre *et al.* 2018). However, this extra challenge of isomer identification is a drawback to this type of analysis and requires high-resolution instruments, high quality samples and time to analyze fragmentation data.

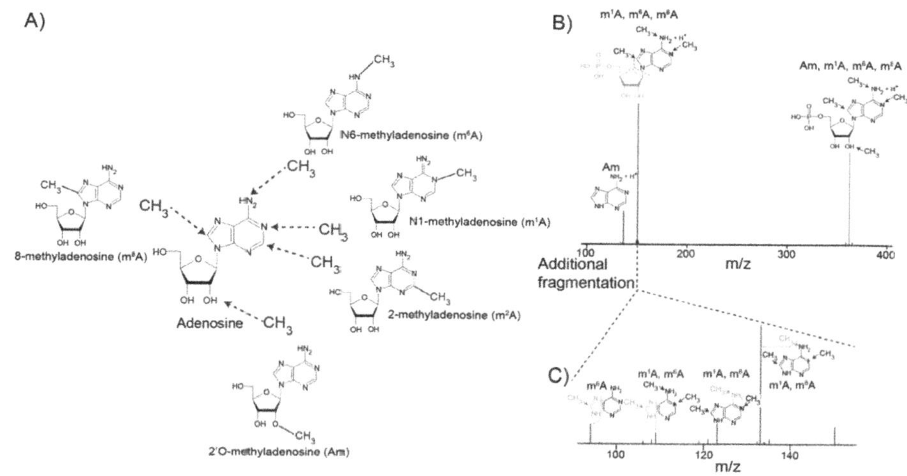

Figure 4.3. *Mass spectrometry distinguishes isobaric species of monomethyladenosine. (A) Different methyladenosine species. (B) A mass spectrum showing methyladenosine, methyladenine and adenine, indicating that ribonucleotide was 2'O-methylated. (C) A mass spectrum following additional fragmentation of methyladenine showing the detection of distinct adenine modifications. The mass spectra shown are included with permission from Will McIntyre (McIntyre* et al. *2018)*

4.2.2. *Sequencing methods for PTM detection*

Sequencing-based platforms are the newest means of detecting RNA modifications and offer some unique benefits compared to chromatography and mass spectrometry. In particular, sequencing demonstrates the context in which the modification occurs, which can aid in the description of consensus sequence sites, such as the well-defined DRACH sequence for m^{6}A placement (Dominissini *et al.* 2012; Liu *et al.* 2014). However, the major drawback of sequencing methods is the analysis of one epitranscriptomic mark at a time. While mass spectrometry and chromatography approaches allow for the detection of many or all known modifications during one analysis, the current sequencing methods only permit the detection of a single RNA modification in one analysis. This limitation is, in part, the result of the limited availability of chemical and molecular reagents to modify and/or facilitate the detection of a handful of RNA modifications as well as the complicated methods involved during library preparation.

Sequencing methods used for RNA modification detection may be broadly separated into two strategies, namely employing an antibody that specifically recognizes an RNA modification or chemical modification of PTMs (Figure 4.4). In both strategies, total RNA is segmented into 100–200 nucleotide oligonucleotides. For identification using antibodies, RNA segments containing the specific RNA modification are immunoprecipitated, RNA-seq libraries are prepared and next generation sequence analysis of immunoprecipitated and non-immunoprecipitated RNAs is compared (Figure 4.4A). Although there are over 160 modifications that have been described in a biological context, to date, there are only 14 modifications for which antibodies are available. The RNA modification-specific antibodies presently available recognize 1-methyladenosine (m^1A), 2-methyladenosine (m^2A), N6-methyladenosine (m^6A), 8-methyladenosine (m^8A), N6,N6-dimethyladenosine ($m^6{}_6A$), 4-methylcytidine (m^4C), 5-methylcytidine (m^5C), N4-acetylcytidine (ac^4C), N2,N2-dimethylguanosine ($m^2{}_2G$), N2,N2,7-trimethylguanosine ($m^{2,2}{}_7G$), 5-hydroxymethylcytidine (hm^5C), 5-formylcytidine (f^5C), 7-methylguanosine (m^7G), inosine (I) and pseudouridine (ψ) (Feederle and Schepers 2017; Abcam; Diagenode).

There are drawbacks to this method, such as cross-reactivity of the antibody to two different RNA modifications. For example, the anti-m^6A antibody detects m^6A and m^6Am sites (Tan *et al.* 2017; Grozhik and Jaffrey 2018). Antibodies can recognize small regions of the nucleotide but detecting both the ribose and nucleoside base alterations at the same time is a challenge. This makes identification of structurally similar modifications difficult and ambiguous.

Some newer methods have eliminated the reliance on antibodies. One of these methods, developed by Hong and colleagues (2018), incorporates an atom-specific replacement of oxygen with either sulfur (S) or selenium (Se) into the 4-position of deoxythymidine triphosphate, thus producing 4S-dTTP or 4Se-dTTP. This weakens the hydrogen-bonding capability of the m^6A pairing while maintaining the A–T pairing (Hong *et al.* 2018). The benefit of this alteration allows for changes that can be carried through reverse transcription, where m^6A preferentially binds with 4S-T. This also sometimes causes aborted transcripts. Both the sequence changes to 4S-T and the sequence terminations can be detected using sequencing. This new and exciting method would benefit from further testing of how diverse modifications may react under the same treatment.

A newer sequencing-based method, called Site-specific Cleavage And Radioactive-labeling, followed by Ligation-assisted Extraction and Thin-layer chromatography (SCARLET) has been applied to determine the precise location of m^6A modifications, a significant improvement over the antibody affinity methods that extrapolate the position of the m^6A modification based on the sequence motif in a 100–200 nucleotide region (Recasens *et al.* 2010). SCARLET uses a combination of cleavage, radiolabeling, enzymatic digestion and thin-layer chromatography to detect m^6A sites (Zheng *et al.* 2013). Here, chimeric oligos with 2'O-methylated/2'H bind to the region of interest in isolated RNA or poly-adenylated RNA. This is then digested using RNase H to produce a cleavage product that is directly 5' of the site of interest. The newly exposed 5' end is then ^{32}P-labeled and bound to a 116-nucleotide splint oligo using DNA ligase. Next, samples are treated with RNase T1 and RNase A to remove all RNA remaining. The ^{32}P-labeled site of interest remains intact in the DNA sequence as either DNA-^{32}P-m^6A(C)p or DNA-^{32}P-A-(C)p. These fragments can then be separated via denaturing gel electrophoresis, and the fragments excised and nuclease P1 digested into mononucleotides. Finally, the presence of m^6A or A at the site in the product is resolved using thin-layer chromatography (Recasens *et al.* 2010).

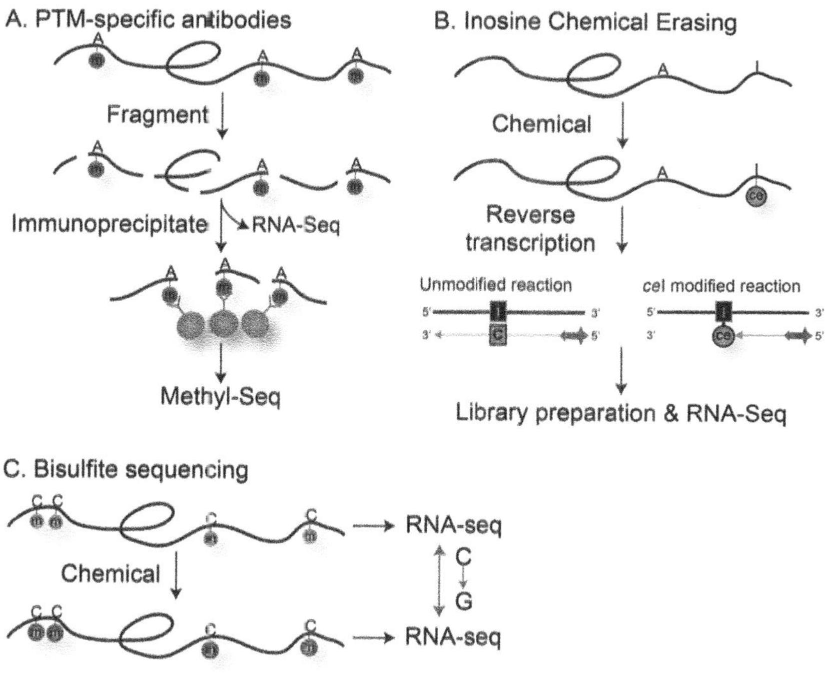

Figure 4.4. *Schematic of example methods to detect individual RNA modifications. (A) Antibodies specific to a PTM immunoprecipitate RNA containing the individual PTM and the PTM is detected by RNA-seq analysis. (B, C) Chemical labeling of individual RNA modifications. (B) Overview of cyanoethylation, primer extension and RNA-seq analysis to identify inosines. Schematic adapted from Sakurai* et al. *2010. (C) Chemical modification and RNA-seq analysis to identify m^5C and hm^5C*

Similar to antibody-based isolation, chemical-based techniques have been applied to identify specific RNA modifications (Figure 4.4B and C). Nucleotide resolution of the RNA modification results from either termination of the primer extension reaction (Figure 4.4B) or a sequence read of a different nucleotide (Figure 4.4C). For example, a cyanoethylation chemical reaction will modify inosines but not adenosines, where subsequent reverse transcription reactions are inhibited by the modified inosine, and accurate inosine identification is determined by sequence analysis of unmodified and chemically modified RNAs (Figure 4.4B) (Sakurai *et al.* 2010). Bisulfite sequencing was initially developed to identify 5-methyldeoxycytosine in DNA, but more recently has also been applied to detect 5-methylcytosine (m^5c) in RNA (Schaefer *et al.* 2009; Amort *et al.* 2017). Here, a bisulfite treatment of RNA under acidic conditions deaminates cytosine to uracil, while m^5C and 5-hydroxymethylcytosine (hm^5C) are protected from this alteration such that U–A and C–G incorporations can be compared between treated and untreated sequences (Figure 4.4C) (Schaefer *et al.* 2009; Amort and Lusser 2017). Notably, bisulfite sequencing does not provide a clear differentiation between m^5C and hm^5C, nor does this approach determine whether the nucleotide was 4-methylcytosine (m^4C). While chemical labeling of RNA can provide nucleotide resolution, the limited number of chemical reagents has hindered analysis of all RNA modifications.

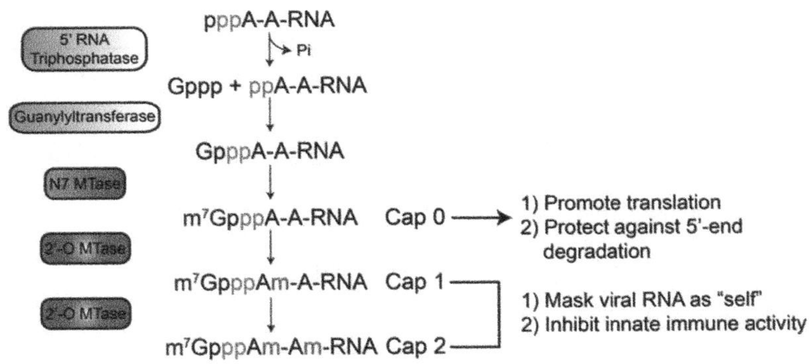

Figure 4.5. *Schematic of the enzymes involved in installing the 5'-cap on cellular mRNAs. The functions of Cap 0, Cap 1 and Cap 2 are also depicted*

Both antibody and chemical labeling approaches to detecting RNA modifications rely on the generation of cDNA, where information on the natural RNA is lost during reverse transcription. More recently, the development of nanopore sequencing has enabled sequencing of a nascent strand of RNA by ionic current, where deviations from the signal for any of the four canonical bases reveal possible RNA modifications

(Smith *et al.* 2019). While the analysis and calling of bases is in its infancy, this new technology method offers sequence context, does not create errors during cDNA copying and also allows direct visualization of which RNA modifications are present.

4.3. RNA modifications deposited by viral enzymes

In eukaryotic cells, the 5' end of cellular mRNAs is predominantly co-transcriptionally modified in the nucleus by the addition of an N7-methylguanosine (m^7G) cap (Wei *et al.* 1975), although RNAs in the cytoplasm that have lost the m^7G cap can also be re-capped (Otsuka *et al.* 2009; Trotman *et al.* 2017). A typical 5'-cap structure is added to mRNAs via three consecutive reactions (Figure 4.5) (Topisirovic *et al.* 2011). First, a 5' RNA triphosphatase cleaves the γ-phosphate on a newly synthesized RNA (pppNp-RNA, where N is the first nucleotide of the RNA, and p denotes the phosphate groups) to form a diphosphate RNA (ppNp-RNA) intermediate (Figure 4.5). Next, GTP is covalently bound to a guanylyltransferase via the γ-phosphate to produce a GTPase-Gp intermediate, which results in the addition of the GMP molecule to the 5' diphosphate RNA via a 5'–5' linkage and results in GpppNp-RNA, also known as the Cap 0 structure (Figure 4.5). Third, donor guanine-N7-methyltransferase (N7 MTase) together with the methyl donor S-adenosyl-L-methionine (SAM) deposits a methyl group on the guanosine at the N7 position to produce m^7GpppNp (Figure 4.5). Finally, the capping reaction is completed by a nucleoside-2'O-methyltransferase (2'O MTase) that methylates the ribose at the 2'O position on the first, and sometimes second, nucleotide to generate the Cap 1 or Cap 2 structure (Figure 4.5). In the nucleus, the 5'-cap protects mRNAs from 5' exoribonuclease degradation, directs pre-mRNA splicing and aids mRNA transport out of the nucleus (Topisirovic *et al.* 2011). In the cytoplasm, the 5'-cap ensures efficient translation and facilitates mRNA stability by protecting the mRNA from 5' exoribonuclease digestion (Topisirovic *et al.* 2011). For those RNAs harboring a 2'O-methyl modification on the penultimate nucleotide, the RNA is recognized as "self", whereas RNAs without this modification activate the innate immune response (Hyde and Diamond 2015).

4.3.1. *Capping of 5' end of viral RNA by viral methyltransferases*

Considering the numerous functions attributed to the cap structure on cellular mRNAs, it is not surprising that many viruses similarly modify the 5' end of viral RNA. Interestingly, the m^7G modified nucleotide protecting the 5' end was first identified on reovirus and vaccinia virus mRNAs, and these studies laid the groundwork to identifying 5'-caps on cellular mRNAs (Furuichi *et al.* 1975a, 1975b; Rose 1975; Wei and Moss 1975). Viruses with a nuclear component in the infectious cycle may subvert the cellular nuclear capping machinery to cap viral RNA. Other viruses, however, "snatch" caps from cellular mRNAs or encode viral capping enzymes (Dong *et al.* 2008; Decroly *et al.* 2012; Ramanathan *et al.* 2016; Decroly and Canard 2017). We refer readers to the outstanding reviews by Decroly *et al.* (2012, 2017) and Ramanathan *et al.* (2016) for a complete overview of the mechanisms and functions of viral RNA caps. Notably, viruses that encode methyltransferase proteins may employ different pathways to install the guanosine at the 5' end of viral RNA and deposit a methyl group at position 7 on the guanosine. Interestingly, the canonical m^7G caps on some alphaviruses, a flavivirus and human immunodeficiency virus type 1 (HIV-1) RNA, have also been shown to be di- and tri-methylated guanosine caps, although the pathway and functional significance of these alternate cap structures have not been thoroughly investigated (Hsuchen and Dubin 1976; van Duijn *et al.* 1986; Dong *et al.* 2008; Yedavalli and Jeang 2010). Regardless of the capping mechanism, m^7G caps on viral RNA promote the efficient translation of viral proteins and protect the viral RNAs from cellular 5'-exonuclease degradation (Decroly *et al.* 2012; Ramanathan *et al.* 2016).

4.3.2. *2'O-methylation of viral RNA*

4.3.2.1. *2'O-methylation at the 5' end of viral RNA*

Methylation of the ribose 2'-hydroxyl group (2'OH) can occur on all four ribonucleotides and on non-canonical nucleotides such as pseudouridine and inosine in different cellular and viral RNAs (Ayadi *et al.* 2019). While the biological function of internal 2'O-methylated nucleotides is incomplete, modification of the first and/or second nucleotides of cellular mRNAs with a 2'O-methylation moiety to produce a Cap 1 or Cap 2 structure (Figure 4.5) masks cellular mRNAs as "self". The function, however, of this same modification on nucleotides in an RNA is still opaque (Hyde and Diamond 2015). To limit recognition by the host, viruses include a Cap 1 structure on the viral RNA by either "snatching" caps from cellular mRNAs or by encoding a viral 2'O-methyltransferase. In particular, viruses in the *Corona-*, *Arteri-*, *Rhabdo-*, *Filo-*, *Paramyxo-*, *Pox-* and *Reoviridae* families and some viruses in the *Flaviviridae* family encode their own methyltransferase enzyme. This enzyme modifies the first nucleotide of the viral RNA by N7-methylation of the guanosine cap and 2'O-methylates the ribose 2'OH of the penultimate nucleotide (Decroly *et al.* 2012). Daffis and colleagues (2010) determined that the addition of the 2'O-methyl modification on the penultimate nucleotide of the viral RNA camouflaged the viral RNA as a cellular mRNA and thus avoided the activation of the innate immune response (Daffis *et al.* 2010). The innate immune response pathway is an important cellular process that functions to restrict pathogens. Cellular sensors recognize particular molecular signatures in pathogens such as double-stranded RNA, 5'-triphosphate RNA or non-2'O-methylated RNA, which activate signaling pathways to induce type I interferon and cytokine expression (McFadden *et al.* 2017). In turn, the release of interferon-α (IFN-α) and IFN-β binds the type I interferon receptor (IFNAR) and initiates signaling in neighboring cells through the Jak kinase/STAT pathway resulting, in the expression of interferon-stimulated genes (ISGs) with specific antiviral activities. For example, viral gene expression may be potently inhibited by the ISGs encoding protein kinase R (PKR), ISG56 (IFIT1) and ISG54 (IFIT2), which function to restrict cellular translation.

The function of the 2'O-methylation modifications as a mark of "self" was first demonstrated in West Nile virus (WNV). WNV, a virus in the *Flaviviridae* family of viruses, has a positive-sense RNA genome. The viral NS5 protein has both methyltransferase and RNA-dependent RNA polymerase activities. The methyltransferase domain in NS5 introduces and methylates the guanosine cap and ribose 2'OH moiety of the first nucleotide to create a Cap 1 structure (Zhou *et al.* 2007). Notably, WNV with deficient 2'O-methyltransferase activity was found to be attenuated in wild-type mice, but was virulent in mice lacking *Ifnar,* the gene encoding the IFN receptor, thus presenting a role for this RNA modification in innate immune sensing (Daffis *et al.* 2010). However, embryonic fibroblasts isolated from *Ifnar*-deficient mice expressed the same amount of IFN mRNA when infected with either wild-type or 2'O-methyltransferase-deficient viruses. These data revealed that the intracellular sensors recognizing the viral RNA and inducing IFN expression remained intact and suggested that an IFN-induced gene might act on non-2'O-methylated WNV RNA. Indeed, when cells were pretreated with IFN, the 2'O-methyltransferase-deficient virus was more affected compared to the wild-type virus (Daffis *et al.* 2010). Daffis and colleagues (2010) showed that the ISGs encoding IFIT (Interferon-Induced protein with Tetratricopeptides) proteins recognized non-2'O-methylated RNA genomes from three different viruses, namely WNV, poxvirus and coronavirus, and restricted the viral infection. The mechanism by which the IFIT proteins are able to discern "self" and "non-self" RNA was revealed by recent X-ray crystallography. In particular, tetratricopeptide subdomains of IFIT assemble around an internal RNA binding tunnel (Abbas *et al.* 2017), with a space constraint of only being able to accommodate a m^7G capped RNA and non-2'O-methylated penultimate nucleotides. The cytoplasmic RNA senor, Retinoic Acid Inducible Gene I (RIG-I), initiates innate immune signaling when uncapped RNAs containing a triphosphate at the 5' end of RNA are recognized (Luo *et al.* 2012). Like IFIT, however, RIG-I can also differentiate between a capped RNA without and with the penultimate 2'O-methylated nucleotide (Schuberth-Wagner *et al.* 2015; Devarkar *et al.* 2016; Ramanathan *et al.* 2016). Thus, IFIT and RIG-I are able to distinguish between cellular mRNA and pathogen RNAs based on the presence or absence of 2'O-methylated nucleotides at the 5' end of RNA (Schuberth-Wagner *et al.* 2015; Devarkar *et al.* 2016; Ramanathan *et al.* 2016; Abbas *et al.* 2017). IFIT was shown to limit viral gene expression by restricting the translation of viral proteins. Such translational inhibition may occur by, first, occluding the interaction of cellular cap-binding enzymes for the m^7G-cap (Habjan *et al.* 2013; Kimura *et al.* 2013; Kumar *et al.* 2014), and, secondly, by interacting with the eukaryotic initiation factor eIF3 to inhibit the translation of viral mRNA (Hinnebusch 2006). In particular, the binding of IFIT with eIF3 can inhibit the assembly of the ternary complex containing eIF2, GTP and the methionine initiator tRNA (Met-tRNAi) and/or the preinitiation complex composed of the ternary complex, mRNA, and the 40S ribosome (Hui *et al.* 2003, 2005; Terenzi *et al.* 2006). Thus, the 2'O-methyl modification ensures the efficient translation of viral proteins by evading antiviral factors induced by type I interferon in response to viral infection (Figure 4.6).

The cytoplasmic innate immune response RNA sensor Melanoma Differentiation-Associated protein 5 (MDA5) recognizes non-2'O-methylated viral RNA as "non-self" and induces the innate immune response pathway (Figure 4.6) (Züst *et al.* 2011). Züst and colleagues dissected this pathway using wild-type and mutant coronaviruses deficient in 2'O-methyltransferase activity. The 5' end of the coronavirus positive-sense RNA contains a Cap 1 structure that is installed by the viral Nsp16 protein, an N7-methylguanosine and ribose 2'O-methyltransferase (Decroly *et al.* 2008; Chen *et al.* 2009; Bouvet *et al.* 2010). Infection studies comparing wild-type viruses and mutant coronaviruses with deficient 2'O-methyltransferase activity revealed that the replication kinetics were decreased with the mutant viruses, yet these viruses induced higher IFN-β levels (Züst *et al.* 2011). Infection of *Ifnar*-deficient macrophages with the Nsp16 mutant viruses similarly induced high expression levels of IFN-β. Together, these data indicated that the secondary IFN-signaling pathway was intact. Instead, the increased IFN production following infection with the Nsp16 mutant coronaviruses was likely in response to a cytosolic RNA sensor (Züst *et al.* 2011). Coronavirus infection in macrophages induces IFN production via the RNA sensor MDA5 (Roth-Cross *et al.* 2008; Zalinger *et al.* 2015). Therefore, as anticipated, the infection of *MDA5*-deficient macrophages with wild-type and 2'O-methyltransferase-deficient coronaviruses failed to induce IFN expression. Further support for the role of MDA5 in sensing non-2'O-methylated RNA was demonstrated when it was shown that 2'O-methyltransferase-deficient coronaviruses in mice deficient in the RNA sensors MDA5 and Toll-like receptor 7 (TLR7) replicated to the same levels as in *Ifnar*-deficient mice (Züst *et al.* 2011). Therefore, installation of a 2'O-methyl group on the first nucleotide of viral RNA camouflages the viral RNA as "self" to thwart recognition by RNA sensors MDA5, TLR7 and RIG-I, as well as by the IFN-induced translation inhibitors, IFIT1 and IFIT2 (Figure 4.6).

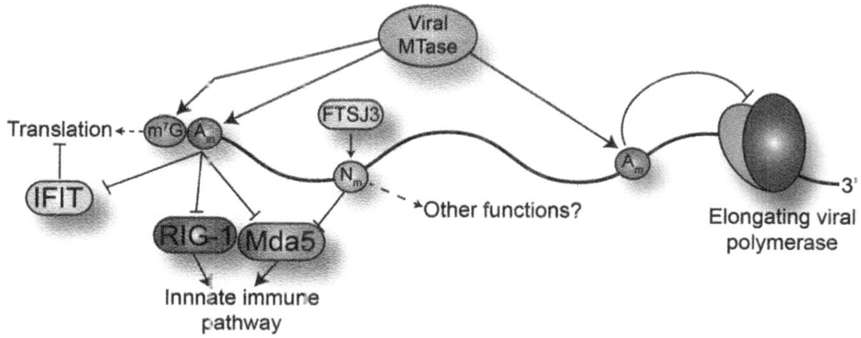

Figure 4.6. *Biological implication of capping the 5' end and 2'O-methylating viral RNA. Viral methyltransferase (MTase) installs a 5'-cap structure and 2'O-methylates the penultimate nucleotide to (1) inhibit sensing by the cytosolic RNA sensor MDA5 and prevent transcription of ISGs, and (2) block an interaction with IFIT to downregulate translation. Deposition of internal 2'O-methyl groups on adenosines (Am) by the viral 2'O-MTase inhibits the elongation of the viral RNA-dependent RNA polymerase during replication. Internal 2'O-methyl marks added by the cellular FTSJ3 2'O-methyltransferase prevent sensing by MDA5*

4.3.2.2. *2'O-methylation in the viral RNA*

In addition to 2'O-methylating the penultimate adenosine nucleotide of the viral genome, the NS5 protein of WNV and the closely related dengue virus (DENV) was also found to methylate internal adenosines (Dong *et al.* 2012). This activity was demonstrated by measuring the *in vitro* methylation of uncapped RNA, where the NS5 protein, or the methyltransferase (MTase) domain of NS5 alone, was found to specifically 2'O-methylate oligonucleotides containing polyAs, but not polyGs, polyCs or polyUs. Interestingly, a 12-nucleotide oligomer containing just adenosine or N6,N6-dimethyladenosine (m^6_6A) was 2'O-methylated by NS5 MTase, but not an oligomer containing 2'O-methyladenosines (Am) (Dong *et al.* 2012). NS5 MTase was found to deposit 2'O-methyl groups on homogeneous oligonucleotides, an uncapped and truncated segment of viral RNA, and 18S and 28S rRNA, suggesting that this activity lacked sequence specificity. Notably, the authors identified an estimated 3.4 ± 0.5 2'O-methylated adenosines per genome on viral genomes isolated from infected *Aedes albopictus* cell culture supernatants (Dong *et al.* 2012). Viral translation from an *in vitro* transcribed viral RNA containing internal

2'O-methyladenosines was modestly decreased following transfection compared to an unmodified viral RNA. Notably, however, the internal 2'O-methyladenosines in the viral RNA significantly decreased replication, particularly the elongation step of the viral RNA-dependent RNA polymerase (Figure 4.6) (Dong *et al.* 2012). Short RNAs containing 2'O-methylated uridine (Um), guanosine (Gm) and adenosine (Am) strongly inhibit cytokine and IFN production from an activated endosomal immune sensor TLR7 (Robbins *et al.* 2007; Petes *et al.* 2017). Therefore, internal 2'O-methyl modifications in viral RNA may alter gene expression. While such a function remains to be demonstrated, it is also possible that 2'O-methyl marks may promote subversion of the host by restricting innate sensing during virus entry, which for WNV and DENV occurs via the endosome (Cruz-Oliveira *et al.* 2015). Interestingly, the internal 2'O-methyltransferase activity is not unique to WNV and DENV NS5 protein. Recently, the polymerase or L (large) protein of Ebola virus, a negative-sense RNA virus, was also shown to 2'O-methylate adenosines in the Ebola virus RNA. This RNA modification activity was virus-specific, as human metapneumovirus (hMPV), another negative-strand RNA virus but in a different virus family to Ebola virus, lacked this activity (Martin *et al.* 2018). At present, the functional consequence of these internal modifications is unknown (Martin *et al.* 2018).

Internal 2'O-methylated nucleotides in human immunodeficiency virus type 1 (HIV-1) and murine leukemia virus (MLV) RNA were recently described (McIntyre *et al.* 2018; Courtney *et al.* 2019a, 2019b; Ringeard *et al.* 2019). In the HIV-1 genomes, the cellular 2'O-methyltransferase FTSJ3 was identified as the 2'O-methyl writer enzyme (Ringeard *et al.* 2019). The interaction of the viral Tat protein with an RNA structure at the 5' end of the viral genome known as the transactivation response (TAR) element facilitates the recruitment cellular RNA polymerase II to promote efficient transcription (Ott *et al.* 2011). Interestingly, Ringeard and colleagues (2019) found that the cellular TAR RNA binding protein (TRBP), which binds HIV-1 RNA, also forms a complex with MAT2A, an S-adenosyl-L-methionine (SAM) metabolism protein and FTSJ3 (Ringeard *et al.* 2019). Moreover, FTSJ3 isolated from cells, or recombinantly expressed, was shown to *in vitro* 2'O-methylate radiolabeled m^{7}GpppG to form m^{7}GpppGm in the context of m^{7}G-capped HIV-1 RNA. This same activity was demonstrated with an oligonucleotide containing 27-adenosines or uridines or guanosines but was absent when an oligonucleotide containing 2'O-methylated adenosines was used. The authors used RiboMeth-Seq together with primer extension validation analyses and identified seventeen 2'O-methylated nucleotides deposited by FTSJ3 in HIV-1 RNA that was isolated from viral particles (Ringeard *et al.* 2019). The function of these internal 2'O-methylated nucleotides was elegantly determined by first extracting viral RNA from virions produced in FTSJ3 knockout cells, and then transfecting this RNA into a pro-monocytic cell line. IFN-α and IFN-β mRNA levels were dramatically upregulated in response to activated MDA5 cytosolic RNA sensor by HIV-1 RNA without (or having a reduced number of) 2'O-methylated nucleotides (Ringeard *et al.* 2019). Thus, internal 2'O-methylations may also enable viruses to escape immune sensing (Figure 4.6). Internal 2'O-methyladenosines in DENV RNA limit elongation by the viral RNA-dependent RNA polymerase (Dong *et al.* 2012). It is therefore possible that 2'O-methylated nucleotides in the HIV-1 genome might similarly hinder reverse transcriptase activity, although such a function remains to be shown.

Recently, mass spectrometry analyses of isolated viral RNAs from DENV, Zika virus (ZIKV), hepatitis C virus (HCV), poliovirus and HIV-1-infected cells showed that 2'O-methyladenosine, 2'O-methylguanosine, 2'O-methylcytosine and 2'O-methyluridine modifications were present on the viral genomes (Lichinchi *et al.* 2016; McIntyre *et al.* 2018). It is likely that a fraction of the 2'O-methyladenosines detected on the DENV and ZIKV genomes arise from the ability of viral methyltransferase to modify internal adenosines. Since the DENV methyltransferase does not 2'O-methylate guanosines, cytosine or uracil (Dong *et al.* 2012), these other 2'O-methylated nucleotides are likely deposited by cellular 2'O-methyltransferases. That 2'O-methylated nucleotides are present on isolated HCV and poliovirus RNA from infected cells and virions is also curious (McIntyre *et al.* 2018), particularly as the single-stranded positive-sense RNA genomes of both viruses are uncapped and the viral open reading frame does not encode a methyltransferase (Lindenbach *et al.* 2013). These 2'O-methyl marks on the viral RNA are therefore likely to be deposited by cellular 2'O-methyltransferases.

Fibrillarin 2'O-methylates ribosomal RNA (rRNA) in the nucleolus (Reichow *et al.* 2007). This cellular 2'O-methyltransferase was found by Deffrasnes and colleagues (2016) to be required by Hendra and Nipah viruses, as well as by other viruses within the same *Paramyxoviridae* family, such as the measles and mumps viruses (Deffrasnes *et al.* 2016). Specifically, the knockdown of fibrillarin using target-specific siRNAs decreased viral titers, which could be restored by expressing an siRNA-resistant form of fibrillarin. Data from an assay using a minigenome and single-cycle infection assay revealed that fibrillarin affected the synthesis of viral RNA. At present, it is unknown whether fibrillarin 2'O-methylates Hendra and Nipah virus mRNAs or negative-sense genomes to impact viral replication by affecting translation, transcription and/or replication by the viral polymerase protein. Since 2'O-methylation of rRNA is important for pre-rRNA processing, an alternate explanation for the effect on viral replication might be a consequence of altering the abundance, 2'O-methylation status or composition of functional ribosomes following fibrillarin depletion. Such changes in ribostasis could impact the translation of viral proteins and, ultimately, viral RNA synthesis (Tollervey *et al.* 1991; Mauro and Edelman 2002; Erales *et al.* 2017; Shi *et al.* 2017). Deffrasnes *et al.* (2016), however, showed that the depletion of fibrillarin had no effect on influenza virus infection, suggesting that the effect on Hendra virus was not a result of ribosome biogenesis or translation deficiencies.

The 2'O-methyl modification at the 5' end of RNA is an important marker of "self" (Hyde and Diamond 2015). The function of internal 2'O-methylation in cellular and viral RNAs is more elusive, particularly as reader enzymes for the 2'O-mark are unknown and installing a 2'O-methyl moiety on the ribose would not predictably affect the RNA structure by changing the base pairing (Prusiner *et al.* 1974). RNA containing 2'O-methylated nucleotides is resistant to hydrolytic cleavage, and the modification can stabilize nucleotide conformation to restrict the shape and flexibility of the RNA strand (Lane and Tamaoki 1967; Trim and Parker 1970). Thus, 2'O-methylated nucleotides in viral RNA could change interactions with cellular components, viral polymerase processivity or avoid immune sensing (Figure 4.6). Additionally, actions required for efficient replication such as RNA unwinding or changes in RNA conformations required for specific steps in the infectious cycle (Iglesias and Gamarnik 2011; Nicholson and White 2015) might be impacted by a 2'O-methyl moiety. A number of high-throughput sequencing approaches have been developed to identify 2'O-methylated nucleotides (Marchand *et al.* 2016; Dai *et al.* 2017; Incarnato *et al.* 2017 *et al.* 2017), which, together with known RNA structures and RNA–protein interactions, might help unravel the role of this modified nucleotide on viral gene expression.

4.4. Editing of viral RNA by cellular enzymes

The post-transcriptional process that modifies the sequence of the original RNA is known as RNA editing. Two such RNA editing events are the isomerization of uridine-to-5-ribosyl uracil (or pseudouridine; Figure 4.7) and the deamination of adenosine-to-inosine (Figure 4.8). These RNA editing events are catalyzed by known writer enzymes (Slotkin and Nishikura 2013; George *et al.* 2014; Spenkuch *et al.* 2014; Li *et al.* 2016). Whether these RNA editing events can be undone remains to be established. Indeed, the eraser enzymes that would convert pseudouridine-to-uridine and inosine-to-adenosine are unknown. Therefore, the editing of viral RNA compared to the reversible addition/removal of chemical moieties has distinct consequences on viral gene expression.

4.4.1. *Editing of uridine-to-pseudouridine (Ψ)*

Pseudouridine is the most abundant RNA modification and is often referred to as the fifth nucleotide (Cohn and Volkin 1951). Pseudouridine is installed in RNA by pseudouridine synthases (PUS), an enzyme that isomerizes uridine to form 5-ribosyl uracil (Figure 4.7) (Cohn 1959). In converting uridine to pseudouridine, the Watson–Crick base-pairing with adenosine is maintained and expanded as pseudouridine is also able to base pair with other nucleotides. Pseudouridine was originally discovered in rRNA and tRNA. More recently, however, transcriptome-wide analysis of pseudouridine by chemically labeling with CMCT (cyclohexyl-N'-(2-morpholinoethyl) carbodiimide metho-p-toluene-sulfonate) and next generation sequence analyses identified the presence of pseudouridine in non-coding RNAs (ncRNAs) and mRNAs (Carlile *et al.* 2014; Lovejoy *et al.* 2014; Schwartz *et al.* 2014; Li *et al.* 2015). To date, there are only a few reports identifying pseudouridine in viral RNAs and the role of pseudouridine in modulating viral gene expression is mostly unknown.

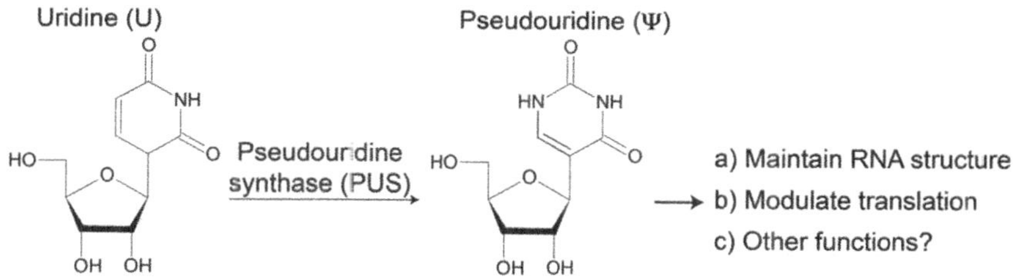

Figure 4.7. *Pseudouridine synthase (PUS) isomerizes uridine to form pseudouridine (5-ribosyl uracil) to significantly affect mRNA function*

A "tRNA-like" domain is present in the last 82 nucleotides of the genomic RNA of plant turnip yellow mosaic virus (TYMV), a virus in the *Tymoviridae* family. One of the stem-loop structures in this "tRNA-like" domain resembles the T-arm of tRNA that contains pseudouridine (Becker 1998). Becker (1998) demonstrated that the TYMV RNA fragment was pseudouridylated following incubation with either yeast extract or purified PUS1 and PUS4 enzymes. Another plant virus in the *Bromoviridae* family, namely brome mosaic virus (BMV), encodes three RNA segments. Interestingly, the intergenic region (IGR) of the BMV RNA3 segment contains a similar T-pseudouridine stem-loop structure which is pseudouridylated at a position corresponding to the same modification in tRNA (Baumstark and Ahlquist 2001). An analysis of the RNA modification profile on the genomes of positive-sense RNA viruses isolated from infected cells and virions by mass spectrometry revealed that pseudouridine is an abundant modification on viral RNAs (McIntyre *et al.* 2018). Interestingly, a CRISPR screen for antiviral host factors modulating HCV and DENV infections identified PUS proteins (Marceau *et al.* 2016). While the PUS proteins were not the leading proteins uncovered in the screen, these data do highlight that PUS proteins and/or the modulation of pseudouridine likely impact viral infection.

In this chapter, our focus is on viral PTMs; however, RNA modifications on cellular RNAs have also been shown to affect viral gene expression. For example, the cellular $tRNA^{Lys3}$ binds with the primer binding site in the HIV genome to initiate reverse transcription, a critical step in the HIV-1 infectious cycle (Mak and Kleiman 1997). Notably, the pseudouridine in the cellular tRNA promotes a stable interaction between the HIV RNA and the $tRNA^{Lys3}$ (Bilbille *et al.* 2009). Secondly, the translation of viral proteins is impacted by cellular tRNAs. For example, the putative RNA-dependent polymerase of tobacco mosaic virus (TMV), a virus in the *Virgaviridae* family, is synthesized via read-through of a weak UAG stop codon between two open reading frames (Zerfass and Beier 1992). In tobacco and yeast, $tRNA^{Tyr}$ contains a pseudouridine in the anticodon loop (GΨA). Unexpectedly, the binding of $tRNA^{Tyr}$ with the UAG stop codon promotes the termination of translation and, hence, termination of the synthesis of the RNA-dependent polymerase. Zerfass and Beier (1992) showed that the pseudouridine in the GΨA anticodon sequence was responsible for the translation termination.

The levels of pseudouridine have been shown to change in response to stress (Carlile *et al.* 2014; Li *et al.* 2015). Because virus infections induce stress responses in cells (Lloyd 2013), it is perhaps not surprising that the pseudouridine and epitranscriptomic landscape is different between mock- and virus-infected cells (McIntyre *et al.* 2018). Beyond changes in the pseudouridine landscape or knowledge of pseudouridine distribution on mRNA (Carlile *et al.* 2014), we have little understanding of how pseudouridine might affect viral gene expression. Pseudouridines in tRNAs impact RNA structure and have roles in frameshifting, misreading or suppressing stop or nonsense codons (Davis 1995; Newby and Greenbaum 2002). Notably, synthetic mRNA containing pseudouridine is camouflaged from immune sensing (Karikó *et al.* 2008), stabilizes mRNAs (Karikó *et al.* 2008; Anderson *et al.* 2011) and affects translation. This modified synthetic mRNA also inhibited the activation of double-stranded RNA protein kinase (Anderson *et al.* 2010; Aspden and Jackson 2010), as well as the stress- and IFN-induced kinases that phosphorylate the translation initiation factor eIF2α and inhibit cellular translation during stress (Dalet *et al.* 2015). If pseudouridine in viral RNA had similar functions, this epitranscriptomic mark would significantly affect viral gene expression. Because high-throughput approaches are available to identify pseudouridines at nucleotide resolution, and the PUS enzymes are known, a broader understanding of the role of this fifth nucleotide in different viral systems is likely on the horizon.

4.4.2. *Editing of adenosine-to-inosine*

Inosine is an abundant RNA modification (Li *et al.* 2009; Danecek *et al.* 2012; Pfaller *et al.* 2018) that is installed by the enzyme Adenosine Deaminase Acting on RNA (ADAR). ADAR catalyzes the deamination of adenine (A) at the C6 position to produce inosine in double-stranded RNA (dsRNA) (Figure 4.8) (Samuel 2011; Slotkin and Nishikura 2013; George *et al.* 2014). While inosine can form base pairs with other nucleotides, it preferentially pairs with cytosine. Thus, the misinterpretation of inosines as guanosine (G) by the translation machinery or viral RNA-dependent RNA polymerases can change the coding potential of mRNAs or viral genomes, alter mRNA regulatory domains, such as splice sites, and change the RNA structure (Figure 4.8). In mammalian cells, three genes encode four ADAR enzymes (ADAR1 p110, ADAR1 p150, ADAR2 and ADAR3); ADAR-1 p110 is constitutively expressed, while interferon induces ADAR-1 p150 expression (Samuel 2011; Savva *et al.* 2012). Notably, only ADAR1 and ADAR2 harbor deaminase activity (Chen *et al.* 2000; Savva *et al.* 2012), and ADAR1 p110 and ADAR2 are localized in the nucleus (Desterro *et al.* 2003; Sansam *et al.* 2003), while ADAR1 p150 shows cytoplasmic localization (Patterson and Samuel 1995; Poulsen *et al* 2001). In addition to promoting A-to-I editing, ADAR1 also functions with other proteins synthesized in response to IFN signaling to modulate viral infection (Samuel 2011).

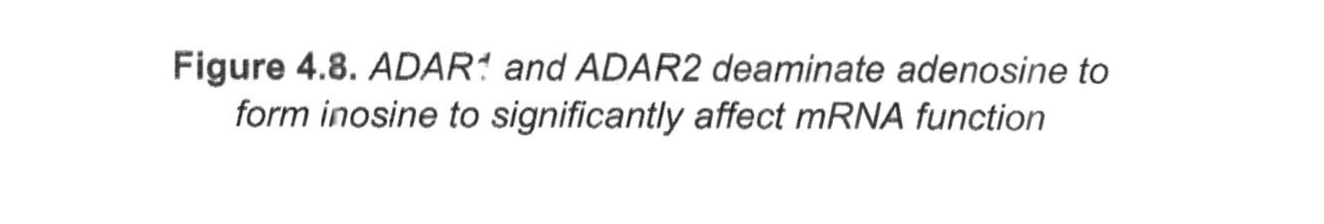

Figure 4.8. *ADAR1 and ADAR2 deaminate adenosine to form inosine to significantly affect mRNA function*

4.4.2.1. *Role of A-to-I editing in paramyxovirus gene expression*

The measles virus, a single-stranded negative-strand RNA virus in the *Paramyxoviridae* family, typically causes an acute infection. Acute measles infection can, in some instances, establish a persistent infection in the central nervous system known as Subacute Sclerosing Panencephalitis (SSPE) (Griffin *et al.* 2018). In these patients, the titers against all measles virus proteins are high, except against the non-glycosylated viral membrane protein or matrix (M) protein (Hall *et al.* 1979). Similarly, M protein expression, when analyzed by immunofluorescence analysis, was found to be low in SSPE brain material (Liebert *et al.* 1986). Even though measles virus replication does not require M, when M expression is low or absent, viral nucleocapsids do not co-localize at the cell membrane with the viral glycoproteins. Because the nucleocapsids remain in the cytoplasm and nucleus when M expression is low, virus spread is limited and virus persistence is promoted (Patterson *et al.* 2001). M genes, cloned from brain autopsies of SSPE patients and analyzed by sequencing, showed biased hypermutations where half of the uridines (U) in the M gene were identified as cytosine (C) (Cattaneo *et al.* 1988). These biased hypermutations were proposed to contribute to the decrease in M expression by promoting frameshift mutations and introducing a termination codon (Cattaneo *et al.* 1988). These groups of nucleotide transitions (A-to-I [G] or U-to-C, depending on RNA strand polarity) during measles virus transcription and replication were attributed to ADAR (Bass *et al.* 1989; Pfaller *et al.* 2014, 2015).

Other paramyxovirus RNA genomes, such as human respiratory syncytial virus (hRSV), parainfluenza virus 5 (PIV5) and human metapneumovirus (hMPV) are also edited by ADAR (Martínez and Melero 2002; Samuel 2011; Rima *et al.* 2014; van den Hoogen *et al.* 2014). Sequence analysis of the hRSV glycoprotein (G) revealed similar biased hypermutations which altered amino acid sequences to affect an antibody epitope (Rueda *et al.* 1994; Martínez *et al.* 1997). It is curious that paramyxovirus genomes are edited by ADAR, particularly as the nucleocapsid proteins coating the viral RNA (Gutsche *et al.* 2015) and double-stranded RNA structures, the substrates for ADAR-1 activity, are unlikely to form and/or undergo A-to-I editing. However, such editing may occur if the nucleocapsid–RNA genome interactions were perturbed (Pfaller *et al.* 2014, 2015, 2018). Recently, the presence of inosine in mRNA was found to camouflage RNA, thereby preventing the activation of the cytoplasmic sensor MDA5 and the innate immune response (Mannion *et al.* 2014; Liddicoat *et al.* 2015). Thus, in addition to introducing mutations in viral proteins, the insertion of inosines in viral RNA might also modulate the innate immune response (Pfaller *et al.* 2018).

4.4.2.2. *Role of A-to-I editing on hepatitis δ virus (HDV) gene expression*

The HDV infectious cycle is intimately regulated by A-to-I editing. In HBV patients, the risk of severe liver damage is increased with coinfection by HDV and HBV (Lempp *et al.* 2016). HDV, the only virus in the *Deltaviridae* family, has an

RNA genome that encodes a single protein, the hepatitis delta antigen (HDAg). HDV is a subviral pathogen that is transmitted by encapsidating the HDV RNA in the HBV virion (Rizzetto *et al.* 1980). In the HDV infectious cycle, replication and assembly are directed by distinct forms of HDAg. In particular, HDV replication is promoted by the small form of the protein HDAg-S, while HDAg-L, the large form, modulates the switch from replication to virus assembly (Kuo *et al.* 1989; Ryu *et al.* 1992). This pivotal transition in the infectious cycle is directed by A-to-I editing. The HDV genome, a circular negative-sense RNA, is replicated through an antigenome, an RNA intermediate that functions as a template for transcription of viral mRNA. Interestingly, the antigenomes have very specific uridine-to-cytidine substitution (Casey *et al.* 1992). Indeed, inosine was identified on the HDV RNA when a double-stranded RNA deaminase, isolated from *Xenopus laevis* eggs, was incubated with radiolabeled *in vitro* transcribed antigenome RNA, and analyzed by thin-layer chromatography (Polson *et al.* 1996). Moreover, an editing assay, together with site-directed mutagenesis studies, showed that adenosine 1012 on the antigenome was edited to inosine (Casey and Gerin 1995; Polson *et al.* 1998). This editing event converts the UAG amber stop codon of HDAg-S to UIG (or "UGG", because inosine, like guanosine, base pairs with cytosine), a tryptophan codon, which results in the synthesis of the elongated protein HDAg-L. ADAR installs inosines in regions of double-stranded RNA (Bass and Weintraub 1987; Rebagliati and Melton 1987). Thus, an important clue to this replication-to-assembly switch came from HDV RNA structure studies. In particular, the HDV genome has extensive complementarity in the circular RNA and forms a putative rod-like dsRNA structure, the extent of A-to-I editing either stabilizes or destabilizes the secondary structure (Jayan and Casey 2005). Inosine was found to be installed during the synthesis of the antigenome, which is the replication step (Sato *et al.* 2004). Hence, HDAg-L expression and, thus, virus production are modulated by RNA secondary structure and the extent of editing.

Using non-replicating HDV antigenome reporter and overexpression assays, both ADAR1 and ADAR2 were shown to edit the HDV antigenome (Sato *et al.* 2001). Subsequently, however, the differential knockdown of either ADAR2 or ADAR1 p150 demonstrated that ADAR2 only modestly, if at all, edited HDV RNA. In contrast, the knockdown of both ADAR1 p110 and ADAR1 p150 significantly affected A-to-I editing of the HDV RNA. Additionally, ADAR-1 knockdown and differential overexpression of ADAR1 p110 or ADAR1 p150 showed that ADAR1 p110 edited the antigenome more efficiently than ADAR1 p150 (Wong and Lazinski 2002). Notably, ADAR1 p150 is induced upon interferon production. Even though the amount of ADAR1 p150 that is produced is a fraction of the ADAR1 p110 form, the HDV RNA is also edited by ADAR1 p150 to promote the expression of

HDAg-L (and downregulation of HDV replication). As such, one antiviral treatment of HDV has been pegylated interferon-α, although newer antivirals, directed to other steps in the infectious cycles, and putatively fewer adverse effects, are being developed (Koh *et al.* 2019; Mentha *et al.* 2019). In summary, A-to-I editing inhibits HDAg-L replication, to exquisitely maintain the balance between replication and virus production (Sato *et al.* 2004).

4.4.2.3. *Role of A-to-I editing in HIV-1 gene expression*

HIV-1 belongs to the *Retroviridae* family of viruses that contain positive-sense RNA genomes that are reverse-transcribed into double-stranded DNA proviruses, which integrate into the host genome (Hu and Hughes 2012). HIV-1 subverts the cellular RNA polymerase II to transcribe viral mRNAs. These HIV-1 mRNAs may contain inosines (Phuphuakrat *et al.* 2008; Doria *et al.* 2009; McIntyre *et al.* 2018). Inosines in HIV-1 RNA were first demonstrated in the stem structure of the transactivation response element (TAR). These inosines were identified on *in vitro* transcribed TAR RNA that was injected into the nucleus of *Xenopus* oocytes and analyzed by thin-layer chromatography (Sharmeen *et al.* 1991). The position of the edited adenosine was identified using deletion and mutagenesis studies, which showed that editing was dependent on Tat, the activator of HIV-1 gene expression (Sharmeen *et al.* 1991). Interestingly, HIV-1 protein levels increased when ADAR1 and ADAR2 were overexpressed (Phuphuakrat *et al.* 2008; Doria *et al.* 2009). Conversely, HIV-1 protein levels decreased when ADAR was knocked down using target-specific siRNAs. Furthermore, the impact of ADAR on HIV-1 protein levels was found to be independent of the A-to-I editing function. These data suggested that ADARs affected translation by inhibiting PKR, the RNA-activated protein kinase that recognizes double-stranded RNA (Doria *et al.* 2009). Moreover, ADAR effects on HIV-1 protein levels were stronger than the impact of ADAR with editing functions.

ADAR1 levels were found to affect the amount of unspliced HIV-1 RNA in the cytoplasm. Because immunoprecipitation studies showed that ADAR interacted with both unspliced and spliced HIV-1 RNAs (Doria *et al.* 2009), it was not unexpected to find A-to-I editing in HIV-1 transcripts. Such changes occurred in the 5' untranslated regions including in the TAR and *Rev* response element (RRE), and Tat coding sequence of all HIV-1 RNAs (Phuphuakrat *et al.* 2008; Doria *et al.* 2009). A-to-G substitutions in the RRE were found to change the codon which increased the amount of capsid (p24) protein and unspliced RNA. A-to-G substitutions also increased the number of HIV-1 virions although these particles had low infectivity. The genome of Rous-associated virus type 1, another retrovirus, was also found to contain inosine (Felder *et al.* 1994). In particular a high number of A-to-G (I) substitutions were identified in the U3 sequence, in the long untranslated repeat. A-to-G conversions did not change the viral transcription or replication; however, a change in the polyadenylation signal caused transcriptional read-through to include sequences from a cellular gene sequences (Felder *et al.* 1994).

4.4.2.4. *Role of A-to-I editing in single-stranded positive-sense RNA virus gene expression*

The genomes of viruses in the *Flaviviridae* family such as HCV, ZIKV and DENV, and a *Picornaviridae* family member, poliovirus, have been shown to contain inosine (Taylor *et al.* 2005; McIntyre *et al.* 2018). In HCV, a role for A-to-I editing in antiviral functions has been described (Taylor *et al.* 2005). In particular, the levels of HCV RNA and proteins decreased in the hepatoma cell line, stably expressing an HCV replicon following interferon treatment. The decrease of HCV protein levels was likely due to interferon treatment modulating PKR expression to affect cap-independent or internal ribosomal entry site (IRES)-mediated translation. However, a role for ADAR was presented when Taylor and colleagues (2005) expressed an RNA that inhibits ADAR activity, which then restored IRES-mediated activity and HCV replicon RNA (Taylor *et al.* 2005). Moreover, increased levels of inosine were demonstrated following sequence analysis of the HCV replicon RNA isolated from replicon cells treated with interferon, as well as by thin-layer chromatography analysis of total RNA isolated from the replicon cells (Taylor *et al.* 2005). These data together suggested that ADAR might edit HCV RNA and that HCV gene expression was promoted when ADAR activity was inhibited. Bioinformatic analyses of the RNA genomes from African versus Asian ZIKV lineages showed more guanosine nucleotides, which may be the result of A-to-I editing (Khrustalev *et al.* 2017; Piontkivska *et al.* 2017). The putative inosines were identified in regions of predicted secondary structure, and on both the positive and negative RNA strands (Khrustalev *et al.* 2017). Because the immune response is altered and apoptosis increases in mouse embryonic fibroblasts in which ADAR has been knocked out (Mannion *et al.* 2014; Liddicoat *et al.* 2015), installation of inosines in ZIKV genomes has been proposed to be one possible evolutionary driver and contributor to the neuropathologies associated with the outbreak of ZIKV in the Americas (Lazear and Diamond 2016).

4.4.2.5. *Role of A-to-I editing in polyomavirus gene expression*

Mouse polyomavirus (MPyV) is a double-stranded (ds) DNA virus in the *Polyomaviridae* family. Genes from the 5.3 kb circular genome are temporally expressed (Huang and Carmichael 2009). In particular, bidirectional non-coding control regions direct the expression of early and late genes, and the functional outcome of these genes is post-transcriptionally regulated. Both early and late genes are expressed immediately after infection, where early genes are preferentially expressed and late transcripts are targeted for degradation in the nucleus (Hyde-Deruyscher and Carmichael 1990). Early and late transcripts are similarly expressed at later times during infection; however, RNA editing on early viral mRNAs favors the expression of the late mRNAs (Kumar and Carmichael 1997). The mode by which early transcript expression is regulated late during infection is unique. Specifically, in the late phase of infection, the termination of late gene transcription is inefficient, such that the cellular RNA polymerase elongates on the circle MPyV DNA genomes a number of times (Acheson 1978, 1981). These elongated multi-genome precursors are spliced and polyadenylated (Treisman and Kamen 1981; Acheson 1984). However, those that are not post-transcriptionally processed amass in the nucleus to high levels. Because these retained transcripts are derived from the circular MPyV genome, parts of the RNAs are complementary (or antisense) to early viral transcripts. The extended base pairing between the elongated transcripts and early mRNAs creates a substrate for ADAR editing. Indeed, an examination of early transcripts that were present during MPyV infection determined that approximately half of the adenosines on the early transcripts had been edited to inosine (Kumar and Carmichael 1997). Moreover, these inosine-edited transcripts were stable and accumulated in the nucleus. By sequestering these early transcripts in the nucleus, early viral mRNAs are not exported into the cytoplasm to be translated and are inert for gene expression. A subsequent study identified a multi-protein complex consisting of an inosine-specific RNA binding protein $p54^{nrb}$, the splicing factor PSF and matrin3, an inner nuclear matrix structural protein (Zhang and Carmichael 2001), that was responsible for trapping hyper-A-to-I edited RNAs in the nucleus. In summary, MPyV uniquely subverts post-transcriptional modifications to temporally regulate viral gene expression.

Bioinformatic analyses of A-to-I (or A-to-G) substitutions in sequencing data have been conducted to identify A-to-I editing in RNA. From such analyses, inosine on a number of viral RNAs has been identified (Samuel 2011). To date, the outcome of A-to-I editing has only been investigated for a few viruses. However, with orthogonal validation approaches and next generation sequence analyses (Sakurai *et al.* 2010; Cattenoz *et al.* 2013; Knutson *et al.* 2018), and in-depth functional studies, the function of this abundant post-transcriptional modification in other viruses will be revealed.

4.5. Deposition of RNA modifications on viral RNA by cellular enzymes

4.5.1. *Role of N6-methyladenosine (m^6A) on viral gene expression*

m^6A is an abundant RNA modification (Dominissini *et al.* 2012). This, together with the availability of an anti-m^6A antibody and advances in next generation sequencing, and that the writer, eraser and reader enzymes are known, has contributed to the broad interest in this particular RNA modification (Shi, Wei and He 2019). METTL3 and METTL14, together with Wilms tumor 1-associated protein (WTAP) and KIAA1429 deposit m^6A on RNA (Meyer and Jaffrey 2017; Shi *et al.* 2019), and the demethylases FTO (fat mass and obesity-associated protein) and ALKBH5 (α-ketoglutarate-dependent dioxygenase AlkB homology 5) erase the methyl group (Figure 4.9) (Jia *et al.* 2011; Zheng *et al.* 2013). The YTH-domain family of RNA binding proteins (YTHDC1, YTHDC2, YTHDF1, YTHDF2 and YTHDF3) are localized in the nucleus and cytoplasm and facilitate (or read) the functions of m^6A (Figure 4.9) (Shi *et al.* 2019). In manipulating the levels of the writer, eraser and reader enzymes, m^6A in cellular RNAs contributes to splicing, localization, translation and RNA stability. Considering the broad impact of this single RNA modification, it is hardly surprising that recent studies have been undertaken investigating how m^6A might modulate RNA and DNA virus infections (Gokhale and Horner 2017; Gonzales-van Horn and Sarnow 2017; Kennedy *et al* 2017; Roundtree *et al.* 2017; Shi *et al.* 2019).

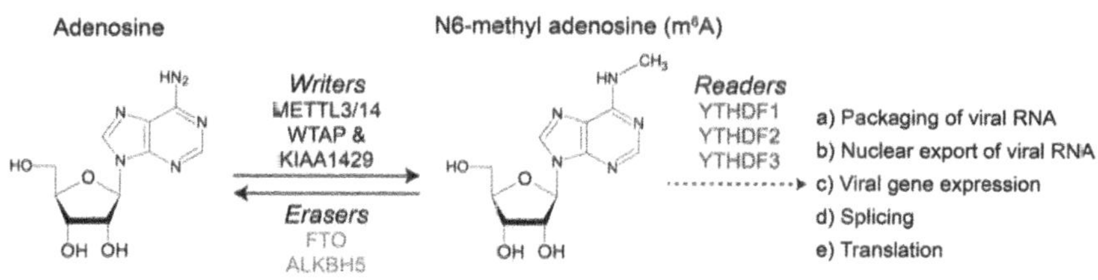

Figure 4.9. *Writer, eraser and reader enzymes modulate m^6A function on RNA*

4.5.1.1. *m^6A modifications and RNA viruses that replicate in the nucleus*

4.5.1.1.1. Influenza A virus (IAV)

Influenza virus, a virus in the *Orthomyxoviridae* family, harbors eight segments of single-stranded negative-sense RNA (Bouvier and Palese 2008). Early studies showed that 55–60% of the complementary RNA or mRNA of influenza virus contained methyladenosine modifications. A third of these methyladenosine modifications were m^6A, which were internally localized in the viral RNA (Krug *et al.* 1976) and that the m^{6}A modifications on different influenza virus mRNAs were differentially distributed (Narayan *et al.* 1987). Courtney *et al.* (2017) recently showed that when the abundance of the METTL3 writer was depleted with siRNAs, the levels of influenza virus protein and mRNA and virus titers decreased, and virus gene expression was promoted by overexpression of the YTHDF reader proteins. Putative m^{6}A sites were identified in both the viral mRNAs and negative-sense genomic RNA, and that, similar to m^{6}A topology in cellular mRNAs, most of the m^{6}A modifications were at the 3' end of the viral RNA (Courtney *et al.* 2017). Notably, Courtney and colleagues showed that the kinetics of viral infection and viral pathogenicity decreased when the m^{6}A sites were mutated, suggesting that m^{6}A on the viral RNA promotes influenza virus gene expression.

4.5.1.1.2. Retroviruses

Early studies investigating RNA methylation with avian sarcoma virus (ASV), a virus in the *Retroviridae* family, first defined the consensus motif recognized by m^{6}A writer enzymes (Dimock and Stoltzfus 1977). Specifically, methyl-moieties on B77 ASV RNA radiolabeled with [methyl-^{3}H]methionine identified methylation of the Cap 1 structure, and 14–16 methyl modifications were internal m^{6}As (Stoltzfus and Dimock 1976). Additional enzymatic digest and chromatography showed that adenosine in the sequence (G>A)-m^{6}A-C was modified (Dimock and Stoltzfus 1977). More recently, RNA-seq studies have refined the m^{6}A consensus motif as DRm^{6}ACH (where D may be G, A or U, R represents the purines G and A, and H may be U, C or A) (Dominissini *et al.* 2012; Liu *et al.* 2014). Analysis of the Rous sarcoma virus (RSV), another retrovirus, similarly revealed 10–15 m^{6}A modifications, with a significant number of the modifications installed on the 500–4,000 nucleotides of the 3' end of the RNA genome (Beemon and Keith 1977). RNA-seq of the m^{6}A landscape with RNA showed that m^{6}A modifications are dominant in long exons in the 5' untranslated region (UTR) and near termination sites in the 3' UTR (Dominissini *et al.* 2012; Meyer *et al.* 2012; Zhou *et al.* 2015). Subsequent investigation of m^{6}A in RSV RNA suggested a role in splicing. Support for this function was parsed from data showing that the number of m^{6}A-sites in RSV RNA varied, the modified nucleotides were upstream of defined splice acceptor sites and the levels of unspliced RSV RNA increased following treatment of cells with an inhibitor of internal methylation namely cycloleucine (Dimock and Stoltzfus 1978; Stoltzfus and Dane 1982; Kane and Beemon 1985).

m^{6}A has been known to decorate the RNAs from both RNA and DNA viruses since the 1970s (Hashimoto and Green 1976; Krug *et al.* 1976; Beemon and Keith 1977; Canaani *et al.* 1979). However, the studies from three labs investigating the role of m^{6}A in HIV-1 gene expression prompted renewed interest in viral epitranscriptomics (Kennedy *et al.* 2016; Lichinchi *et al.* 2016b; Tirumuru *et al.* 2016; Lu *et al.* 2018). This viral epitranscriptomic renaissance was facilitated by the availability of an anti-m^{6}A antibody for immunoprecipitations of modified RNA and RNA-seq (Figure 4.4A). These studies demonstrated numerous m^{6}A modifications on HIV-1 RNA, particularly in the 5' and 3' long terminal repeats (LTRs) (Kennedy *et al.* 2016; Tirumuru *et al.* 2016), in genes encoding the envelope (*Env*) and reverse transcriptase (*Rev*), and in the *Rev* response element (RRE) (Kennedy *et al.* 2016; Lichinchi *et al.* 2016b; Tirumuru *et al.* 2016). Modulation of m^{6}A writer, eraser and reader proteins by siRNA depletion or overexpression showed that m^{6}A on HIV-1 RNA affected HIV-1 gene expression at early and late stages of the infectious cycle (Lichinchi *et al.* 2016b; Tirumuru *et al.* 2016; Lu *et al.* 2018). Tirumuru and colleagues (2016) modulated the levels of reader proteins and quantified early and late reverse transcription (RT) products and showed deposition of m^{6}A on HIV-1 RNA affected events before the transport of cDNA into the nucleus and integration into host DNA. A role for m^{6}A in viral RNA stability was also proposed when reader proteins were overexpressed, which resulted in more HIV-1 genomic RNA associated with the readers, yet, at early times after infection, the overall levels of viral RNA decreased (Tirumuru *et al.* 2016). While further studies are required to establish the mechanism by which m^{6}A affects the stability of viral RNA, it is possible that the modification alone or the m^{6}A–reader interaction inhibits reverse transcription which promotes turnover of RT products. Alternatively, the recognition of m^{6}A moieties by reader proteins might target viral RNA for degradation by RNA exo- or endonucleases.

The RRE, a critical *cis*-element in HIV-1 RNA, facilitates export of the viral RNA to the cytoplasm following binding by the Rev protein. Lichinchi and colleagues (2016) used m^{6}A-immunoprecipitation and RNA-seq, and an orthogonal validation approach, to demonstrate two conserved m^{6}A sites in the RRE loop. The function of RRA m^{6}A sites was elucidated by altering the levels of the writer and eraser enzymes, and hence m^{6}A levels on the HIV-1 genome. Indeed, the amount of Rev binding to the RRE was changed. Subsequent mutation of the m^{6}A consensus motifs showed that the amount of genomic RNA and nuclear export decreased when the second m^{6}A site was perturbed (Lichinchi *et al.* 2016b). m^{6}A modifications were also identified in the HIV-1 5' UTR at or in the close range of the primer-binding site (PBS) and dimer initiation sequence (DIS), respectively. New rounds of reverse transcription require the association of a tRNA^{Lys} to the PBS, and the specific encapsidation of two HIV-1 genomes is directed by the DIS. The levels of newly synthesized HIV-1 proteins were modestly altered by the mutation of either or both m^{6}A sites. These mutations, however, had a notable effect on subsequent HIV-1 infections. These data, together with assays showing that m^{6}A reader proteins bound unprocessed Gag protein in an HIV-1 RNA-dependent manner, raised the possibility that the assembly of new HIV-1 particles might also be affected by m^6A modifications on the viral RNA (Lu *et al.* 2018). Recently, 20 putative m^6A modifications were described on the MLV RNA (Courtney *et al.* 2019b). Courtney and colleagues showed that the levels of intracellular and virion proteins decreased when three of the m^6A sites in the *Env* gene decreased, viral RNA levels when expressed from the proviral expression plasmid remained constant and viral gene expression was reduced in subsequent infections. A role for m^6A in MLV gene expression was further supported when YTHDF2 reader levels were modulated, which increased MLV protein abundance (Courtney *et al.* 2019b).

4.5.1.2. *m^6A modifications and RNA viruses that replicate in the cytoplasm*

Because rRNA, tRNAs and other small RNAs are extensively post-transcriptionally modified in the nucleus, it had been assumed that m^6A modifications on the RNAs of viruses with a cytoplasmic infectious cycle would be absent. However, m^6A modifications were recently identified on the RNA genomes of viruses in the *Flaviviridae* and *Picornaviridae* families (Gokhale *et al.* 2016; Lichinchi *et al.* 2016a; McIntyre *et al.* 2018; Hao *et al.* 2019). A mechanistic study of m^6A during HCV infection demonstrated that modulation of the levels of the writer and eraser proteins affected the abundance of intracellular HCV protein, virus titers and the amount of HCV RNA in the extracellular media, indicating a role in new virus particle production (Gokhale *et al.* 2016). Notably, HCV replication was not affected by the levels of the writer, eraser and reader proteins; however, the reader proteins, together with core protein, co-localized at lipid droplets, the sites of virus assembly. These data supported a role for m^6A regulatory proteins in virus assembly (Gokhale *et al.* 2016; Vieyres and Pietschmann 2019). Fourteen putative m^6A modifications were distributed at 14 positions with no apparent bias with respect to the viral genes or positions along the HCV genome. While disruption of four putative m^6A sites in the envelope E1 gene did not alter HCV protein levels, translation or replication of the HCV genome, virus titers dramatically increased. Predictably, less HCV RNA were associated with the YTHDF2 reader when m^6A sites in the E1 gene were mutated. Conversely, however, the amount of HCV RNA immunoprecipitated by the core protein increased when m^6A sites in E1 were mutated (Gokhale *et al.* 2016). In summary, the binding of reader proteins to m^6A moieties in HCV RNA modulates the amount of readers associated with core protein, and thus the assembly of new HCV particles. In addition to HCV, m^6A marks have been identified on the RNA genomes of other viruses in the *Flaviviridae* family, such as yellow fever virus (YFV), WNV, DENV and ZIKV (Gokhale *et al.* 2016; Lichinchi *et al.* 2016a; McIntyre *et al.* 2018). Additional studies show that ZIKV and enterovirus 71 (EV71) infections were affected when the levels of the writer, eraser and reader proteins were altered, indicating a role for m^{6}A in viral gene expression (Lichinchi *et al.* 2016a; Hao *et al.* 2019).

4.5.1.3. *m^{6}A modifications and DNA viruses*

4.5.1.3.1. Simian virus 40 (SV40)

Simian virus 40 (SV40) is a member of the *Polyomaviridae* family and has been extensively studied due to its ability to cause tumors in mammalian cells. Early research examining RNA modifications on SV40 RNA identified internal m^{6}A as well as m^{7}G caps on the viral RNAs (Aloni *et al.* 1979). In particular, Aloni *et al.* (1979) labeled RNA with tritiated-labeled S-adenosylmethionine (SAM), the methyl donor for the methylation of RNA, and analyzed the viral RNA following segmentation into five fragments. Radiolabeling on three RNA segments corresponded to the late (L) transcript and two segments from the early (E) transcript. The E transcripts contained m^{7}G caps, and the L transcripts contained 80% of all m^{6}A present in viral RNA, with the remaining 20% present on the early transcripts. Aloni *et al.* (1979) posited that m^{6}A functioned as a binding site or a signal for enzymatic cleavage to carry out the excision and ligation processes during the viral lifecycle.

Canaani *et al.* (1979) expanded these initial studies and established the position and consensus sequence for m^{6}A in SV40 RNA (Canaani *et al.* 1979). In particular, they examined the segmented SV40 RNA and then performed strategic digestions using RNase A, which cleaves after C and U, and RNase T1, which cleaves following G residues. They determined that m^{6}A-containing fragments in the highest abundance were m^{6}ApC and Gpm^{6}A. This study was the first to describe a consensus sequence for m^{6}A, namely A/Gpm^{6}ApC, which was later validated in methyl-seq studies (Fu *et al.* 2014). Lastly, Canaani *et al.* (1979) demonstrated that SV40 had about the same amount of m^{6}A residues per 1,000 bases as cellular mRNA.

Finkel and Groner (1983) provided new clues as to the function of m^{6}A on SV40 RNA. For these studies, they used cycloleucine, an inhibitor of adenosine methylation (Dimock and Stoltzfus 1978; Stoltzfus and Dane 1982). They determined that the inhibition of adenosine methylation changed the amount of E and L transcripts and the proportion of viral RNA in the nucleus versus cytoplasm, with more viral RNA accumulating in the nucleus. These findings suggested that the deposition of m^{6}A modification on SV40 mRNA may play a role in RNA processing and nucleo-cytoplasmic shuttling of viral RNAs.

The role of m^{6}A on SV40 RNA was recently revisited (Tsai *et al.* 2018). To determine the function of m^{6}A during SV40 infection, Tsai *et al.* (2018) altered the abundance of the YTHDF2 enzyme, a reader that facilitates nuclear export of modified RNAs, and METTL3/METTL14, the writer enzymes responsible for depositing m^{6}A on RNA (Meyer and Jaffrey 2017). Manipulation of YTHDF2 levels altered SV40 protein levels and viral titers. Using antibodies directed directly to the m^{6}A modification and the reader enzymes combined with RNA-seq, Tsai *et al.* (2018) cross-validated the positions of m^{6}A in the SV40 RNA which were enriched in the VP1 gene. SV40 viruses that were modified at putative m^{6}A sites did not replicate as well as wild-type viruses, indicating a proviral role for m^{6}A in SV40 infection. Consistent with the studies undertaken by Aloni *et al.* (1979) and Finkel and Groner (1983), Tsai and colleagues (2018) reported that E and L transcripts were differentially m^{6}A modified and postulated that these epitranscriptomic marks function to alter the shuttling of cellular and viral RNAs into the nucleus. In particular, L transcripts had more m^{6}A moieties than E transcripts, and thus were more likely to be shuttled into the cytoplasm. Editing of adenosine-to-inosine on mouse polyoma virus (MPyV) similarly regulates nucleo-cytoplasmic shuttling of transcripts (Kumar and Carmichael 1997), suggesting that these viruses might employ epitranscriptomic marks to modulate the expression of early and late genes.

4.5.1.3.2. Adenovirus

Adenoviruses, in the family *Adenoviridae*, are dsDNA viruses that replicate in the nucleus. Viral DNA, similar to cellular DNA, is transcribed by RNA polymerase II into mRNA which is then modified with a 5' m^{7}G cap and a 200 nucleotide poly(A) tail and exported into the cytoplasm for translation (Sommer *et al.* 1976). Sommer and colleagues (1976) investigated RNA modifications during adenovirus infections by ^{3}H-methionine-labeling infected cells, harvesting at 14–18 hours post-infection and isolating cytoplasmic and nuclear fractions of RNA. They found that both mRNA fractions contained m^{7}G caps and m^{6}A; however, the proportion of m^{6}A modifications was greater on nuclear viral RNAs compared to cytoplasmic viral RNAs (Sommer *et al.* 1976). The viral mRNAs contained approximately one m^{7}G cap per molecule, of either the Cap 1 or Cap 2 structure, and the internal methylations, m^{6}A or m^{5}C, were primarily located in the 5' two-thirds of the mRNA molecules. The location of these modifications could plausibly lend some evidence to their function.

4.5.1.3.3. Herpesviruses

Herpes simplex, in the family *Herpesviridae*, is a large dsDNA virus that replicates in the nuclei of infected cells. Viral messages are synthesized by RNA polymerase II and capped and polyadenylated at the 5' and 3' ends of the viral RNAs, respectively. Internal methylations on RNA, including m^{6}A, were detected in infected cells that were first labeled both ^{3}H-methionine and ^{32}P-orthophosphate, and then mRNAs were separated from non-mRNAs via poly(A) selection. These early experiments showed that viral mRNAs contained, on average, one m^{7}GpppNmpNmpNp (Cap 2) or m^{7}GpppNmpNp (Cap 1) per 2,780 nucleotides and this cap was confirmed to be RNase-resistant, protecting the viral RNA from host RNA degradation pathways in the cell (Moss *et al.* 1977). The majority of internal methylations identified were m^{6}As, which was thought to be the most common internal modification in cells and on many viral mRNAs, although the function of m^{6}A on RNA had yet to be determined (Moss *et al.* 1977).

More recent research with Kaposi's sarcoma-associated herpesvirus (KSHV) has been undertaken to examine the role of m^{6}A in this herpesvirus infection (Tan *et al.* 2017; Ye *et al.* 2017; Hesser *et al.* 2018). KSHV has been associated with a number of malignancies such as Kaposi's sarcoma and primary effusion lymphoma (Cesarman *et al.* 1996; Said *et al.* 1996). Following initial infection, herpesviruses enter a latent phase, and upon stimulation, the virus can be reactivated into a lytic infection. Hesser *et al.* (2018) investigated m^{6}A marks by first isolating polyadenylated transcripts from the KSHV-positive renal carcinoma cell line that had been induced into the lytic cycle and then the mononucleotides were analyzed by LC-MS/MS. Notably, the level of m^{6}A induced in the lytic KSHV cells was threefold higher than uninduced latent KSHV-infected cells (Ye *et al.* 2017; Hesser *et al.* 2018). m^{6}A-seq analysis identified sites enriched in m^{6}A; in particular, 10 viral transcripts displayed m^{6}A modifications, with more m^{6}A marks present in coding regions. Analysis of the whole host and viral transcriptome revealed that host mRNAs saw an overall reduction in m^{6}A levels during lytic infection compared to that during latent infection, whereas viral transcripts increased in methylation under the same conditions. These data suggested that a switch in m^{6}A modifying targets occurs during lytic infection of KSHV (Hesser *et al.* 2018). Functional studies examining the effects of the METTL3 writer and nuclear YTHDF2 reader on viral infection revealed cell line-specific results. In both the iSLK.219 and iSLK.BAC16 cell lines, depletion of m^{6}A enzymes hampered the production of virions; however, the method of depletion differed. In iSLK.219, the depletion of writers and readers did not impact the transcription of ORF50, which is a critical to lytic infection. In iSLK.BAC16 cells, the depletion of writers and readers caused a subsequent loss of lytic replication and lower levels of ORF50. The results demonstrate that, though m^{6}A clearly impacts KSHV infection, it can do so via different mechanisms in different cell lines. Tan and colleagues similarly showed that KSHV transcripts in latent and lytic cycles were abundantly m^{6}A modified. Notably, depletion of the reader enzyme inhibited the turnover of KSHV mRNA to promote the lytic cycle. Moreover, KSHV changed the m^{6}A landscape on cellular mRNAs (Tan *et al.* 2017).

In addition to affecting the stability of KSHV transcripts, m^6A marks were also found to facilitate pre-mRNA splicing. Thus, in the complex infectious cycle of KSHV, m^6A functions at distinct times to promote viral gene expression.

4.5.1.4. *m^6A modifications and hepatitis B viruses*

Hepatitis B virus (HBV), a member of the *Hepadnaviridae* family, is associated with chronic liver infections that can result in liver failure and hepatocellular carcinoma (Seeger and Mason 2000). HBV has a partially double-stranded relaxed circular DNA (rcDNA) genome which undergoes a number of steps to form a covalently closed circular DNA (cccDNA). HBV employs an RNA intermediate known as the pregenomic RNA (pgRNA) to complete the infectious cycle (Seeger and Mason 2000). Depletion of the m^6A methyltransferase METTL3 increased virus protein production but decreased reverse transcription of the pgRNA. One m^6A site was identified in an RNA stem-loop structure (Imam *et al.* 2018). This stem-loop structure or epsilon stem-loop was present in the 3' UTR of all HBV mRNAs, as well as at the 5' and 3' ends of pgRNA. Notably, the fate of the RNA depended on which pgRNA termini contained the m^6A modification. In particular, the deposition of m^6A at the 5'-terminus of the epsilon stem-loop was required for efficient reverse transcription, while m^6A installation at the 3' end affected the stability of all HBV RNAs (Imam *et al.* 2018). Together, these data show that in HBV, m^6A spatiotemporally regulates the HBV life cycle.

4.5.2. *Role of 5-methylcytosine (m^5C) in viral gene expression*

Cytosine in DNA and different species of RNA may be methylated at position 5 of the nucleobase. In tRNA and rRNA, m^5C has been shown to affect the structure and function (Trixl and Lusser 2019). m^5C in RNA may be identified by classical analytical approaches such as chromatography and mass spectrometry, and more recent high-throughput sequencing technologies (Trixl and Lusser 2019). In contrast to m^6A, no clear RNA motif in mRNAs for m^5C has been identified. In mRNA, this modified nucleotide is enriched in regions that are immediately downstream of the AUG start codon, in 3' UTRs and CG-rich regions (Yang *et al.* 2017). The m^5C writer enzymes belonging to either DNMT2 or NOL/NOP2/sun (NSUN) methyltransferase subgroups use S-adenosyl-L-methionine (SAM) as the methyl donor group (Figure 4.10) (Bujnicki *et al.* 2004; Motorin *et al.* 2009). Seven NSUN proteins have been identified in mammals, of which NSUN2 was shown to install m^5C on tRNAs and mRNAs (Squires *et al.* 2012). An m^5C eraser enzyme to date has not been described. However, Aly/REF export factor (ALYREF or THOC4), an RNA transport adaptor protein, functions as a putative reader enzymes that recognize m^5C marks on mRNA (Yang *et al.* 2017).

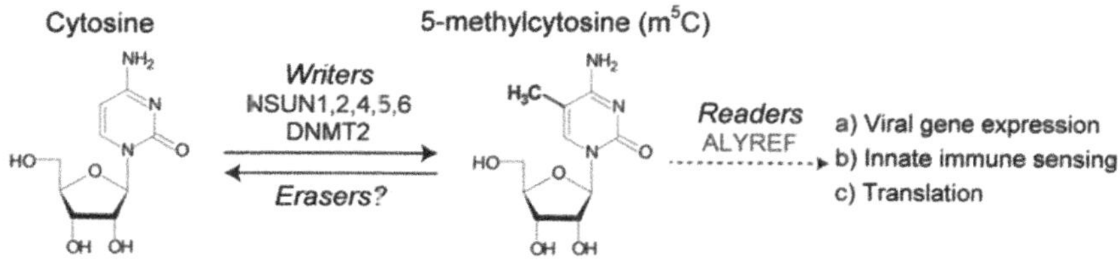

Figure 4.10. *Writer and reader enzymes modulate the m^{5}C function on viral RNA*

Sindbis virus is an alphavirus in the *Togaviridae* family. Early methylation studies using electrophoresis to separate ^{14}C-uridine-labeled RNA and ^{3}H-methylated chemical groups revealed that Sindbis virus RNA was heavily methylated (Dubin and Stollar 1975; Dubin *et al.* 1977). Dubin and colleagues showed that m^{5}C made up one-fourth to one-half of all the internal methyl marks on intracellular viral RNA (Dubin *et al.* 1977). Notably, m^{5}C was the only methyl mark identified on the subgenome that encodes the structural proteins required for the assembly of new virus particles (Dubin and Stollar 1975), although m^{5}C marks were absent on Sindbis RNA in viral particles. At present, the position and function of m^{5}C in Sindbis RNA is unknown. Interestingly, Durdevic and colleagues suggested that Mt2 (or Dnmt), an m^{5}C methyltransferase, affected innate immune responses in *Drosophila* (Durdevic *et al.* 2013). In particular, when *Dnmt2* mutant flies were infected with *Drosophila* C virus (DCV), a positive-sense RNA virus in the *Dicistroviridae* family, viral RNA and protein significantly increased, and viral RNA interacted with Dnmt, suggesting that the DCV RNA was m^{5}C-modified (Durdevic *et al.* 2013). These studies were recently supported by Bhattacharya *et al.* (2017) who showed that the infection of *Drosophila* with Sindbis virus upregulated *Mt2* expression (Bhattacharya *et al.* 2017). This study aimed to determine the cellular mechanism by which *Wolbachia pipientis,* an intracellular endosymbiont, controlled viral infection in arthropods. As a model for Sindbis virus infection in mosquitoes, the authors used *Drosophila* flies and a cell line and showed that Sindbis virus protein and RNA levels and virus infectivity were decreased in the presence of a *Wolbachia* infection. On comparing *Drosophila* immune response genes in mock-infected flies, with and without *Wolbachia*, *Mt2* levels were notably increased. Furthermore, when knocked-out or shRNA-depletion approaches were used to decrease the levels of *Mt2*, Sindbis virus infectivity increased (Bhattacharya *et al.* 2017). Considering that cytoplasmic Sindbis RNA contains m^{5}C marks, but not viral RNA encapsidated in virions, it is possible that m^{5}C functions directly to spatiotemporally regulate virus gene expression or indirectly to modulate the innate immune response.

The m^5C RNA landscape in cells changes following infection with DENV, ZIKV, HCV, poliovirus and HIV-1, and analyses of viral RNA isolated from infected cells and virions show m^5C on the viral RNA (McIntyre *et al.* 2018; Courtney *et al.* 2019a). The role of m^5C in HIV-1 infection was recently dissected by Courtney and colleagues (2019a) using mass spectrometry, m^5C-seq analyses and CRISPR knockout cell lines of the m^5C-methyltransferase NSUN2. By mass spectrometry, they identified 10.5 m^5C marks or 0.645–1.422% of HIV-1 gRNA depending on the cell line the virus was propagated in. Notably, this level of m^5C modification is greater than that on mRNA (~0.046%). By m^5C-seq analyses, they identified ~19 m^5C marks, likely due to non-specificity of the antibody. Using NSUN2 knockout cells, the levels of nano-luciferase expression from a reporter genome or p24 and p55 protein produced from HIV-1 infections were decreased, yet viral RNA levels did not change. Similar to APOBEC3G, NSUN2 appears to be packaged into virions, although the mechanism or consequence of packaging the writer enzyme was not explored. Knockout of NSUN2 decreased the number of ribosomes associated with HIV-1 mRNA, while cellular mRNAs that lack m^5C marks are bound by similar numbers of ribosomes in the absence or presence of NSUN2. Thus, NSUN2 knockout likely decreases HIV-1 translation (Courtney *et al.* 2019a). NSUN2 is also known to modify tRNAs; hence, the effects of NSUN2 knockout on HIV-1 translation could also be due to tRNA deficiencies, rather than the absence of m^5C marks. Courtney *et al.* (2019a) also showed that the modulation of m^5C marks affected alternative splicing at one specific site in the *Vif* gene of HIV-1 when NSUN2 was absent. To support this observation, alternative splicing was shown to be decreased in HIV-1 genomes harboring mutations at putative m^5C sites. The absence of NSUN2 affects HIV-1 and installation of m^5C marks on the viral genome, although the effects on HIV-1 gene expression may be direct or indirect.

Courtney *et al.* (2019b) also investigated different epitranscriptomic marks (including m^5C) on the MLV genomic RNA (gRNA) isolated from cells and virions. To dissect the function of these putative m^5C modification sites, four putative sites of m^5C modifications in the polymerase gene were mutated to prevent m^5C installation (Courtney *et al.* 2019b) and then the absence of the four m^5C marks on the mutant MLV gRNA was validated by RNA-seq. Unexpectedly, the mutation of the four cytosines resulted in an overall decrease of m^5C marks on the mutant viral genome. Expression of m^5C mutant MLV from a proviral expression vector had no effect on the gRNA levels, and only modestly affected MLV Gag p65 and p30 protein levels compared to wild-type MLV. In contrast, the examination of proteins and RNA levels in MLV particles released from cells transfected with the provirus DNA plasmid showed a 2-fold reduction of Gag protein levels without affecting the level of gRNA. Even with this decrease of Gag proteins in the virion, the assembly of virus particles was not affected; however, the absence (and overall decrease) of m^5C modifications negatively affected the infectivity of these viruses (Courtney *et al.* 2019b). Knockdown of the NSUN2 writer enzyme with target-specific siRNAs modestly decreased MLV Gag protein levels, while overexpression of Flag-tagged NSUN2 had no effect (Courtney *et al.* 2019b). The small effects on MLV gene expression might, in part, be a result of the discrepancy between the mass spectrometry and m^5C-seq analyses, which identified 13 and 40 m^5C marks, respectively, on the MLV gRNA. It is feasible that some of the m^5C sites identified by m^5C-seq analyses as a result of antibody non-specificity might represent 5-hydroxylmethyl cytosine (hm^5C) or 5,2'O-dimethylcytosine (m^5Cm) which would indicate regulation by other methyltransferases. Regardless, the data show that the presence or absence of m^5C marks impacts MLV infection (Courtney *et al.* 2019b). That the subsequent round of infection from the newly produced mutant virions decreased infectivity is interesting. Possible explanations include that decreased levels of Gag in the mutant virion affects early steps in the MLV infectious cycle such as membrane fusion and entry. Alternatively, m^5C marks, similar to ribose 2'O-methylation and inosine (Mannion *et al.* 2014; Hyde and Diamond 2015; Liddicoat *et al.* 2015), conceal the viral RNA from innate sensing. m^5C marks have been identified on a number of RNA virus genomes (McIntyre *et al.* 2018). Because molecular approaches are available to map m^5C sites, and the m^5C writer and reader enzymes are known, additional studies will elucidate the mechanisms by which m^5C on viral RNA affects viral gene expression.

4.5.3. *The viral epitranscriptome*

Studies investigating the role of an m^6A, m^5C, 5'-cap, 2'O-methylation, pseudouridine and inosine during virus infection have been aided by the availability of antibodies and reactive chemicals to these epitranscriptomic marks, together with recent advances in RNA-seq. With more than 160 known RNA modifications, the epitranscriptomic code modulating viral infection is likely far more sophisticated than we currently understand (Cantara *et al.* 2011; Boccaletto *et al.* 2018). A recent study by McIntyre and colleagues (2018) provided the first insight into the breadth and diversity on viral RNA genomes. These authors used the high sensitivity and agnostic analysis of mass spectrometry together with RNA affinity captures to identify the epitranscriptomic landscape on HIV-1, poliovirus, HCV, DENV and ZIKV genomes. This systems-wide analysis revealed 30–45 different RNA modifications on the viral RNA genomes. McIntyre *et al.* (2018) showed that the abundance of m^6A on the viral RNAs identified by mass spectrometry was comparable to the number of m^6A sites determined by immunoprecipitation and

RNA-seq analysis. The viral genomes were also modified by other notable epitranscriptomic marks, such as 4-acetylacytosine (ac^4C), 2'O-methylated nucleotides, 5-methylcyosine (m^5C) and 1-methyladenosine (m^1A) (McIntyre *et al.* 2018). Indeed, the recent study by Courtney *et al.* (2019a) examining m^5C marks on HIV-1 corroborated the mass spectrometry data (McIntyre *et al.* 2018). McIntyre *et al.* (2018) also identified N7-methylguanosine (m^7G) on HIV-1, poliovirus, HCV, DENV and ZIKV genomes. m^7G marks on HCV and poliovirus RNA genomes were unexpected, particularly as both viruses lack 5'-cap structures (McIntyre *et al.* 2018). Support for these internal m^7G marks was recently demonstrated using a new chemical-assisted sequencing approach to show a role for internal m^7G in mRNA translation (Zhang *et al.* 2019). The mass spectrometric data included in the Courtney *et al.* (2019a, 2019b) studies similarly show m^7G marks on HIV-1 and MLV genomes.

4.6. Conclusion

Viral RNAs are decorated with a plethora of epitranscriptomic marks (McIntyre *et al.* 2018; Courtney *et al.* 2019a, 2019b). Investigations undertaken so far on m^6A alone highlight the significant regulation that an epitranscriptomic mark can have on viral gene expression. However, with more than 160 known RNA modifications (Cantara *et al.* 2011; Boccaletto *et al.* 2018), the next challenge is to decrypt and elucidate the function of this RNA code. At present, the epitranscriptomic field is hindered by the dearth of reagents to all RNA modifications, and while mass spectrometry can detect all RNA modifications, analysis of viral RNAs longer than 300 nucleotides in a milieu of other cellular RNAs has yet to identify RNA modification sites at nucleotide resolution. Additionally, the effects of an epitranscriptomic mark are typically evaluated by changing the abundance of the writer, eraser and reader enzymes. This approach presents challenges, first because these enzymes for many RNA modifications are unknown. Second, the outcomes of overexpressing or depleting these enzymes need to be carefully examined, as the effects on viral gene expression might be an indirect rather than direct outcomes resulting from effects on cellular RNA homeostasis. Deciphering the epitranscriptomic code will expand our knowledge of virus–cell interactions, and aid identification of new antiviral targets which will contribute to the development of novel therapeutics. Viruses have also been shown to be exquisite cellular and molecular biology explorers; thus, interpreting the viral epitranscriptome will likely also illuminate the functions of the epitranscriptome in cellular RNA processes and broaden our understanding of numerous health conditions linked to RNA malfunction.

4.7. References

Abbas, Y.M., Laudenbach, B.T., Martínez-Montero, S., Cencic, R., Habjan, M., Pichlmair, A., Damha, M.J., Pelletier, J., and Nagar B. (2017). Structure of human IFIT1 with capped RNA reveals adaptable mRNA binding and mechanisms for sensing N1 and N2 ribose 2'-O methylations. *Proceedings of the National Academy of Sciences of the United States of America*, 114(11), E2106–E2115.

Acheson, N.H. (1978). Polyoma virus giant RNAs contain tandem repeats of the nucleotide sequence of the entire viral genome. *Proceedings of the National Academy of Sciences of the United States of America*, 75(10), 4754–4758.

Acheson, N.H. (1981). Efficiency of processing of viral RNA during the early and late phases of productive infection by polyoma virus. *Journal of Virology*, 37(2), 628–635. Available at: http://www.ncbi.nlm.nih.gov/pubmed/6261010 [Accessed 22 January 2020]

Acheson, N.H. (1984). Kinetics and efficiency of polyadenylation of late polyomavirus nuclear RNA: Generation of oligomeric polyadenylated RNAs and their processing into mRNA. *Molecular and Cellular Biology*, 4(4), 722–729.

Aloni, Y., Dhar, R., and Khoury, G. (1979). Methylation of nuclear simian virus 40 RNAs. *Journal of Virology*, 32(1), 52–60.

Amort, T. and Lusser, A. (2017). Detection of 5-methylcytosine in specific poly(A) RNAs by bisulfite sequencing. *Methods in Molecular Biology*, 107–121.

Amort, T., Rieder, D., Wille, A., Khokhlova-Cubberley, D., Riml, C., Trixl, L., Jia, X.-Y., Micura, R., and Lusser, A. (2017). Distinct 5-methylcytosine profiles in poly(A) RNA from mouse embryonic stem cells and brain. *Genome Biology*, 18, 1.

Anderson, B.R., Muramatsu, H., Nallagatla, S.R., Bevilacqua, P.C., Sansing, L.H., Weissman, D., and Karikó, K. (2010). Incorporation of pseudouridine into mRNA enhances translation by diminishing PKR activation. *Nucleic Acids Research*, 38, 5884–5892.

Anderson, B.R., Muramatsu, H., Jha, B.K., Silverman, R.H., Weissman, D., and Karikó, K. (2011). Nucleoside modifications in RNA limit activation of 2'-5'- oligoadenylate synthetase and increase resistance to cleavage by RNase L. *Nucleic Acids Research*, 39, 9329–9338.

Aspden, J.L. and Jackson, R.J. (2010). Differential effects of nucleotide analogs on scanning-dependent initiation and elongation of mammalian mRNA translation *in vitro*. *RNA*, 16(6), 1130–1137.

Ayadi, L., Galvanin, A., Pichot, F., Marchand, V., and Motorin, Y. (2019). RNA ribose methylation (2'-O-methylation): Occurrence, biosynthesis and biological functions. *Biochimica et Biophysica Acta - Gene Regulatory Mechanisms*, 1862, 253–269.

Bass, B.L. and Weintraub, H. (1987). A developmentally regulated activity that unwinds RNA duplexes. *Cell*, 48(4), 607–613.

Bass, B.L., Weintraub, H., Cattaneo, R., and Billeter, M.A. (1989). Biased hypermutation of viral RNA genomes could be due to unwinding/modification of double-stranded RNA. *Cell*, 56, 331.

Baumstark, T. and Ahlquist, P. (2001). The brome mosaic virus RNA3 intergenic replication enhancer folds to mimic a tRNA TpsiC-stem loop and is modified *in vivo*. *RNA*, 7(11), 1652–1670. Available at: http://www.ncbi.nlm.nih.gov/pubmed/11720293 [Accessed 1 July 2019].

Becker, H. (1998). Pseudouridine and ribothymidine formation in the tRNA-like domain of turnip yellow mosaic virus RNA. *Nucleic Acids Research*, 26(17), 3991–3997.

Beemon, K. and Keith, J. (1977). Localization of N6-methyladenosine in the Rous sarcoma virus genome. *Journal of Molecular Biology*, 113(1), 165–179.

Bhattacharya, T., Newton, I.L.G., and Hardy, R.W. (2017). Wolbachia elevates host methyltransferase expression to block an RNA virus early during infection. *PLoS Pathogens*, 13(6), e1006427.

Bilbille, Y., Vendeix, F.A.P., Guenther, R., Malkiewicz, A., Ariza, X., Vilarrasa, J., and Agris, P.F. (2009). The structure of the human tRNALys3 anticodon bound to the HIV genome is stabilized by modified nucleosides and adjacent mismatch base pairs. *Nucleic Acids Research*, 37, 3342–3353.

Boccaletto, P., Machnicka, M.A., Purta, E., Pitkowski, P., Baginski, B., Wirecki, T.K., De Crécy-Lagard, V., Ross, R., Limbach, P.A., Kotter, A., Helm, M., and Bujnicki, J.M. (2018). MODOMICS: A database of RNA modification pathways. 2017 update. *Nucleic Acids Research,* 46, D303–D307.

Bouvet, M., Debarnot, C., Imbert, I., Selisko, B., Snijder, E.J., Canard, B., and Decroly, E. (2010). *In vitro* reconstitution of SARS-coronavirus mRNA cap methylation. *PLoS Pathogens*, 6, e1000863.

Bouvier, N.M. and Palese, P. (2008). The biology of influenza viruses. *Vaccine*, 26(4), D49–D53.

Bujnicki, J.M., Feder, M., Ayres, C.L., and Redman, K.L. (2004). Sequence-structure-function studies of tRNA:m^{5}C methyltransferase Trm4p and its relationship to DNA:m^{5}C and RNA:m^{5}U methyltransferases. *Nucleic Acids Research*, 32(8), 2453–2463.

Canaani, D. *et al.* (1979). Identification and mapping of N6-methyladenosine containing sequences in simian virus 40 RNA. *Nucleic Acids Research*, 6(8), 2879–2899.

Cantara, W.A. *et al.* (2011). The RNA modification database, RNAMDB: 2011 update. *Nucleic Acids Research*, 39(1), D195–D201.

Carlile, T.M. *et al.* (2014). Pseudouridine profiling reveals regulated mRNA pseudouridylation in yeast and human cells. *Nature*, 515(7525), 143–146.

Casey, J.L. and Gerin, J.L. (1995). Hepatitis D virus RNA editing: Specific modification of adenosine in the antigenomic RNA. *Journal of Virology*, 69(12), 7593–7600.

Casey, J.L. *et al.* (1992). Structural requirements for RNA editing in hepatitis delta virus: Evidence for a uridine-to-cytidine editing mechanism. *Proceedings of the National Academy of Sciences*, 89(15), 7149–7153.

Cattaneo, R. *et al.* (1988). Biased hypermutation and other genetic changes in defective measles viruses in human brain infections. *Cell*, 55(2), 255–265.

Cattenoz, P.B. *et al.* (2013). Transcriptome-wide identification of A > I RNA editing sites by inosine specific cleavage. *RNA*, 19(2), 257–270.

Cesarman, E. *et al.* (1996). Kaposi's sarcoma-associated herpesvirus in non-AIDS-related lymphomas occurring in body cavities [Online]. *American Journal of Pathology*, 149(1), 53–57. Available: http://www.ncbi.nlm.nih.gov/pubmed/8686762 [Accessed 27 January 2020].

Chan, C.T.Y. *et al.* (2010). A quantitative systems approach reveals dynamic control of tRNA modifications during cellular stress. *PLoS Genetics*, 6(12), 1–9.

Chen, C.X. *et al.* (2000). A third member of the RNA-specific adenosine deaminase gene family, ADAR3, contains both single- and double-stranded RNA binding domains. *RNA*, 6(5), 755–767.

Chen, Y. *et al.* (2009). Functional screen reveals SARS coronavirus nonstructural protein nsp14 as a novel cap N7 methyltransferase. *Proceedings of the National Academy of Sciences*, 106(9), 3484–3489.

Clemmer, D.E. and Jarrold, M.F. (1997). Ion mobility measurements and their applications to clusters and biomolecules. *Journal of Mass Spectrometry*, 577–592.

Cohn, W.E. (1959). 5-Ribosyl uracil, a carbon-carbon ribofuranosyl nucleoside in ribonucleic acids. *Biochimica et Biophysica Acta*, 32, 569–571.

Cohn, W.E. and Volkin, E. (1951). Nucleoside-5′-phosphates from ribonucleic acid. *Nature*, 167(4247), 483–484.

Courtney, D.G. *et al.* (2017). Epitranscriptomic enhancement of influenza A virus gene expression and replication. *Cell Host & Microbe*, 22(3), 377–386.

Courtney, D.G. *et al.* (2019a). Epitranscriptomic addition of m^5C to HIV-1 transcripts regulates viral gene expression. *Cell Host & Microbe*, 26(2), 217–227.

Courtney, D.G. *et al.* (2019b). Extensive epitranscriptomic methylation of A and C residues on murine leukemia virus transcripts enhances viral gene expression. *mBio*, 10(3), e01209 19.

Cruz-Oliveira, C. *et al.* (2015). Receptors and routes of dengue virus entry into the host cells. *FEMS Microbiology Reviews*, 39(2), 155–170.

Daffis, S. *et al.* (2010). 2′-O methylation of the viral mRNA cap evades host restriction by IFIT family members. *Nature*, 468(7322), 452–456.

Dai, Q. *et al.* (2017). Nm-seq maps 2′-O-methylation sites in human mRNA with base precision. *Nature Methods*, 14(7), 695–698.

Dalet, A., Gatti, E., and Pierre, P. (2015). Integration of PKR-dependent translation inhibition with innate immunity is required for a coordinated anti-viral response. *FEBS Letters*, 589(14), 1539–1545.

Danecek, P. *et al.* (2012). High levels of RNA-editing site conservation amongst 15 laboratory mouse strains. *Genome Biology*, 13(4), R26.

Davis, D.R. (1995). Stabilization of RNA stacking by pseudouridine. *Nucleic Acids Research*, 23(24), 5020–5026.

Decroly, E. and Canard, B. (2017). Biochemical principles and inhibitors to interfere with viral capping pathways. *Current Opinion in Virology*, 24, 87–96.

Decroly, E. *et al.* (2008). Coronavirus nonstructural protein 16 is a Cap-0 binding enzyme possessing (nucleoside-2’O)-methyltransferase activity. *Journal of Virology*, 82(16), 8071–8084.

Decroly, E. *et al.* (2012). Conventional and unconventional mechanisms for capping viral mRNA. *Nature Reviews Microbiology*, 10(1), 51–65.

Deffrasnes, C. *et al.* (2016). Genome-wide siRNA screening at biosafety level 4 reveals a crucial role for fibrillarin in henipavirus infection, *PLoS Pathogens*, 12(3), e1005478.

Desterro, J.M.P. *et al.* (2003). Dynamic association of RNA-editing enzymes with the nucleolus. *Journal of Cell Science*, 116(9), 1805–1818.

Devarkar, S.C. *et al.* (2016). Structural basis for m7G recognition and 2′-O-methyl discrimination in capped RNAs by the innate immune receptor RIG-I. *Proceedings of the National Academy of Sciences of the United States of America*, 113(3), 596–601.

Dimock, K. and Stoltzfus, C.M. (1977). Sequence specificity of internal methylation in B77 avian sarcoma virus RNA subunits. *Biochemistry*, 16(3), 471–478.

Dimock, K. and Stoltzfus, C.M. (1978). Cycloleucine blocks 5'-terminal and internal methylations of avian sarcoma virus genome RNA. *Biochemistry*, 17(17), 3627–3632.

Dominissini, D. *et al.* (2012). Topology of the human and mouse m^6A RNA methylomes revealed by m^6A-seq. *Nature*, 485(7397), 201–206.

Dong, H. *et al.* (2008a). Separate molecules of West Nile virus methyltransferase can independently catalyze the N7 and 2′-O methylations of viral RNA cap. *Virology*, 377(1), 1–6.

Dong, H., Zhang, B., and Shi, P.Y. (2008b). Flavivirus methyltransferase: A novel antiviral target. *Antiviral Research*, 80(1), 1–10.

Dong, H. *et al.* (2012). 2′-O methylation of internal adenosine by flavivirus NS5 methyltransferase. *PLoS Pathogens*, 8(4), e1002642.

Doria, M. *et al.* (2009). Editing of HIV-1 RNA by the double-stranded RNA deaminase ADAR1 stimulates viral infection. *Nucleic Acids Research*, 37(17), 5848–5858.

Dubin, D.T. and Stollar, V. (1975). Methylation of sindbis virus "26S" messenger RNA, *Biochemical and Biophysical Research Communications*, 66(4), 1373–1379.

Dubin, D.T. *et al.* (1977). Sindbis virus messenger RNA: The 5′-termini and methylated residues of 26 and 42 S RNA, *Virology*, 77(2), 457–470.

van Duijn, L.P. *et al.* (1986). Additional methylation at the N(2)-position of the cap of 26S Semliki Forest virus late mRNA and initiation of translation. *Virus Research*, 5(1), 61–66. Available at: http://www.ncbi.nlm.nih.gov/pubmed/3751287 [Accessed 11 August 2019].

Durdevic, Z. *et al.* (2013). Efficient RNA virus control in *Drosophila* requires the RNA methyltransferase Dnmt2. *EMBO Reports*, 14(3), 269–275.

Edupuganti, R.R. *et al.* (2017). N6-methyladenosine (m6A) recruits and repels proteins to regulate mRNA homeostasis. *Nature Structural & Molecular Biology*, 24(10), 870–878.

Erales, J. *et al.* (2017). Evidence for rRNA 2′-O-methylation plasticity: Control of intrinsic translational capabilities of human ribosomes. *Proceedings of the National Academy of Sciences*, 114(49), 12934–12939.

Feederle, R. and Schepers, A. (2017). Antibodies specific for nucleic acid modifications. *RNA Biology*, 14(9), 1089–1098.

Felder, M.-P. *et al.* (1994). Functional and biological properties of an avian variant long terminal repeat containing multiple A to G conversions in the U3 sequence [Online]. *Journal of Virology*, 68(8), 4759–4767. Available: http://www.ncbi.nlm.nih.gov/pubmed/8035477 [Accessed 29 July 2019].

Fontecave, M., Atta, M., and Mulliez, E. (2004). S-adenosylmethionine: Nothing goes to waste. *Trends in Biochemical Sciences*, 29(5), 243–249.

Fu, Y. *et al.* (2014). Gene expression regulation mediated through reversible m^6A RNA methylation. *Nature Reviews Genetics*, 293–306.

Furuichi, Y., *et al.* (1975a). Blocked, methylated 5'-terminal sequence in avian sarcoma virus RNA. *Nature*, 257(5527), 618–620.

Furuichi, Y., *et al.* (1975b). Reovirus messenger RNA contains a methylated, blocked 5'-terminal structure: m-7G(5')ppp(5')G-MpCp-. *Proceedings of the National Academy of Sciences*, 72(1), 362–366.

George, C.X., John, L., and Samuel, C.E. (2014). An RNA editor, adenosine deaminase acting on double-stranded RNA (ADAR1). *Journal of Interferon & Cytokine Research*, 34(6), 437–446.

Gokhale, N.S. and Horner, S.M. (2017). RNA modifications go viral. *PLOS Pathogens*, 13(3), e1006188.

Gokhale, N.S. *et al.* (2016). N6-Methyladenosine in Flaviviridae viral RNA genomes regulates infection. *Cell Host & Microbe*, 20(5), 654–665.

Gonzales-van Horn, S.R. and Sarnow, P. (2017). Making the mark: The role of adenosine modifications in the life cycle of RNA viruses. *Cell Host & Microbe*, 21(6), 661–669.

Griffin, D.E., Lin, W.-H.W., and Nelson, A.N. (2018). Understanding the causes and consequences of measles virus persistence. *F1000Research*, 7, 237.

Grozhik, A.V. and Jaffrey, S.R. (2018). Distinguishing RNA modifications from noise in epitranscriptome maps. *Nature Chemical Biology*, 14(3), 215–225.

Gutsche, I. *et al.* (2015). Near-atomic cryo-EM structure of the helical measles virus nucleocapsid. *Science*, 348(6235), 704–707.

Habjan, M. *et al.* (2013). Sequestration by IFIT1 impairs translation of 2'O-unmethylated capped RNA. *PLoS Pathogens*, 9(10), e1003663.

Hall, W.W., Lamb, R.A., and Choppin, P.W. (1979). Measles and subacute sclerosing panencephalitis virus proteins: Lack of antibodies to the M protein in patients with subacute sclerosing panencephalitis. *Proceedings of the National Academy of Sciences*, 76(4), 2047–2051.

Hao, H. *et al.* (2019). N6-methyladenosine modification and METTL3 modulate enterovirus 71 replication. *Nucleic Acids Research*, 47(1), 362–374.

Hashimoto, S.I. and Green, M. (1976). Multiple methylated cap sequences in adenovirus type 2 early mRNA [Online]. *Journal of Virology*, 20(2), 425–435. Available: http://www.pubmedcentral.nih.gov/articlerender.fcgi?artid=PMC355010 [Accessed 12 August 2019].

Hesser, C.R. *et al.* (2018). N6-methyladenosine modification and the YTHDF2 reader protein play cell type specific roles in lytic viral gene expression during Kaposi's sarcoma-associated herpesvirus infection. *PLOS Pathogens*, 14(4), e1006995.

Hinnebusch, A.G. (2006). eIF3: A versatile scaffold for translation initiation complexes. *Trends in Biochemical Sciences*, 31(10), 553–562.

Hong, T. *et al.* (2018). Precise antibody-independent m^{6}A identification via 4SedTTP-involved and FTO-assisted strategy at single-nucleotide resolution'. *Journal of the American Chemical Society*, 140(18), 5886–5889.

van den Hoogen, B.G. *et al.* (2014). Excessive production and extreme editing of human metapneumovirus defective interfering RNA is associated with type I IFN induction. *Journal of General Virology*, 95(8), 1625–1633.

Hsuchen, C.C. and Dubin, D.T. (1976). Di- and trimethylated congeners of 7-methylguanine in Sindbis virus mRNA. *Nature*, 264(5582), 190–191.

Hu, W.-S. and Hughes, S.H. (2012). HIV-1 reverse transcription. *Cold Spring Harbor Perspectives in Medicine*, 2(10), a006882–a006882.

Huang, Y. and Carmichael, G.G. (2009). RNA processing in the polyoma virus life cycle. *Frontiers in Bioscience*, 14(13), 4968–4977.

Hui, D.J. *et al.* (2003). Viral stress-inducible protein p56 inhibits translation by blocking the interaction of eIF3 with the ternary complex eIF2·GTP· Met-tRNAi. *Journal of Biological Chemistry*, 278(41), 39477–39482.

Hui, D.J. *et al.* (2005). Mouse p56 blocks a distinct function of eukaryotic initiation factor 3 in translation initiation. *Journal of Biological Chemistry*, 280(5), 3433–3440.

Hyde, J.L. and Diamond, M.S. (2015). Innate immune restriction and antagonism of viral RNA lacking 2'-O methylation. *Virology*, 479–480, 66–74.

Hyde-Deruyscher, R.P. and Carmichael, G.G. (1990). Polyomavirus late pre-mRNA processing: DNA replication-associated changes in leader exon multiplicity suggest a role for leader-to-leader splicing in the early-late switch. *Journal of Virology*, 64(12), 5823–5832. Available at: http://www.ncbi.nlm.nih.gov/pubmed/2173769 [Accessed 11 January 2020].

Iglesias, N.G. and Gamarnik, A.V. (2011). Dynamic RNA structures in the dengue virus genome. *RNA Biology*, 8(2), 249–57.

Imam, H. *et al.* (2018). N6-methyladenosine modification of hepatitis B virus RNA differentially regulates the viral life cycle. *Proceedings of the National Academy of Sciences of the United States of America*, 115(35), 8829–8834.

Incarnato, D. *et al.* (2017). High-throughput single-base resolution mapping of RNA 2-O-methylated residues. *Nucleic Acids Research*, 45(3), 1433–1441.

Jayan, G.C. and Casey, J.L. (2005). Effects of conserved RNA secondary structures on hepatitis delta virus genotype I RNA editing, replication, and virus production. *Journal of Virology*, 79(17), 11187–11193.

Jia, G. *et al.* (2011). N6-Methyladenosine in nuclear RNA is a major substrate of the obesity-associated FTO. *Nature Chemical Biology*, 7(12), 885–887.

Kane, S.E. and Beemon, K. (1985). Precise localization of m^6A in Rous sarcoma virus RNA reveals clustering of methylation sites: Implications for RNA processing. *Molecular and Cellular Biology*, 5(9), 2298–2306.

Karikó, K. *et al.* (2008). Incorporation of pseudouridine into mRNA yields superior nonimmunogenic vector with increased translational capacity and biological stability. *Molecular Therapy*, 16(11), 1833–1840.

Kellner, S., Burhenne, J., and Helm, M. (2010). Detection of RNA modifications. *RNA Biology*, 7(2), 237–247.

Kennedy, E.M. *et al.* (2016). Posttranscriptional m^6A editing of HIV-1 mRNAs enhances viral gene expression. *Cell Host & Microbe*, 19(5), 675–685.

Kennedy, E.M. *et al.* (2017). Viral epitranscriptomics. *Journal of Virology*, 91(9), e02263-16.

Khrustalev, V.V. *et al.* (2017). Mutational pressure in zika virus: Local ADAR-editing areas associated with pauses in translation and replication. *Frontiers in Cellular and Infection Microbiology*, 7, 44.

Kimura, T. *et al.* (2013). Ifit1 inhibits Japanese encephalitis virus replication through binding to 5' capped 2'-O unmethylated RNA. *Journal of Virology*, 87(18), 9997–10003.

Knutson, S.D., Ayele, T.M., and Heemstra, J.M. (2018). Chemical labeling and affinity capture of inosine-containing RNAs using acrylamidofluorescein. *Bioconjugate Chemistry*, 29(9), 2899–2903.

Koh, C., Da, B.L., and Glenn, J.S. (2019). HBV/HDV coinfection: A challenge for therapeutics. *Clinics in Liver Disease*, 23(3), 557–572.

Krug, R.M., Morgan, M.A., and Shatkin, A.J. (1976). Influenza viral mRNA contains internal N6-methyladenosine and 5'-terminal 7-methylguanosine in cap structures [Online]. *Journal of Virology*, 20(1), 45–53. Available: http://www.ncbi.nlm.nih.gov/pubmed/1086370 [Accessed 18 July 2019].

Kruse, S. *et al.* (2011). A novel synthesis and detection method for cap-associated adenosine modifications in mouse mRNA. *Scientific Reports*, 1.

Kumar, M. and Carmichael, G.G. (1997). Nuclear antisense RNA induces extensive adenosine modifications and nuclear retention of target transcripts. *Proceedings of the National Academy of Sciences*, 94(8), 3542–3547.

Kumar, P. *et al.* (2014). Inhibition of translation by IFIT family members is determined by their ability to interact selectively with the 5'-terminal regions of cap0-, cap1- and 5'ppp- mRNAs. *Nucleic Acids Research*, 42(5), 3228–45.

Kuo, M.Y., Chao, M., and Taylor, J. (1989). Initiation of replication of the human hepatitis delta virus genome from cloned DNA: Role of delta antigen [Online]. *Journal of Virology*, 63(5), 1945–50. Available: http://www.pubmedcentral.nih.gov/articlerender.fcgi?artid=PMC250607 [Accessed 3 August 2019].

Lane, B.G. and Tamaoki, T. (1967). Studies of the chain termini and alkali-stable dinucleotide sequences in 16 s and 28 s ribosomal RNA from L cells. *Journal of Molecular Biology*, 27(2), 335–348.

Lazear, H.M. and Diamond, M.S. (2016). Zika virus: New clinical syndromes and its emergence in the Western Hemisphere. *Journal of Virology*, 90(10), 4864–4875.

Lempp, F.A., Ni, Y., and Urban, S. (2016). Hepatitis delta virus: Insights into a peculiar pathogen and novel treatment options. *Nature Reviews Gastroenterology & Hepatology*, 13(10), 580–589.

Li, J.B. *et al.* (2009). Genome-wide identification of human RNA editing sites by parallel DNA capturing and sequencing. *Science*, 324(5931), 1210–1213.

Li, X. *et al.* (2015). Chemical pulldown reveals dynamic pseudouridylation of the mammalian transcriptome. *Nature Chemical Biology*, 11(8), 592–597.

Li, X., Ma, S., and Yi, C. (2016). Pseudouridine: The fifth RNA nucleotide with renewed interests. *Current Opinion in Chemical Biology*, 33, 108–116.

Lichinchi, G. *et al.* (2016a). Dynamics of human and viral RNA methylation during Zika virus infection. *Cell Host & Microbe*, 20(5), 666–673.

Lichinchi, G. *et al.* (2016b). Dynamics of the human and viral m^{6}A RNA methylomes during HIV-1 infection of T cells. *Nature Microbiology*, 1(4), 16011.

Liddicoat, B.J. *et al.* (2015). RNA editing by ADAR1 prevents MDA5 sensing of endogenous dsRNA as nonself. *Science*, 349(6252), 1115–1120.

Liebert, U.G. *et al.* (1986). Restricted expression of Measles virus proteins in brains from cases of subacute sclerosing panencephalitis. *Journal of General Virology*, 67(11), 2435–2444.

Lindenbach, B.D. *et al.* (2013). Flaviviridae. In *Fields Virology*, 6th edition, Knipe, D.M. and Howley, P.M. (eds). Lippincott Williams & Wilkins, Philadelphia, 712–746.

Liu, J. *et al.* (2014). A METTL3-METTL14 complex mediates mammalian nuclear RNA N6-adenosine methylation. *Nature Chemical Biology*, 10(2), 93–95.

Lloyd, R.E. (2013). Regulation of stress granules and P-bodies during RNA virus infection. *Wiley Interdisciplinary Reviews. RNA*, 4(3), 317–331.

Lovejoy, A.F., Riordan, D.P., and Brown, P.O. (2014). Transcriptome-wide mapping of pseudouridines: Pseudouridine synthases modify specific mRNAs in *S. cerevisiae*. *PLoS ONE*, 9(10), e110799.

Lu, W. *et al.* (2018). *N*^{6}-Methyladenosine–binding proteins suppress HIV-1 infectivity and viral production. *Journal of Biological Chemistry*, 293(34), 12992–13005.

Luo, D. *et al.* (2012). Visualizing the determinants of viral RNA recognition by innate immune sensor RIG-I. *Structure*, 20(11), 1983–1988.

Mak, J. and Kleiman, L. (1997). Primer tRNAs for reverse transcription [Online]. *Journal of Virology*, 71(11), 8087–8095. Available: http://www.pubmedcentral.nih.gov/articlerender.fcgi?artid=PMC192263 [Accessed 2 July 2019].

Mannion, N.M. *et al.* (2014). The RNA-editing enzyme ADAR1 controls innate immune responses to RNA. *Cell Reports*, 9(4), 1482–1494.

Marceau, C.D. *et al.* (2016). Genetic dissection of Flaviviridae host factors through genome-scale CRISPR screens. *Nature*, 535(7610), 159–163.

Marchand, V. *et al.* (2016). Illumina-based RiboMethSeq approach for mapping of 2′-O-Me residues in RNA. *Nucleic Acids Research*, 44(16), e135.

Martin, B. *et al.* (2018). The methyltransferase domain of the Sudan ebolavirus L protein specifically targets internal adenosines of RNA substrates, in addition to the cap structure. *Nucleic Acids Research*, 46(15), 7902–7912.

Martínez, I. and Melero, J.A. (2002). A model for the generation of multiple A to G transitions in the human respiratory syncytial virus genome: Predicted RNA secondary structures as substrates for adenosine deaminases that act on RNA. *Journal of General Virology*, 83(6), 1445–1455.

Martínez, I., Dopazo, J., and Melero, J.A. (1997). Antigenic structure of the human respiratory syncytial virus G glycoprotein and relevance of hypermutation events for the generation of antigenic variants. *Journal of General Virology*, 78(10), 2419–2429. doi

Mauro, V.P. and Edelman, G.M. (2002). The ribosome filter hypothesis. *Proceedings of the National Academy of Sciences of the United States of America*, 99(19), 12031–12036.

McFadden, M.J., Gokhale, N.S., and Horner, S.M. (2017). Protect this house: Cytosolic sensing of viruses. *Current Opinion in Virology*, 22, 36–43.

McIntyre, W. *et al.* (2018). Positive-sense RNA viruses reveal the complexity and dynamics of the cellular and viral epitranscriptomes during infection. *Nucleic Acids Research*, 46(11), 5776–5791.

McLuckey, S.A., Van Berkel, G.J., and Glish, G.L. (1992). Tandem mass spectrometry of small, multiply charged oligonucleotides. *Journal of the American Society for Mass Spectrometry*, 3(1), 60–70.

Mentha, N. *et al.* (2019). A review on hepatitis D: From virology to new therapies. *Journal of Advanced Research*, 17, 3–15.

Meyer, K.D. and Jaffrey, S.R. (2017). Rethinking m^{6}A readers, writers, and erasers. *Annual Review of Cell and Developmental Biology*, 33.

Meyer, K.D. *et al.* (2012). Comprehensive analysis of mRNA methylation reveals enrichment in 3' UTRs and near stop codons. *Cell*, 149(7), 1635–1646.

Moss, B. *et al.* (1977). 5'-Terminal and internal methylated nucleosides in herpes simplex virus type 1 mRNA [Online]. *Journal of Virology*, 23(2), 234–239. Available: http://www.pubmedcentral.nih.gov/articlerender.fcgi?artid=PMC515825 [Accessed 27 September 2019].

Motorin, Y., Lyko, F., and Helm, M. (2009). 5-methylcytosine in RNA: Detection, enzymatic formation and biological functions. *Nucleic Acids Research*, 38(5), 1415–1430.

Narayan, P. *et al.* (1987). Unequal distribution of N6-methyladenosine in influenza virus mRNAs [Online]. *Molecular and Cellular Biology*, 7(4), 1572–1575. Available: http://www.pubmedcentral.nih.gov/articlerender.fcgi?artid=PMC365250 [Accessed 26 January 2019].

Newby, M.I. and Greenbaum, N.L. (2002). Investigation of Overhauser effects between pseudouridine and water protons in RNA helices. *Proceedings of the National Academy of Sciences*, 99(20), 12697–12702.

Nicholson, B.L. and White, K.A. (2015). Exploring the architecture of viral RNA genomes. *Current Opinion in Virology*, 12, 66–74.

Otsuka, Y., Kedersha, N.L., and Schoenberg, D.R. (2009). Identification of a cytoplasmic complex that adds a cap onto 5'-monophosphate RNA. *Molecular and Cellular Biology*, 29(8), 2155–2167.

Ott, M., Geyer, M., and Zhou, Q. (2011). The control of HIV transcription: Keeping RNA polymerase II on track. *Cell Host & Microbe*, 10(5), 426–435.

Patterson, J.B. and Samuel, C.E. (1995). Expression and regulation by interferon of a double-stranded-RNA-specific adenosine deaminase from human cells: Evidence for two forms of the deaminase. *Molecular and Cellular Biology*, 15(10), 5376–5388.

Patterson, J.B. *et al.* (2001). Evidence that the hypermutated M protein of a subacute sclerosing panencephalitis measles virus actively contributes to the chronic progressive CNS disease. *Virology*, 291(2), 215–225.

Petes, C., Odoardi, N., and Gee, K. (2017). The toll for trafficking: Toll-like receptor 7 delivery to the endosome. *Frontiers in Immunology*, 8(SEP), 1075.

Pfaller, C.K. *et al.* (2014). Measles virus C protein impairs production of defective copyback double-stranded viral RNA and activation of protein kinase R. *Journal of Virology*, 88(1), 456–468.

Pfaller, C.K. *et al.* (2015). Measles virus defective interfering RNAs are generated frequently and early in the absence of C protein and can be destabilized by adenosine deaminase acting on RNA-1-like hypermutations. *Journal of Virology*, 89(15), 7735–7747.

Pfaller, C.K. *et al.* (2018). Extensive editing of cellular and viral double-stranded RNA structures accounts for innate immunity suppression and the proviral activity of ADAR1 p150. *PLoS Biology*, 16(11), e2006577.

Phuphuakrat, A. *et al.* (2008). Double-stranded RNA adenosine deaminases enhance expression of human immunodeficiency virus type 1 proteins. *Journal of Virology*, 82(21), 10864–10872.

Piontkivska, H. *et al.* (2017). RNA editing by the host ADAR system affects the molecular evolution of the Zika virus. *Ecology and Evolution*, 7(12), 4475–4485.

Polson, A.G., Bass, B.L., and Casey, J.L. (1996). RNA editing of hepatitis delta virus antigenome by dsRNA-adenosine deaminase. *Nature*, 380(6573), 454–456.

Polson, A.G. *et al.* (1998). Hepatitis delta virus RNA editing is highly specific for the amber/W site and is suppressed by hepatitis delta antigen. *Molecular and Cellular Biology*, 18(4), 1919–1926.

Poulsen, H. *et al.* (2001). CRM1 mediates the export of ADAR1 through a nuclear export signal within the Z-DNA binding domain. *Molecular and Cellular Biology*, 21(22), 7862–7871.

Prusiner, P., Yathindra, N., and Sundaralingam, M. (1974). Effect of ribose O(2′)-methylation on the conformation of nucleosides and nucleotides. *BBA Section Nucleic Acids And Protein Synthesis*, 366(2), 115–123.

Ramanathan, A., Robb, G.B., and Chan, S.H. (2016). mRNA capping: Biological functions and applications. *Nucleic Acids Research*, 44(16), 7511–7526.

Rebagliati, M.R. and Melton, D.A. (1987). Antisense RNA injections in fertilized frog eggs reveal an RNA duplex unwinding activity. *Cell*, 48(4), 599–605.

Recasens, M. *et al.* (2010). Proliferation promotion and pro-collagen synthesis stimulation by infrared radiation on human fibroblasts. *RNA*, 19, 1–9.

Reichow, S.L. *et al.* (2007). The structure and function of small nucleolar ribonucleoproteins. *Nucleic Acids Research*, 35(5), 1452–1464.

Rima, B.K. *et al.* (2014). Stability of the parainfluenza virus 5 genome revealed by deep sequencing of strains isolated from different hosts and following passage in cell culture. *Journal of Virology*, 88(7), 3826–3836.

Ringeard, M. *et al.* (2019). FTSJ3 is an RNA 2′-O-methyltransferase recruited by HIV to avoid innate immune sensing. *Nature*, 565(7740), 500–504.

Rizzetto, M. *et al.* (1980). Delta Agent: Association of delta antigen with hepatitis B surface antigen and RNA in serum of delta-infected chimpanzees. *Proceedings of the National Academy of Sciences*, 77(10), 6124–6128.

Robbins, M. *et al.* (2007). 2′-O-methyl-modified RNAs act as TLR7 antagonists. *Molecular Therapy*, 15(9), 1663–1669.

Rose, J.K. (1975). Heterogneeous 5’-terminal structures occur on vesicular stomatitis virus mRNAs [Online]. *The Journal of Biological Chemistry*, 250(20), 8098–8104. Available: http://www.ncbi.nlm.nih.gov/pubmed/170280 [Accessed 6 July 2019].

Rose, R.E. *et al.* (2015). Profiling ribonucleotide modifications at full-transcriptome level: A step toward MS-based epitranscriptomics. *RNA*, 21(7), 1361–1374.

Rose, R.E. *et al.* (2016). Global epitranscriptomics profiling of RNA post-transcriptional modifications as an effective tool for investigating the epitranscriptomics of stress response. *Molecular & Cellular Proteomics*, 15(3), 932–944.

Roth-Cross, J.K., Bender, S.J., and Weiss, S.R. (2008). Murine coronavirus mouse hepatitis virus is recognized by MDA5 and induces type I interferon in brain macrophages/microglia. *Journal of Virology*, 82(20), 9829–9838.

Roundtree, I.A. *et al.* (2017). Dynamic RNA modifications in gene expression regulation. *Cell*, 169(7), 1187–1200.

Rueda, P., García-Barreno, B., and Melero, J.A. (1994). Loss of conserved cysteine residues in the attachment (G) glycoprotein of two human respiratory syncytial virus escape mutants that contain multiple A-G substitutions (hypermutations) [Online]. *Virology*, 198(2), 653–662. Available: http://www.ncbi.nlm.nih.gov/pubmed/7507282 [Accessed 13 July 2019].

Ryu, W.S., Bayer, M., and Taylor, J. (1992). Assembly of hepatitis delta virus particles [Online]. *Journal of Virology*, 66(4), 2310–2315. Available: http://www.ncbi.nlm.nih.gov/pubmed/1548764 [Accessed 3 August 2019].

Said, J.W. *et al.* (1996). Primary effusion lymphoma in women: Report of two cases of Kaposi's sarcoma herpes virus-associated effusion-based lymphoma in human immunodeficiency virus-negative women. *Blood*, 88(8), 3124–3128.

Sakurai, M. *et al.* (2010). Inosine cyanoethylation identifies A-to-I RNA editing sites in the human transcriptome. *Nature Chemical Biology*, 6(10), 733–740.

Samuel, C.E. (2011). Adenosine deaminases acting on RNA (ADARs) are both antiviral and proviral. *Virology*, 411(2), 180–193.

Sansam, C.L., Wells, K.S., and Emeson, R.B. (2003). Modulation of RNA editing by functional nucleolar sequestration of ADAR2. *Proceedings of the National Academy of Sciences of the United States of America*, 100(2), 14018–14023.

Sato, S., Wong, S.K., and Lazinski, D.W. (2001). Hepatitis delta virus minimal substrates competent for editing by ADAR1 and ADAR2. *Journal of Virology*, 75(18), 8547–8555.

Sato, S., Cornillez-Ty, C., and Lazinski, D.W. (2004). By inhibiting replication, the large hepatitis delta antigen can indirectly regulate amber/W editing and its own expression. *Journal of Virology*, 78(15), 8120–8134.

Savva, Y.A., Rieder, L.E., and Reenan, R.A. (2012). The ADAR protein family. *Genome Biology*, 252.

Schaefer, M. *et al.* (2009). RNA cytosine methylation analysis by bisulfite sequencing. *Nucleic Acids Research*, 37(2).

Schuberth-Wagner, C. *et al.* (2015). A conserved histidine in the rna sensor RIG-I controls immune tolerance to N1-2'O-methylated self RNA. *Immunity*, 43(1), 41–51.

Schwartz, S. *et al.* (2014). Transcriptome-wide mapping reveals widespread dynamic-regulated pseudouridylation of ncRNA and mRNA. *Cell*, 159(1), 148–162.

Seeger, C. and Mason, W.S. (2000). Hepatitis B virus biology. *Microbiology and Molecular Biology Reviews*, 64(1), 51–68.

Sharmeen, L. *et al.* (1991). Tat-dependent adenosine-to-inosine modification of wild-type transactivation response RNA. *Proceedings of the National Academy of Sciences*, 88(18), 8096–8100.

Shi, H., Wei, J., and He, C. (2019). Where, when, and how: Context-dependent functions of RNA methylation writers, readers, and erasers. *Molecular Cell*, 74(4), 640–650.

Shi, Z. *et al.* (2017). Heterogeneous ribosomes preferentially translate distinct subpools of mRNAs genome-wide. *Molecular Cell*, 67(1), 71–83.

Slotkin, W. and Nishikura, K. (2013). Adenosine-to-inosine RNA editing and human disease. *Genome Medicine*, 5(11), 105.

Smith, A.M. *et al.* (2019). Reading canonical and modified nucleobases in 16S ribosomal RNA using nanopore native RNA sequencing. *PLoS ONE*, 14(5), 1–15.

Sommer, S. *et al.* (1976). The methylation of adenovirus-specific nuclear and cytoplasmic RNA. *Nucleic Acids Research*, 3(3), 749–765.

Song, J. and Yi, C. (2017). Chemical modifications to RNA: A new layer of gene expression regulation. *ACS Chemical Biology*, 316–325.

Spenkuch, F., Motorin, Y., and Helm, M. (2014). Pseudouridine: Still mysterious, but never a fake (uridine)! *RNA Biology*, 11(12), 1540–1554.

Squires, J.E. *et al.* (2012). Widespread occurrence of 5-methylcytosine in human coding and non-coding RNA. *Nucleic Acids Research*, 40(11), 5023–5033.

Stoltzfus, C.M. and Dane, R.W. (1982). Accumulation of spliced avian retrovirus mRNA is inhibited in S-adenosylmethionine-depleted chicken embryo fibroblasts [Online]. *Journal of Virology*, 42(3), 918–931. Available: http://www.pubmedcentral.nih.gov/articlerender.fcgi?artid=PMC256926 [Accessed 12 August 2019].

Stoltzfus, C.M. and Dimock, K. (1976). Evidence of methylation of B77 avian sarcoma virus genome RNA subunits [Online]. *Journal of Virology*, 18(2), 586–595. Available: http://www.ncbi.nlm.nih.gov/pubmed/178899 [Accessed 6 August 2019].

Tan, B. *et al.* (2017). Viral and cellular N^6-methyladenosine and N^6,2′-O-dimethyladenosine epitranscriptomes in the KSHV life cycle. *Nature Microbiology*, 3(1), 108–120.

Taylor, D.R. *et al.* (2005). New antiviral pathway that mediates hepatitis C virus replicon interferon sensitivity through ADAR1. *Journal of Virology*, 79(10), 6291–6298.

Terenzi, F. *et al.* (2006). Distinct induction patterns and functions of two closely related interferon-inducible human genes, ISG54 and ISG56. *Journal of Biological Chemistry*, 281(45), 34064–34071.

Tirumuru, N. *et al.* (2016). N^6-methyladenosine of HIV-1 RNA regulates viral infection and HIV-1 Gag protein expression. *eLife*, 5, e15528.

Tollervey, D. *et al.* (1991). The small nucleolar RNP protein NOP1 (fibrillarin) is required for pre-rRNA processing in yeast [Online]. *The EMBO Journal*, 10(3), 573–583. Available: http://www.ncbi.nlm.nih.gov/pubmed/1825809 [Accessed 1 July 2019].

Topisirovic, I. *et al.* (2011). Cap and cap-binding proteins in the control of gene expression. *Wiley Interdisciplinary Reviews: RNA*, 2(2), 277–298.

Treisman, R. and Kamen, R. (1981). Structure of polyoma virus late nuclear RNA. *Journal of Molecular Biology*, 148(3), 273–301.

Trim, A.R. and Parker, J.E. (1970). Preparation, purification and analyses of thirteen alkali-stable dinucleotides from ribonucleic acid. *Biochemical Journal*, 116(4), 589–598.

Trixl, L. and Lusser, A. (2019). The dynamic RNA modification 5-methylcytosine and its emerging role as an epitranscriptomic mark. *Wiley Interdisciplinary Reviews: RNA*, 10(1), e1510.

Trotman, J.B. *et al.* (2017). RNA guanine-7 methyltransferase catalyzes the methylation of cytoplasmically recapped RNAs. *Nucleic Acids Research*, 45(18), 10726–10739.

Tsai, K., Courtney, D.G., and Cullen, B.R. (2018). Addition of m^{6}A to SV40 late mRNAs enhances viral structural gene expression and replication. *PLoS Pathogens*, 14(2), e1006919.

Verbeck, G.F. *et al.* (2002). A fundamental introduction to ion mobility mass spectrometry applied to the analysis of biomolecules. *Journal of Biomolecular Techniques*, 13(2), 56–61.

Vieyres, G. and Pietschmann, T. (2019). HCV pit stop at the lipid droplet: Refuel lipids and put on a lipoprotein coat before exit. *Cells*, 8(3), 233.

Wei, C.M. and Moss, B. (1975). Methylated nucleotides block 5'-terminus of vaccinia virus messenger RNA. *Proceedings of the National Academy of Sciences*, 72(1), 318–322.

Wei, C.-M., Gershowitz, A., and Moss, B. (1975). Methylated nucleotides block 5′ terminus of HeLa cell messenger RNA. *Cell*, 4(4), 379–386.

Wong, S.K. and Lazinski, D.W. (2002). Replicating hepatitis delta virus RNA is edited in the nucleus by the small form of ADAR1. *Proceedings of the National Academy of Sciences*, 99(23), 15118–15123.

Yang, X. *et al.* (2017). 5-methylcytosine promotes mRNA export – NSUN2 as the methyltransferase and ALYREF as an m^{5}C reader. *Cell Research*, 27(5), 606–625.

Ye, F., Chen, E.R., and Nilsen, T.W. (2017). Kaposi's sarcoma-associated herpesvirus utilizes and manipulates RNA N^6-adenosine methylation to promote lytic replication. *Journal of Virology*, 91(16), e00466-17.

Yedavalli, V.S.R.K. and Jeang, K.-T. (2010). Trimethylguanosine capping selectively promotes expression of Rev-dependent HIV-1 RNAs. *Proceedings of the National Academy of Sciences*, 107(33), 14787–14792.

Zalinger, Z.B. *et al.* (2015). MDA5 is critical to host defense during infection with murine coronavirus. *Journal of Virology*, 89(24), 12330–12340.

Zerfass, K. and Beier, H. (1992). Pseudouridine in the anticodon G psi A of plant cytoplasmic tRNA(Tyr) is required for UAG and UAA suppression in the TMV-specific context. *Nucleic Acids Research*, 20(22), 5911–5918.

Zhang, Z. and Carmichael, G.G. (2001). The fate of dsRNA in the nucleus: A p54nrb-containing complex mediates the nuclear retention of promiscuously A-to-I edited RNAs. *Cell*, 106(4), 465–476.

Zhang, L.-S. *et al.* (2019). Transcriptome-wide mapping of internal N^7-methylguanosine methylome in mammalian mRNA. *Molecular Cell*, 74(6), 1304–1316.

Zheng, G. *et al.* (2013). ALKBH5 is a mammalian RNA demethylase that impacts RNA metabolism and mouse fertility. *Molecular Cell*, 49(1), 18–29.

Zhou, Y. *et al.* (2007). Structure and function of flavivirus NS5 methyltransferase. *Journal of Virology*, 81(8), 3891–3903.

Zhou, J. *et al.* (2015). Dynamic m^6A mRNA methylation directs translational control of heat shock response. *Nature*, 526(7574), 591–594.

Zhu, Y., Pirnie, S.P., and Carmichael, G.G. (2017). High-throughput and site-specific identification of 2′-O-methylation sites using ribose oxidation sequencing (RibOxi-seq). *RNA*, 23(8), 1303–1314.

Züst, R. *et al.* (2011). Ribose 2'-O-methylation provides a molecular signature for the distinction of self and non-self mRNA dependent on the RNA sensor Mda5. *Nature Immunology*, 12(2), 137–143.

5

Defective Viral Particles

Carolina B. LÓPEZ

Department of Pathobiology, University of Pennsylvania, Philadelphia, USA

5.1. Introduction

Defective viral particles are a subpopulation of viral particles that are unable to complete a full replication cycle in the absence of a co-infecting standard virus. They are generated during the replication of plant, arthropod and mammalian viruses, and contain defective viral genomes (DVGs) that result from genomic alterations introduced during viral replication (Genoyer and Lopez 2019b; Vignuzzi and Lopez 2019). In this chapter, we focus on DVGs that retain functional replication promoters and maintain the potential to be replicated when complemented by a co-infecting standard virus. Defective viral genomes that are able to interfere with the replication of the standard virus by competing for the viral replication machinery are known as defective interfering (DI) genomes. Defective viral particles can induce antiviral immune responses, promote the survival of a subpopulation of infected cells and affect how the virus spreads. These functions are independent of the replication interference properties of DI genomes. For the sake of consistency, in this chapter, we will refer to defective viral genomes as DI genomes only when they have demonstrated interfering activity.

For a color version of all figures in this book, see www.iste.co.uk/saleh/virology.zip.

Virology,
coordinated by María-Carla SALEH and Félix AUGUSTO REY. © ISTE Ltd 2021

5.2. Discovery of defective viral genomes and early research

In the early 1940s, Preben Von Magnus observed that influenza virus infectivity decreased when the virus was passaged at high titers. He postulated that particles of "incomplete" virus produced during high multiplicity of infection interfered with the replication of the standard virus (Henle *et al.* 1943; Henle 1950). His postulate was confirmed when a component with a lower sedimentation rate than the standard virus was identified in the allantoic fluid of embryonated chicken eggs infected with influenza virus. The presence of this component in virus stocks correlated with the lower replication of the standard virus. These components were called "Von Magnus particles" (Gard *et al.* 1952). Von Magnus particles were found to accumulate at different rates in different infections (Fazekas De St. Groth and Graham 1954), and their accumulation depended on the ratio of replicative over non-replicative viral particles present in the virus stock (Beale and Finter 1956). During the late 1950s and 1960s, incomplete interfering particles were described in a number of other viruses, including the *Arenavirus* lymphocytic choriomeningitis virus (LCMV) (Lehmann-Grube *et al.* 1969), the avian influenza virus (Rott and Scholtissek 1963), the *Paramyxovirus* Sendai virus (SeV) (Sokol *et al.* 1964), the *Phenuivirus* Rift Valley fever virus (Mims 1956), *Polyomaviruses* (Uchida *et al.* 1966; Blackstein *et al.* 1969) and the *Rhabdovirus* vesicular stomatitis virus (VSV) (Bellett and Cooper 1959; Huang and Wagner 1966b). In 1956, Ada *et al.* reported that influenza particles with lower infectivity had decreased nucleotide content compared to complete particles (Ada and Perry 1956), and in the late 1960s genomic RNA species shorter than the full-length viral genome were described in incomplete viral particles (Hackett *et al.* 1967; Duesberg 1968). It was then postulated that these altered genomes were the result of errors that occurred during virus replication or during the incorporation of the viral genome into the virion (Pons and Hirst 1969; Kingsbury and Portner 1970).

In 1970, a seminal article by Alice Huang and David Baltimore defined Von Magnus particles and the incomplete and interfering particles described in other viruses as defective interfering particles (DIPs). A DIP contains standard viral structural proteins, a part of the viral genome, and can be propagated in the presence of a helper standard virus. In addition, DIPs interfere with the intracellular replication of a standard virus through a mechanism independent of interferons (Huang and Baltimore 1970). Based on knowledge gathered on the structural and functional properties of DIPs produced during influenza virus and VSV infections *in vitro*, Huang and Baltimore proposed that DIPs determined the course of natural viral infections and the establishment of virus persistence.

DVGs were then described in a large number of viruses and found not only to have interfering activity but also to correlate with the establishment of persistent infections *in vitro* (Lehmann-Grube *et al.* 1969; Kawai *et al.* 1975; Roux and Waldvogel 1981) and *in vivo* (Popescu and Lehmann-Grube 1977; Barrett *et al.* 1986), reduced virulence and pathogenesis in mice (Mims 1956; Rabinowitz and Huprikar 1979; Barrett and Dimmock 1984), and increased production of type I IFNs (interferons) during infections (Marcus and Sekellick 1977; Sekellick and Marcus 1978; Johnston 1981) (Table 5.1).

Despite the strong initial surge of DIP research propelled by the work of Huang and Baltimore, the relevance of DIPs during natural infections was questioned due to a lack of evidence supporting their existence in scenarios other than high-titer virus passages in the laboratory (Mims 1956; Viola *et al.* 1978). Together with a lack of sensitive technologies to distinguish DI genomes from full-length viral genomes during infections *in vivo*, the dismissal of the field led to a decline in DIP research. By the early 2000s, the study of DVGs was mostly limited to their use as antivirals and as tools to investigate the mechanism of virus replication and recombination.

Virus	Type of DVG	Known functions	References
Single-stranded RNA viruses – positive sense			
Arteriviridae			
Porcine reproductive and respiratory syndrome virus; equine arteritis virus	Deletion	Interference	(Molenkamp *et al.* 2000; Han *et al.* 2006; Xiao *et al.* 2011)
Closteroviridae			
Citrus tristeza virus	Deletion	UN (unknown)	(Mawassi *et al.* 1995a, 1995b)
Coronaviridae			
Berne virus; bovine coronavirus; infectious bronchitis virus; mouse hepatitis virus; transmissible gastroenteritis virus	Deletion	Interference Persistence	(Makino *et al.* 1984, 1988; van der Most *et al.* 1991; Snijder *et al.* 1991; Mendez *et al.* 1995; Penzes *et al.* 1995)

Flaviviridae			
Dengue virus; Japanese encephalitis virus; hepatitis C virus; Murray Valley encephalitis virus; tick-borne encephalitis virus; West Nile virus	Deletion	Interference Persistence Modulate virulence	(Schmaljohn and Blair 1977; Brinton 1983; Prince *et al.* 1996; Lancaster *et al.* 1998; Mandl *et al.* 1998; Li *et al.* 2011)
Nodaviridae			
Flock House virus	Deletion	Modulate virulence	(Jaworski and Routh 2017)
Picornaviridae			
Poliovirus; encephalomyocarditis virus; foot-and-mouth disease virus; mengovirus; coxsackievirus B3	Deletion	Interference Modulate virulence	(Cole *et al.* 1971; McClure *et al.* 1980; Radloff and Young 1983; Chapman *et al.* 2008)
Secoviridae			
Tomato black ring virus	Deletion	Interference	(Hasiow-Jaroszewska *et al.* 2018)
Togaviridae			
Rubella virus; Semliki Forest virus; Sindbis virus	Deletion	Interference Persistence Modulate virulence IFN induction	(Bruton and Kennedy 1976; Bruton *et al.* 1976; Kennedy *et al.* 1976; Frey and Hemphill 1988)
Tombusviridae			
Cucumber necrosis virus; tomato bushy stunt virus; turnip crinkle virus; cymbidium ringspot virus	Deletion	Interference Modulate virulence	(Hillman *et al.* 1987; Li *et al.* 1989; Rochon 1991; Havelda *et al.* 1995)
Single-stranded RNA viruses – negative sense			
Arenaviridae			
Lymphocytic choriomeningitis mammarenavirus	Deletion	Interference Modulate virulence Persistence	(Lehmann-Grube *et al.* 1969)

Filoviridae			
Ebola virus	Deletion Copy-back	Persistence	(Calain *et al.* 1999)
Orthomyxoviridae			
Influenza virus	Deletion	Interference IFN induction Persistence Modulate virulence	(Magnus 1947; Isaacs and Edney 1950; Gard *et al.* 1952; Ada and Perry 1956; Rabinowitz and Huprikar 1979; Davis *et al.* 1980; De and Nayak 1980; Chanda *et al.* 1983; Bean *et al.* 1985; Dimmock *et al.* 1986; Chambers and Webster 1987; Morgan and Dimmock 1992; Baum *et al.* 2010; Scott *et al.* 2011; Saira *et al.* 2013; Vasilijevic *et al.* 2017)
Paramyxoviridae			
Human parainfluenza virus 3; parainfluenza virus 5; simian virus; measles virus; mumps virus; Sendai virus	Deletion Copy-back	Interference IFN induction Persistence Modulate virulence	(Sokol *et al.* 1964; Kingsbury and Portner 1970; Kingsbury *et al.* 1970; Portner and Kingsbury 1972; Hall *et al.* 1974; Roux and Holland 1979; Johnston 1981; Roux and Waldvogel 1981; Andzhaparidze *et al.* 1982; Andzhaparidze *et al.* 1983; Re *et al.* 1983; Murphy *et al.* 1987; Calain and Roux 1988; Bellocq *et al.* 1990; Moscona 1991; Calain *et al.* 1992; Engelhorn *et al.* 1993;

			Sidhu *et al.* 1994; Whistler *et al.* 1996; Murphy and Parks 1997; Strahle *et al.* 2006; Yount *et al.* 2006, 2008; Killip *et al.* 2013; Shivakoti *et al.* 2013; Tapia *et al.* 2013; Ho *et al.* 2015; Santak *et al.* 2015; Mura *et al.* 2017; Xu *et al.* 2017)
Peribunyaviridae			
Bunyamwera virus	Deletion	Interference	(Patel and Elliott 1992)
Phenuiviridae			
Rift Valley Fever virus (RVFV); Toscana virus	Deletion	Interference Modulate virulence	(Mims 1956; Marchi *et al.* 1998)
Pneumoviridae			
Human metapneumovirus; human respiratory syncytial virus	Deletion Copy-back	Interference IFN induction Persistence	(Treuhaft and Beem 1982; Treuhaft *et al.* 1983; Sarmiento *et al.* 2002; Valdovinos and Gomez 2003; van den Hoogen *et al.* 2014; Sun *et al.* 2015)
Rhabdoviridae			
Vesicular stomatitis virus; rabies virus	Deletion Copy-back	Interference IFN induction Persistence	(Cooper and Bellett 1959; Huang *et al.* 1966; Huang and Wagner 1966b; Stampfer *et al.* 1969; Huang and Baltimore 1970; Palma and Huang 1974; Kawai *et al.* 1975; Marcus and Sekellick 1977; Wiktor *et al.* 1977; Ramseur and Friedman 1978;

			Schubert *et al.* 1978; Sekellick and Marcus 1978; Andzhaparidze *et al.* 1981; Lazzarini *et al.* 1981; Perrault 1981; Rao and Huang 1982; O'Hara *et al.* 1984; Cave *et al.* 1985; DePolo and Holland 1986; Marcus and Gaccione 1989)
Tospoviridae			
Tomato spotted wilt virus	Deletion	Modulate virulence	(Resende Rde *et al.* 1991)
Double-stranded RNA viruses			
Birnaviridae			
Infectious necrotic pancreatic virus; infectious bursal disease virus	UN	Persistence	(Macdonald and Yamamoto 1978; Muller *et al.* 1986)
Partitiviridae			
Rosellinia necatrix virus	Deletion	Interference	(Chiba *et al.* 2016)
Reoviridae			
Type 3 reovirus; wound tumor virus	Deletion	Interference	(Ahmed and Fields 1981; Anzola *et al.* 1987)
Retroviruses			
Retroviridae			
Human immunodeficiency virus 1	Deletion Hypermutation Frame shift	Persistence	(Tominaga *et al.* 1996; Sanchez *et al.* 1997; Ho *et al.* 2013; Bruner *et al.* 2016)

DNA viruses			
Herpesviridae			
Equid alphaherpesvirus 1; herpes simplex 1 virus	Deletion Rearrangements	Interference Persistence	(Murray *et al.* 1975; Dauenhauer *et al.* 1982; Schroder *et al.* 1984; Baumann *et al.* 1987)

Table 5.1. *RNA viruses with described DVGs*

In the last decade, the study of DVGs has witnessed a renaissance that coincided with the demonstration that DVGs form during natural infections (Li *et al.* 2011; Saira *et al.* 2013; Tapia *et al.* 2013; Sun *et al.* 2015; Xu *et al.* 2015; Vasilijevic *et al.* 2017) (Figure 5.1). The impact of DVGs on natural infections and the mechanisms involved are the subject of active investigation. Whether their activities are dominated by an interfering effect is unclear as new understanding of the mechanism behind the various functions of DVGs is beginning to be unveiled, as discussed in detail later in this chapter.

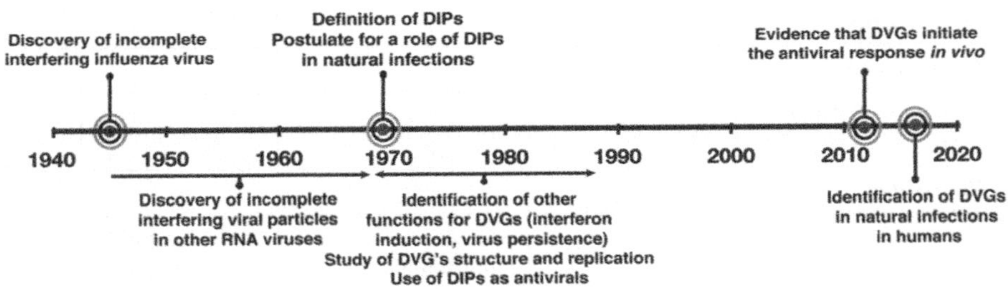

Figure 5.1. *Timeline of key discoveries during defective viral particle research*

5.3. Classes of defective viral genomes

DVGs arise from modifications in the viral genome that render it unable to complete a full replication cycle unless complemented by a standard virus. These modifications range from single nucleotide mutations, hypermutations and frame shifts to large deletions, insertions and complex genomic rearrangements (Figure 5.2). The ability of DVGs to be propagated upon complementation with a standard virus differentiates them from other forms of deleterious genomic alterations, as DVGs retain signals necessary for viral replication.

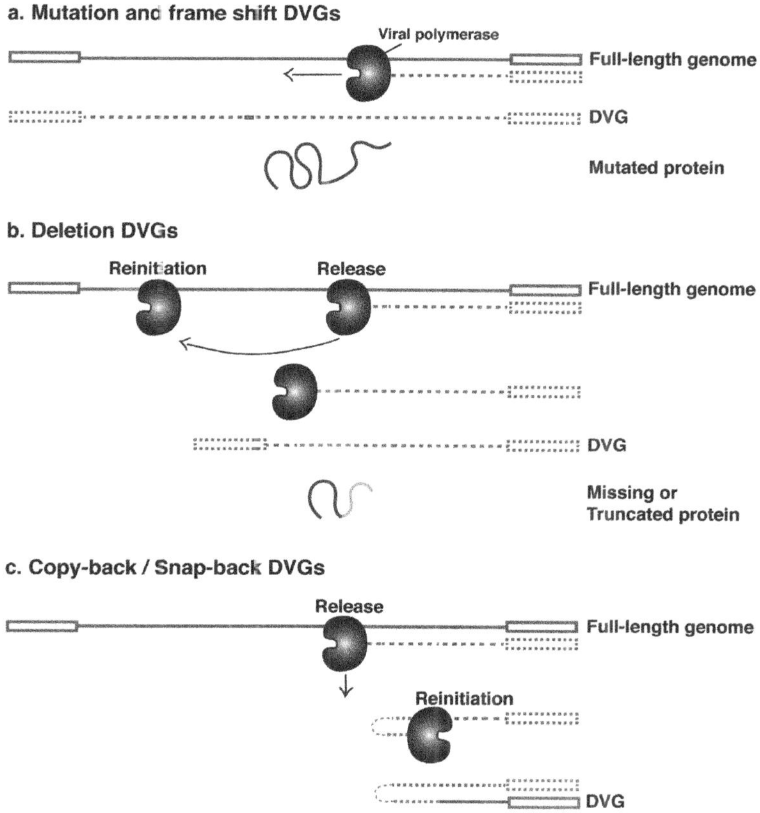

Figure 5.2. *Classes of defective viral genomes (DVGs). (a) Mutations and frame shifts lead to alterations in viral proteins that could affect or eliminate their function. (b) Deletion DVGs occur when the viral polymerase skips part of the genome during replication, leading to genomic truncations and loss or altered viral proteins. (c) Copy-back and snap-back DVGs occur when the polymerase detaches from its template and continues copying the nascent strand, producing drastically altered truncated viral genomes that cannot be transcribed but can be replicated and packaged. To form copy-back DVGs, the polymerase rejoins at a position upstream from where it was released, forming a panhandle theoretical structure. To form snap-back DVGs, the polymerase rejoins at the nucleotides immediately before where it was released, forming a theoretical hairpin structure. Modified from (Genoyer and Lopez (2019b)). The continuous line represents the viral genome. The dotted line represents the antigenome. The boxes at the end of the genomes represent replication promoter sequences*

5.3.1. *Mutations and frame shifts*

These types of DVGs have been mostly described and studied in RNA viruses, but they also exist in DNA viruses. RNA viruses are replicated by a virus-encoded RNA-dependent RNA polymerase that lacks proofreading activity. This property allows for the introduction of mutations in each cycle of replication, allowing fast adaptation of the virus to changing environmental conditions. A large majority of these mutations result in unviable viruses due to the highly constrained nature of viral genome organization (Sanjuan *et al.* 2004; Poirier *et al.* 2015). However, sometimes mutations result in viruses carrying one or more non-functional proteins that can be complemented by a co-infecting standard virus (Figure 5.2a). In some cases, these mutant viruses can compete with the standard virus for the proteins they are missing and therefore interfere with the standard virus replication process, meeting the definition of a DIP (Perales *et al.* 2007; Kupke *et al.* 2019).

5.3.2. *Deletion DVGs*

Deletion DVGs either lack a few nucleotides that result in the loss of activity of one viral gene or contain large truncations that remove one or more viral genes. These DVGs retain the 5’ and 3’ ends of the viral genome and other genomic elements necessary for replication. Deletion DVGs are formed during virus replication when the viral polymerase falls off the template strand and reinitiates further along the same template, or on a new template, to continue replication while skipping internal regions of the genome (Figure 5.2b). Deletion DVGs are predominant in positive strand viruses and orthomyxoviruses. Some deletion DVGs maintain the in-frame coding region that retains the expression of viral proteins, while other deletion DVGs contain out-of-frame genes that are not expressed (Lancaster *et al.* 1998; Sugiyama *et al.* 2009; Liao *et al.* 2014).

The recombination between homologous sequences is a major driver of the generation of deletion DVGs in *Tombusviruses*, *Nodaviruses* and *Alphaviruses* (White and Morris 1994; Routh *et al.* 2012; Poirier *et al.* 2015). In addition, the recombination between deletion DVGs and a standard full-length *Coronavirus* genome has been proposed as a mechanism for DVG diversification and evolution in this virus family (Furuya *et al.* 1993). In *Picornaviruses*, copy-choice recombination in predicted secondary structures of the viral genome has been proposed to mediate the generation of deletion DVGs (Kuge *et al.* 1986). In *Orthomyxoviruses*, tertiary structures in the ribonucleoprotein (Jennings *et al.* 1983), mutations that alter the viral polymerase elongation or multimerization capacity (Fodor *et al.* 2003; Vasilijevic *et al.* 2017), or mutations in the viral protein NS2 (Odagiri and Tobita 1990; Odagiri *et al.* 1994) drive the generation of deletion DVGs. Deletion DVGs are also formed in other RNA viruses, including *Flaviviruses* (Li and Aaskov 2014) and *Bunyaviruses* (Patel and Elliott 1992), but mechanisms leading to their generation are still unknown. Deletions combined with genomic rearrangements have also been described in DNA *Herpesviruses* (Baumann *et al.* 1987).

5.3.3. *Copy-back and snap-back DVGs*

Copy-back and snap-back DVGs are truncated and rearranged viral genomes flanked by a reversed complementary copy of the end region of the viral genome, normally the 5' end of negative-sense RNA viruses. Copy-back and snap-back DVGs form theoretical panhandle or hairpin structures, respectively (Figure 5.2c). These types of DVGs are generated predominantly during the replication of negative-sense single-stranded RNA viruses and occur when the viral polymerase detaches from the template and reinitiates in the nascent complementary strand, copying back the end of the genome (Kolakofsky 1976; Schubert *et al.* 1978). Despite a lack of sequence homology in the sites of polymerase release and reinitiation during the formation of copy-back or snap-back DVGs, analysis of DVGs from hRSV revealed sequences in the virus genome that serve as a hotspot for the generation of copy-back and snap-back DVGs. This hotspot is less than 250 nucleotides long and contains two major polymerase rejoin regions. Single nucleotide modifications of one of these regions can change the population of DVGs present during infection by shifting the rejoin region to any other regions (Sun *et al.* 2019). In VSV, conserved sequences that were six nucleotides long were found at the 5' end of different copy-back DVGs (Meier *et al.* 1984). In addition to genomic determinants, mutations in the non-structural paramyxovirus C proteins (Irie *et al.* 2014) alter the generation rate of copy-back DVGs during infection with measles, SeV or human parainfluenza virus through an unclear mechanism (Boonyaratanakornkit *et al.* 2011; Sanchez-Aparicio *et al.* 2017). It has also been suggested that a mutation in the SeV nucleoprotein that alters the density of the nucleoprotein in the viral genome promotes the formation of SeV copy-back DVGs (Yoshida *et al.* 2018).

5.3.4. *Others*

DVG variants, often referred to as mosaics, contain a mix of deletions, insertions, duplications and/or inversions of parts of the genome. They have been described among deletion and copy-back DVGs in *Arteriviruses* and *Rhabdoviruses* (O'Hara *et al.* 1984; Hillman *et al.* 1987; White and Morris 1999; Molenkamp *et al.* 2000). In addition, editing of viral RNA by cellular enzymes, such as the adenosine deaminase acting on RNA (ADAR), can increase the diversity of the DVG population in a given infection (O'Hara *et al.* 1984; Cattaneo *et al.* 1988; van den Hoogen *et al.* 2014; Pfaller *et al.* 2015).

Viruses containing a segmented genome, such as influenza virus and Rift Valley fever virus, have an additional type of defective particles, referred to as semi-infectious particles. Semi-infectious particles lack one or more genomic segments while retaining other complete unaltered segments, and are likely the result of a failure to transport and/or package all genomic segments into a virion (Brooke *et al.* 2013; Diefenbacher *et al.* 2018). The large majority of influenza virus particles belong to this category as they fail to express one or more gene products (Brooke *et al.* 2013; Jacobs *et al.* 2019). Semi-infectious particles have been proposed to promote genomic reassortment of viral segments (Fonville *et al.* 2015) and to contribute to modulating the gene expression dosage of influenza virus for optimal virus fitness (Brooke *et al.* 2014).

5.4. Impacts on the virus–host interaction

DVGs have several described functions during infection: they can interfere with the replication of the standard virus by competing for the replication machinery, can act as strong immunostimulatory molecules, can promote the establishment of virus persistence and can determine the dynamics of virus spread and transmission (Figure 5.3). As DVGs can be packaged and spread to the neighboring cells in the form of defective virus particles, they have the potential to affect virus replication and the host responses to the infection in large regions of the infected tissue. The size of the DVG and the range and quality of viral proteins expressed from DVGs may have an impact on the efficiency of DVG packaging into defective viral particles (Re and Kingsbury 1988; Ziegler *et al.* 2016). Therefore, different DVGs may spread at different rates in the infected tissue. Importantly, DVGs accumulate at different rates in different cells during an infection, and some infected cells become dominated by full-length viral genomes, while others by DVGs (Genoyer and Lopez 2019a). The heterogeneity in the composition of viral genomes in individual cells associates with distinct functional responses among infected cells. Cells enriched in DVGs strongly express type I and III interferons and better survive the infection, while cells enriched in full-length viral genomes are the primary producers of viral particles (Genoyer and Lopez 2019a).

5.4.1. *Interference with virus replication*

DIPs can interfere with the replication of the standard virus both *in vivo* and *in vitro*. Interference occurs when DI genomes compete with the standard virus for resources needed for genome replication, encapsidation and/or virus budding, leading to a reduction in standard virus replication (Figure 5.3a) (Barrett *et al.* 1981; Calain and Roux 1995). Interference can also occur indirectly through the induction of the antiviral response, which will be discussed in the next section. Copy-back and snap-back DI genomes produced predominantly in negative-sense RNA viruses are generally much shorter than the full-length viral genome and are flanked by the 5' end promoter, known as the trailer promoter. The binding of the viral polymerase to the trailer promoter occurs with greater strength than to the 3' leader promoter. Since the trailer promoter is present at the two ends of the copy back and snap-back DI genomes, while only at the 5' end of the full-length viral genome, copy-back and snap-back genomes have a faster replication rate (Rao and Huang 1982; Calain and Roux 1995). Faster replication, together with an inability to act as templates for transcription to produce viral proteins, leads to deficient quantities of viral polymerase and nucleoprotein necessary to coat the genomes. In these conditions, DI genomes outcompete full-length viral genomes for resources necessary for replication (Portner and Kingsbury 1971; Rao and Huang 1982). This situation leads to interference with the production of infectious particles. Copy-back and snap-back DI genomes can also outcompete deletion DI genomes that retain the weaker 3' end (leader) promoter (Perrault and Semler 1979). The efficiency of copy-back DI genome packaging into virus particles increases with the genome size (Re and Kingsbury 1988), suggesting that while shorter copy-back genomes are better at interfering with other forms of the viral genome replicating within an infected cell, longer copy-back genomes may have a larger impact on the replication of the standard virus as viral particles spread in a tissue. For deletion DI genomes, which retain the replication promoters of the full-length virus, a longer coding sequence and maintenance of a larger pool of functional viral proteins may provide a competitive advantage over shorter DI genomes. In this situation, longer genomes may be more interfering (Cole and Baltimore 1973; Derdeyn and Frey 1995).

5.4.2. *Stimulation of the immune response*

Copy-back DVGs can act as strong pathogen-associated molecular patterns and stimulate the signaling of the intracellular virus sensors retinoic acid inducible gene I (RIG-I)-like receptors (RLRs). The two major RLRs are the RIG-I and the melanoma differentiation-associated protein 5 (MDA5). Signaling by RLRs in response to DVGs triggers a strong antiviral immune response characterized by the expression of type I and III interferons (IFNs), cytokines and chemokines, and a large number of interferon-stimulated genes (ISGs) with antiviral activity (Figure 5.3b) (Marcus and Sekellick 1977; Fuller and Marcus 1980; Johnston 1981; Marcus and Gaccione 1989; Lopez *et al.* 2004; Strahle *et al.* 2006; Strahle *et al.*

2007; Yount *et al.* 2008; Baum *et al.* 2010; van den Hoogen *et al.* 2014; Ho *et al.* 2015; Sun *et al.* 2015; Xu *et al.* 2015). Copy-back DVGs from SeV, MeV, hRSV and others stimulate RIG-I signaling through their exposed 5' triphosphate motif (Mercado-Lopez *et al.* 2013; Ho *et al.* 2015; Xu *et al.* 2015). The potent stimulatory activity of copy-back DVGs cannot be recapitulated by infecting with larger amounts of standard virus, suggesting that specific features of DVGs drive their immunostimulatory activity (Marcus and Sekellick 1977; Yount *et al.* 2006; Patel *et al.* 2013; Tapia *et al.* 2013; Runge *et al.* 2014).

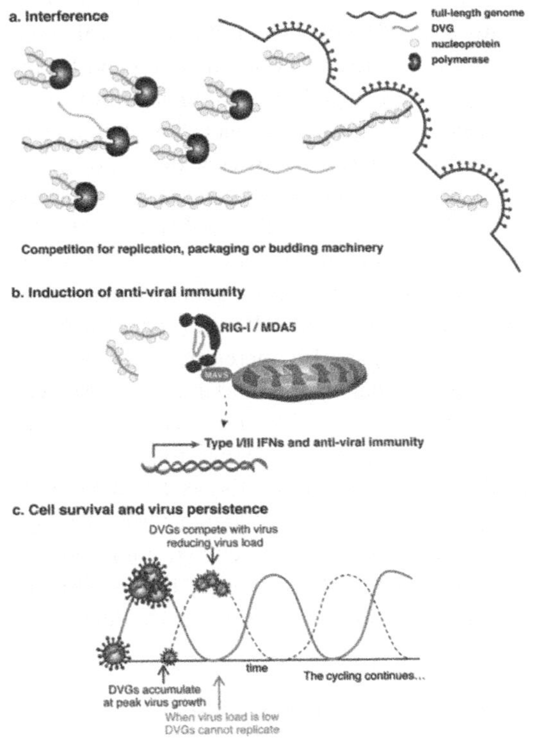

Figure 5.3. *Functions of DVGs. (a) DVGs can interfere with virus production during replication, genome packaging or budding processes. (b) Copy-back DVGs strongly stimulate the intracellular viral sensor proteins RIG-I and MDA5 that signal for the expression of type I and III IFNs and the initiation of the antiviral response. (c) DVGs that accumulate during infection interfere with the replication of the standard virus until the viral proteins necessary for DVG replication are reduced to low levels. At this stage, DVGs cannot be replicated and their ability to interfere with the replication of the standard virus diminishes, allowing replication of the standard virus and reinitiation of the cycle. Figure (c) depicts the model of cyclic DIP and standard virus production put forward by Huang and Baltimore in 1970 (Huang and Baltimore 1970)*

Indeed, an internal stem–loop motif (DVG_{70-114}) formed at the site of polymerase reinitiation enhances the binding of a highly immunostimulatory SeV copy-back DVG to RLRs, and strongly promotes RLR signaling (Xu *et al.* 2015). The immunostimulatory activity of copy-back DVGs is also observed during natural infections in humans. Detection of copy-back DVGs in respiratory samples collected from children infected with hRSV associates with an increase in the number of transcripts of genes encoding antiviral proteins in the samples (Sun *et al.* 2015).

Strong stimulation of the innate immune system by copy-back DVGs leads to the development of potent adaptive immune responses that protect from reinfection with the same pathogen. Copy-back DVGs induce the expression of TNF, IL-6, IL-1β and other pro-inflammatory cytokines and chemokines (Strahle *et al.* 2006; Yount *et al.* 2008; Shivakoti *et al.* 2013; Tapia *et al.* 2013; Sun *et al.* 2015) and maximize the antigen presentation capacity of dendritic cells and the stimulation of T cells (Yount *et al.* 2006; Mercado-Lopez *et al.* 2013). This strong immunostimulatory ability of copy-back DVGs makes them good candidates for vaccine adjuvants and antivirals (Dimmock and Easton 2014; Vasou *et al.* 2017).

Deletion DVGs can also be immunostimulatory. During the infection of mice with influenza virus, the expression of genes with antiviral activity, including interferons, is only detectable when DVGs accumulate to high levels (Tapia *et al.* 2013). In addition, influenza virus isolates with an impaired ability to generate DVGs are highly pathogenic in humans and mice, and show reduced ability to trigger antiviral immunity *in vitro* (Vasilijevic *et al.* 2017). The immunostimulatory activity of deletion DVGs also extends to infections in insects. DVGs from Sindbis virus are processed by Dicer-2 into small interfering (si)RNAs that amplify the antiviral RNAi response in *Drosophila* (Poirier *et al.* 2018).

5.4.3. *Antivirals and vaccines*

The strong interfering property of DIPs has led to an active effort to explore the use of DIPs as antivirals (Marriott and Dimmock 2010; Mercado-Lopez *et al.* 2013; Dimmock and Easton 2014; Dimmock and Easton 2015; Frensing 2015). Supporting a potential antiviral role of DIPs in infections with influenza virus, VSV or Semliki Forest virus DI genomes were shown to interfere with the replication of the co-infecting standard virus, leading to reduced and delayed virus spread and eventually to virus clearance (Holland and Doyle 1973; Rabinowitz and Huprikar 1979; Barrett *et al.* 1984). In addition, synthetically engineered DIPs known as "therapeutic interfering particles (TIPs)" are being developed as a strategy to outcompete the standard virus on infected organisms (Notton *et al.* 2014; Rast *et al.* 2016).

The immunostimulatory ability of DVGs has also been explored as a means to enhance the immune response during vaccination. In mice, vaccines against the RSV and IAV complemented with copy-back DVGs or DVG-derived oligonucleotides that retain the parental DVG immunostimulatory motifs (Xu *et al.* 2015) show improved antibody production, antigen-specific T helper 1 $CD4^+$ T cells and $CD8^+$ T cells, and increased protection from virus challenge (Martinez-Gil *et al.* 2013; Mercado-Lopez *et al.* 2013; Fisher *et al.* 2018). Interestingly, DVGs have been described in effective human live-attenuated vaccines against polio, measles and influenza viruses (McLaren and Holland 1974; Bellocq *et al.* 1990; Xue *et al.* 2016; Singanayagam *et al.* 2017). Although it is not known how DVGs have an impact on the efficacy of these vaccines, it is likely that their strong immunostimulatory ability is an important factor.

5.4.4. *Establishment of virus persistence*

Since early after their discovery, DVGs were shown to aid in the establishment of persistent infections in tissue cultures infected with a number of viruses, including the *Arenavirus* LCMV (Lehmann-Grube *et al.* 1969; Meyer and Southern 1997), the *Filovirus* Ebola (Calain *et al.* 1999), the *Flaviviruses* Japanese encephalitis virus (Schmaljohn and Blair 1977; Yoon *et al.* 2006), Murray Valley encephalitis virus (Lancaster *et al.* 1998) and West Nile virus (Brinton 1982), the *Orthomyxovirus* influenza (De and Nayak 1980; Frensing *et al.* 2013), the *Paramyxoviruses* SeV (Roux and Holland 1979; Roux and Waldvogel 1981), HPIV3 (Moscona 1991; Moscona and Peluso 1993), measles (Viola *et al.* 1978; Baczko *et al.* 1986), mumps (Andzhaparidze *et al.* 1982) and canine distemper virus (ter Meulen and Martin 1976), the *Pneumovirus* hRSV (Sarmiento *et al.* 2002; Valdovinos and Gomez 2003), the Reovirus type 3 (Ahmed and Graham 1977; Ahmed *et al.* 1980), and the *Rhabdoviruses* rabies (Kawai *et al.* 1975; Andzhaparidze *et al.* 1981) and VSV (Wagner *et al.* 1963; Sinarachatanant and Huang 1977; Ramseur and Friedman 1978). DVGs are detected in mice persistently infected with LCMV (Popescu and Lehmann-Grube 1977; Atkinson *et al.* 1986) or Semliki Forest virus (Atkinson *et al.* 1986; Barrett *et al.* 1986) and in humans infected with measles virus (Viola *et al.* 1978; Baczko *et al.* 1986; Sidhu *et al.* 1994) or hepatitis C virus (Yagi *et al.* 2005; Iwai *et al.* 2006; Noppornpanth *et al.* 2007), suggesting that DVGs may also drive persistent infections *in vivo*.

Persistent infections in the presence of DVGs are characterized by asynchronous cycling of full-length viral genomes and DVGs, both *in vitro* and *in vivo* (Figure 5.3c) (Mims 1956; Cave *et al.* 1985; Frensing *et al.* 2013). Since the 1970s, this cycling pattern has been interpreted to be a consequence of DVG interference with the replication of full-length viral genomes (Huang and Baltimore 1970). When enough DVGs accumulate, they interfere with the replication of the standard virus, which, in turn, impairs the accumulation of DVGs because of a lack of viral proteins necessary for replication. Reduced DVG amounts allow for standard virus replication and thus reinitiation of the cycling as long as the host is alive. A molecular mechanism complementing these observations was recently described to explain the role of SeV copy-back DVGs in promoting the establishment of persistent infections *in vitro* (Xu *et al.* 2017). Tumor necrosis factor (TNF) produced by infected cells in response to RLR sensing of DVGs is the predominant cytotoxic factor during SeV infection (Vasou *et al.* 2017; Xu *et al.* 2017). The cellular response to DVGs also involves the upregulation of multiple elements of the TNF prosurvival pathway, including the TNF receptor 2 and its signaling molecule TRAF1 that protect the cells containing DVGs from TNF-mediated death. This extended cell survival is associated with the establishment of a persistent infection *in vitro* (Xu *et al.* 2017). Whether this mechanism applies to other viruses, or to infections *in vivo,* is yet to be determined.

5.4.5. *Impact on virus spread*

DVGs can be packaged into defective viral particles that infect other cells and have an impact on the dynamics of virus spread during infection. In the presence of a co-infecting standard virus, DVGs may be amplified, packaged and released from the cells. As the ratio of defective viral particles to standard virus increases during infection, the impact of DVGs on the infection becomes more apparent. Copy-back DVGs, for example, will overcome the virus-encoded antagonists of the antiviral response that induces the expression of type I and III IFNs. IFNs signal the neighboring cells of the presence of the virus that render them resistant to infection, thereby limiting further spread of the virus (Akpinar *et al.* 2016; Baltes *et al.* 2017). Studies using Semliki forest virus, influenza virus, SeV and hRSV have shown reduced virus titers and morbidity when the content of defective viral particles is larger (Gamboa *et al.* 1976; Rabinowitz and Huprikar 1979; Crouch *et al.* 1982; Barrett and Dimmock 1984; Tapia *et al.* 2013; Sun *et al.* 2015). In contrast, due to the high recombination and reassortment rates of some viruses, such as positive-strand RNA viruses, deletion DVGs can provide a repertoire of mutated genes that may be advantageous to the virus by enhancing virus fitness and spread (Li *et al.* 2018).

5.5. Host factors affecting DVG accumulation and activity

The host species and the cell type infected can affect the generation, accumulation or activity of DVGs. For example, human tissue infected with the same viral stock of hRSV largely differs in the rate of accumulation of DVGs (Duesberg 1968), and the generation of DVGs during infections with influenza virus is enhanced if the host is different from the natural reservoir (Ngunjiri *et al.* 2012; Ngunjiri *et al.* 2013). Similarly, DVGs accumulate at higher rates in cells and mouse strains resistant to West Nile virus infection compared to susceptible strains (Brinton 1983). Despite this evidence, specific host factors that influence DVG accumulation and activity are yet to be identified.

5.6. Conclusion

The study of DVGs has been reinvigorated in the last decade with the aid of new molecular biology and imaging-based tools. This period has revealed that DVGs play a critical role in determining infection outcome by regulating how the virus replicates and spreads. In addition, strong immunostimulatory functions have been described for some DVGs and, in some cases, the immunostimulatory activity associates with the survival of infected cells and the establishment of persistent infections *in vitro*. New knowledge of the mechanisms that mediate the generation of DVGs has been generated. In particular, the discovery of hotspots for the generation of copy-back DVGs, the identification of copy-choice recombination mechanism for the generation of deletion DVGs, and the role of selective packaging in the case of semi-infection particles. Much more remains to be discovered regarding the molecular mechanism involved in the generation and maintenance of DVGs, the evolutionary processes that lead to DVG occurrence, and the impact of different classes of DVGs on infections. It also remains unclear whether the multiple functions ascribed to copy-back DVGs apply to deletion or other types of defective viruses. Multidisciplinary approaches will be required to further advance our understanding of defective viruses.

5.7. References

Acevedo, A., Brodsky, L., and Andino, R. (2014). Mutational and fitness landscapes of an RNA virus revealed through population sequencing. *Nature*, 505(7485), 686–690.

Ada, G.L. and Perry, B.T. (1956). Influenza virus nucleic acid: relationship between biological characteristics of the virus particle and properties of the nucleic acid. *J Gen Microbiol*, 14(3), 623–633.

Ahmed, R. and Fields, B.N. (1981). Reassortment of genome segments between reovirus defective interfering particles and infectious virus: construction of temperature-sensitive and attenuated viruses by rescue of mutations from DI particles. *Virology*, 111(2), 351–363.

Ahmed, R. and Graham, A.F. (1977). Persistent infections in L cells with temperature-sensitive mutants of reovirus. *J Virol*, 23(2), 250–262.

Ahmed, R., Chakraborty, P.R., Graham, A.F., Ramig, R.F., and Fields, B.N. (1980). Genetic variation during persistent reovirus infection: presence of extragenically suppressed temperature-sensitive lesions in wild-type virus isolated from persistently infected L cells. *J Virol*, 34(2), 383–389.

Akpinar, F., Inankur, B., and Yin, J. (2016). Spatial-temporal patterns of viral amplification and interference initiated by a single infected cell. *J Virol*, 90(16), 7552–7566.

Andzhaparidze, O.G., Bogomolova, N.N., Boriskin, Y.S., Bektemirova, M.S., and Drynov, I.D. (1981). Comparative study of rabies virus persistence in human and hamster cell lines. *J Virol*, 37(1), 1–6.

Andzhaparidze, O.G., Boriskin Yu, S., Bogomolova, N.N., and Drynov, I.D. (1982). Mumps virus-persistently infected cell cultures release defective interfering virus particles. *J Gen Virol*, 63(2), 499–503.

Andzhaparidze, O.G., Bogomolova, N.N., Boriskin Yu, S., and Drynov, I.D. (1983). Chronic non-cytopathic infection of human continuous cell lines with mumps virus. *Acta Virol*, 27(4), 318–328.

Anzola, J.V., Xu, Z.K., Asamizu, T., and Nuss, D.L. (1987). Segment-specific inverted repeats found adjacent to conserved terminal sequences in wound tumor virus genome and defective interfering RNAs. *Proc Natl Acad Sci USA*, 84(23), 8301–8305.

Atkinson, T., Barrett, A.D., Mackenzie, A., and Dimmock, N.J. (1986). Persistence of virulent Semliki Forest virus in mouse brain following co-inoculation with defective interfering particles. *J Gen Virol*, 67(Pt 6), 1189–1194.

Baczko, K., Liebert, U.G., Billeter, M., Cattaneo, R., Budka, H., and ter Meulen, V. (1986). Expression of defective measles virus genes in brain tissues of patients with subacute sclerosing panencephalitis. *J Virol*, 59(2), 472–478.

Baltes, A., Akpinar, F., Inankur, B., and Yin, J. (2017). Inhibition of infection spread by co-transmitted defective interfering particles. *PLoS One*, 12(9), e0184029.

Barrett, A.D. and Dimmock, N.J. (1984). Modulation of Semliki Forest virus-induced infection of mice by defective-interfering virus. *J Infect Dis*, 150(1), 98–104.

Barrett, A.D., Crouch, C.F., and Dimmock, N.J. (1981). Assay of defective-interfering semliki forest virus by the inhibition of synthesis of virus-specified RNAs. *J Gen Virol*, 54(Pt 2), 273–280.

Barrett, A.D., Guest, A.R., Mackenzie, A., and Dimmock, N.J. (1984). Protection of mice infected with a lethal dose of Semliki Forest virus by defective interfering virus: modulation of virus multiplication. *J Gen Virol*, 65(Pt 11), 1909–1920.

Barrett, A.D., Cross, A.J., Crow, T.J., Johnson, J.A., Guest, A.R., and Dimmock, N.J. (1986). Subclinical infections in mice resulting from the modulation of a lethal dose of Semliki Forest virus with defective interfering viruses: neurochemical abnormalities in the central nervous system. *J Gen Virol*, 67(Pt 8), 1727–1732.

Baum, A., Sachidanandam, R., and Garcia-Sastre, A. (2010). Preference of RIG-I for short viral RNA molecules in infected cells revealed by next-generation sequencing. *Proc Natl Acad Sci USA*, 107(37), 16303–16308.

Baumann, R.P., Staczek, J., and O'Callaghan, D.J. (1987). Equine herpesvirus type 1 defective-interfering (DI) particle DNA structure: the central region of the inverted repeat is deleted from DI DNA. *Virology*, 159(1), 137–146.

Beale, A.J. and Finter, N.B. (1956). The infectivity of chorio-allantoic membrane influenza virus and incomplete influenza virus by the six-hour soluble antigen production test. *J Hyg (Lond)*, 54(1), 68–78.

Bean, W.J., Kawaoka, Y., Wood, J.M., Pearson, J.E., and Webster, R.G. (1985). Characterization of virulent and avirulent A/chicken/Pennsylvania/83 influenza A viruses: potential role of defective interfering RNAs in nature. *J Virol*, 54(1), 151–160.

Bellett, A.J. and Cooper, P.D. (1959). Some properties of the transmissible interfering component of vesicular stomatitis virus preparations. *J Gen Microbiol*, 21, 498–509.

Bellocq, C., Mottet, G., and Roux, L. (1990). Wide occurrence of measles virus subgenomic RNAs in attenuated live-virus vaccines. *Biologicals*, 18(4), 337–343.

Blackstein, M.E., Stanners, C.P., and Farmilo, A.J. (1969). Heterogeneity of polyoma virus DNA: isolation and characterization of non-infectious small supercoiled molecules. *J Mol Biol*, 42(2), 301–313.

Boonyaratanakornkit, J., Bartlett, E., Schomacker, H., Surman, S., Akira, S., Bae, Y.S., Collins, P., Murphy, B., and Schmidt, A. (2011). The C proteins of human parainfluenza virus type 1 limit double-stranded RNA accumulation that would otherwise trigger activation of MDA5 and protein kinase R. *J Virol*, 85(4), 1495–1506.

Brinton, M.A. (1982). Characterization of West Nile virus persistent infections in genetically resistant and susceptible mouse cells. I. Generation of defective nonplaquing virus particles. *Virology*, 116(1), 84–98.

Brinton, M.A. (1983). Analysis of extracellular West Nile virus particles produced by cell cultures from genetically resistant and susceptible mice indicates enhanced amplification of defective interfering particles by resistant cultures. *J Virol*, 46(3), 860–870.

Brooke, C.B., Ince, W.L., Wrammert, J., Ahmed, R., Wilson, P.C., Bennink, J.R., and Yewdell, J.W. (2013). Most influenza a virions fail to express at least one essential viral protein. *J Virol*, 87(6), 3155–3162.

Brooke, C.B., Ince, W.L., Wei, J., Bennink, J.R., and Yewdell, J.W. (2014). Influenza A virus nucleoprotein selectively decreases neuraminidase gene-segment packaging while enhancing viral fitness and transmissibility. *Proc Natl Acad Sci USA*, 111(47), 16854–16859.

Bruner, K.M., Murray, A.J., Pollack, R.A., Soliman, M.G., Laskey, S.B., Capoferri, A.A., Lai, J., Strain, M.C., Lada, S.M., Hoh, R., Ho, Y.C., Richman, D.D., Deeks, S.G., Siliciano, J.D., and Siliciano, R.F. (2016). Defective proviruses rapidly accumulate during acute HIV-1 infection. *Nat Med*, 22(9), 1043–1049.

Bruton, C.J. and Kennedy, S.I. (1976). Defective-interfering particles of Semliki Forest virus: structural differences between standard virus and defective-interfering particles. *J Gen Virol*, 31(3), 383–395.

Bruton, C.J., Porter, A., and Kennedy, S.I. (1976). Defective-interfering particles of Semliki Forest virus: intracellular events during interference. *J Gen Virol*, 31(3), 397–416.

Calain, P. and Roux, L. (1988). Generation of measles virus defective interfering particles and their presence in a preparation of attenuated live-virus vaccine. *J Virol*, 62(8), 2859–2866.

Calain, P. and Roux, L. (1995). Functional characterisation of the genomic and antigenomic promoters of Sendai virus. *Virology*, 212(1), 163–173.

Calain, P., Curran, J., Kolakofsky, D., and Roux, L. (1992). Molecular cloning of natural paramyxovirus copy-back defective interfering RNAs and their expression from DNA. *Virology*, 191(1), 62–71.

Calain, P., Monroe, M.C., and Nichol, S.T. (1999). Ebola virus defective interfering particles and persistent infection. *Virology*, 262(1), 114–128.

Cattaneo, R., Schmid, A., Eschle, D., Baczko, K., ter Meulen, V., and Billeter, M.A. (1988). Biased hypermutation and other genetic changes in defective measles viruses in human brain infections. *Cell*, 55(2), 255–265.

Cave, D.R., Hendrickson, F.M., and Huang, A.S. (1985). Defective interfering virus particles modulate virulence. *J Virol*, 55(2), 366–373.

Chambers, T.M. and Webster, R.G. (1987). Defective interfering virus associated with A/Chicken/Pennsylvania/83 influenza virus. *J Virol*, 61(5), 1517–1523.

Chanda, P.K., Chambers, T.M., and Nayak, D.P. (1983). *In vitro* transcription of defective interfering particles of influenza virus produces polyadenylic acid-containing complementary RNAs. *J Virol*, 45(1), 55–61.

Chapman, N.M., Kim, K.S., Drescher, K.M., Oka, K., and Tracy, S. (2008). 5' terminal deletions in the genome of a coxsackievirus B2 strain occurred naturally in human heart. *Virology*, 375(2), 480–491.

Cheng, C.P., Pogany, J., and Nagy, P.D. (2002). Mechanism of DI RNA formation in tombusviruses: dissecting the requirement for primer extension by the tombusvirus RNA dependent RNA polymerase *in vitro*. *Virology*, 304(2), 460–473.

Chiba, S., Lin, Y.H., Kondo, H., Kanematsu, S., and Suzuki, N. (2016). A novel betapartitivirus RnPV6 from *Rosellinia necatrix* tolerates host RNA silencing but is interfered by its defective RNAs. *Virus Res*, 219, 62–72.

Cole, C.N. and Baltimore, D. (1973). Defective interfering particles of poliovirus. 3. Interference and enrichment. *J Mol Biol*, 76(3), 345–361.

Cole, C.N., Smoler, D., Wimmer, E., and Baltimore, D. (1971). Defective interfering particles of poliovirus. I. Isolation and physical properties. *J Virol*, 7(4), 478–485.

Cooper, P.D. and Bellett, A.J. (1959). A transmissible interfering component of vesicular stomatitis virus preparations. *J Gen Microbiol*, 21, 485–497.

Crouch, C.F., Mackenzie, A., and Dimmock, N.J. (1982). The effect of defective-interfering Semliki Forest virus on the histopathology of infection with virulent Semliki Forest virus in mice. *J Infect Dis*, 146(3), 411–416.

Dauenhauer, S.A., Robinson, R.A., and O'Callaghan, D.J. (1982). Chronic production of defective-interfering particles by hamster embryo cultures of herpesvirus persistently infected and oncogenically transformed cells. *J Gen Virol*, 60(Pt 1), 1–14.

Davis, A.R., Hiti, A.L., and Nayak, D.P. (1980). Influenza defective interfering viral RNA is formed by internal deletion of genomic RNA. *Proc Natl Acad Sci USA*, 77(1), 215–219.

De, B.K. and Nayak, D.P. (1980). Defective interfering influenza viruses and host cells: establishment and maintenance of persistent influenza virus infection in MDBK and HeLa cells. *J Virol*, 36(3), 847–859.

DePolo, N.J. and Holland, J.J. (1986). Very rapid generation/amplification of defective interfering particles by vesicular stomatitis virus variants isolated from persistent infection. *J Gen Virol*, 67(Pt 6), 1195–1198.

Derdeyn, C.A. and Frey, T.K. (1995). Characterization of defective-interfering RNAs of rubella virus generated during serial undiluted passage. *Virology*, 206(1), 216–226.

Diefenbacher, M., Sun, J., and Brooke, C.B. (2018). The parts are greater than the whole: the role of semi-infectious particles in influenza A virus biology. *Curr Opin Virol*, 33, 42–46.

Dimmock, N.J. and Easton, A.J. (2014). Defective interfering influenza virus RNAs: time to reevaluate their clinical potential as broad-spectrum antivirals? *J Virol*, 88(10), 5217–5227.

Dimmock, N.J. and Easton, A.J. (2015). Cloned defective interfering influenza RNA and a possible pan-specific treatment of respiratory virus diseases. *Viruses*, 7(7), 3768–3788.

Dimmock, N.J., Beck, S., and McLain, L. (1986). Protection of mice from lethal influenza: evidence that defective interfering virus modulates the immune response and not virus multiplication. *J Gen Virol*, 67(Pt 5), 839–850.

Duesberg, P.H. (1968). The RNA of influenza virus. *Proc Natl Acad Sci USA*, 59(3), 930–937.

Engelhorn, M., Stricker, R., and Roux, L. (1993). Molecular cloning and characterization of a Sendai virus internal deletion defective RNA. *J Gen Virol*, 74(Pt 1), 137–141.

Fazekas De St. Groth, S., and Graham, D.M. (1954). The production of incomplete virus particles among influenza strains: experiments in eggs. *Br J Exp Pathol*, 35(1), 60–74.

Fisher, D.G., Coppock, G.M., and Lopez, C.B. (2018). Virus-derived immunostimulatory RNA induces type I IFN-dependent antibodies and T-cell responses during vaccination. *Vaccine*, 36(28), 4039–4045.

Fodor, E., Mingay, L.J., Crow, M., Deng, T., and Brownlee, G.G. (2003). A single amino acid mutation in the PA subunit of the influenza virus RNA polymerase promotes the generation of defective interfering RNAs. *J Virol*, 77(8), 5017–5020.

Fonville, J.M., Marshall, N., Tao, H., Steel, J., and Lowen, A.C. (2015). Influenza virus reassortment is enhanced by semi-infectious particles but can be suppressed by defective interfering particles. *PLoS Pathog*, 11(10), e1005204.

Frensing, T. (2015). Defective interfering viruses and their impact on vaccines and viral vectors. *Biotechnol J*, 10(5), 681–689.

Frensing, T., Heldt, F.S., Pflugmacher, A., Behrendt, I., Jordan, I., Flockerzi, D., Genzel, Y., and Reichl, U. (2013). Continuous influenza virus production in cell culture shows a periodic accumulation of defective interfering particles. *PLoS One*, 8(9), e72288.

Frey, T.K. and Hemphill, M.L. (1988). Generation of defective-interfering particles by rubella virus in Vero cells. *Virology*, 164(1), 22–29.

Fuller, F.J. and Marcus, P.I. (1980). Interferon induction by viruses. IV. Sindbis virus: early passage defective-interfering particles induce interferon. *J Gen Virol*, 48(1), 63–73.

Furuya, T., Macnaughton, T.B., La Monica, N., and Lai, M.M. (1993). Natural evolution of coronavirus defective-interfering RNA involves RNA recombination. *Virology*, 194(1), 408–413.

Gamboa, E.T., Harter, D.H., Duffy, P.E., and Hsu, K.C. (1976). Murine influenza virus encephalomyelitis. III. Effect of defective interfering virus particles. *Acta Neuropathol (Berl)*, 34(2), 157–169.

Garcia-Arriaza, J., Domingo, E., and Escarmis, C. (2005). A segmented form of foot-and-mouth disease virus interferes with standard virus: a link between interference and competitive fitness. *Virology*, 335(2), 155–164.

Gard, S., Von Magnus, P., Svedmyr, A., and Birch-Andersen, A. (1952). Studies on the sedimentation of influenza virus. *Arch Gesamte Virusforsch*, 4(5), 591–611.

Genoyer, E. and Lopez, C.B. (2019a). Defective viral genomes alter how Sendai virus interacts with cellular trafficking machinery, leading to heterogeneity in the production of viral particles among infected cells. *J Virol*, 93(4).

Genoyer, E. and Lopez, C.B. (2019b). The impact of defective viruses on infection and immunity. *Annu Rev Virol*. 6, 547–566.

Grande-Perez, A., Lazaro, E., Lowenstein, P., Domingo, E., and Manrubia, S.C. (2005). Suppression of viral infectivity through lethal defection. *Proc Natl Acad Sci USA*, 102(12), 4448–4452.

de Groot, R.J., van der Most, R.G., and Spaan, W.J. (1992). The fitness of defective interfering murine coronavirus DI-a and its derivatives is decreased by nonsense and frameshift mutations. *J Virol*, 66(10), 5898–5905.

Hackett, A.J., Schaffer, F.L., and Madin, S.H. (1967). The separation of infectious and autointerfering particles in vesicular stomatitis virus preparations. *Virology*, 31(1), 114–119.

Hall, W.W., Martin, S.J., and Gould, E. (1974). Defective interfering particles produced during the replication of measles virus. *Med Microbiol Immunol*, 160(2–3), 155–164.

Han, J., Burkhart, K.M., Vaughn, E.M., Roof, M.B., and Faaberg, K.S. (2006). Replication and expression analysis of PRRSV defective RNA. *Adv Exp Med Biol*, 581, 445–448.

Hasiow-Jaroszewska, B., Minicka, J., Zarzynska-Nowak, A., Budzynska, D., and Elena, S.F. (2018). Defective RNA particles derived from tomato black ring virus genome interfere with the replication of parental virus. *Virus Res*, 250, 87–94.

Havelda, Z., Dalmay, T., and Burgyan, J. (1995). Localization of cis-acting sequences essential for cymbidium ringspot tombusvirus defective interfering RNA replication. *J Gen Virol*, 76(Pt 9), 2311–2316.

Henle, W. (1950). Interference phenomena between animal viruses; a review. *J Immunol*, 64(3), 203–236.

Henle, W. and Henle, G. (1943). Interference of inactive virus with the propagation of virus of influenza. *Science*, 98(2534), 87–89.

Hillman, B.I., Carrington, J.C., and Morris, T.J. (1987). A defective interfering RNA that contains a mosaic of a plant virus genome. *Cell*, 51(3), 427–433.

Ho, Y.C., Shan, L., Hosmane, N.N., Wang, J., Laskey, S.B., Rosenbloom, D.I., Lai, J., Blankson, J.N., Siliciano, J.D., and Siliciano, R.F. (2013). Replication-competent noninduced proviruses in the latent reservoir increase barrier to HIV-1 cure. *Cell*, 155(3), 540–551.

Ho, T.H., Kew, C., Lui, P.Y., Chan, C.P., Satoh, T., Akira, S., Jin, D.Y., and Kok, K.H. (2015). PACT- and RIG-I-dependent activation of type I Interferon production by a defective interfering RNA derived from measles virus vaccine. *J Virol*, 90(3), 1557–1568.

Holland, J.J. and Doyle, M. (1973). Attempts to detect homologous autointerference *in vivo* with influenza virus and vesicular stomatitis virus. *Infect Immun*, 7(4), 526–531.

van den Hoogen, B.G., van Boheemen, S., de Rijck, J., van Nieuwkoop, S., Smith, D.J., Laksono, B., Gultyaev, A., Osterhaus, A.D., and Fouchier, R.A. (2014). Excessive production and extreme editing of human metapneumovirus defective interfering RNA is associated with type I IFN induction. *J Gen Virol*, 95(Pt 8), 1625–1633.

Huang, A.S. and Baltimore, D. (1970). Defective viral particles and viral disease processes. *Nature*, 226(5243), 325–327.

Huang, A.S. and Wagner, R.R. (1966a). Comparative sedimentation coefficients of RNA extracted from plaque-forming and defective particles of vesicular stomatitis virus. *J Mol Biol*, 22(2), 381–384.

Huang, A.S. and Wagner, R.R. (1966b). Defective T particles of vesicular stomatitis virus. II. Biologic role in homologous interference. *Virology*, 30(2), 173–181.

Huang, A.S., Greenawalt, J.W., and Wagner, R.R. (1966). Defective T particles of vesicular stomatitis virus. I. Preparation, morphology, and some biologic properties. *Virology*, 30(2), 161–172.

Irie, T., Okamoto, I., Yoshida, A., Nagai, Y., and Sakaguchi, T. (2014). Sendai virus C proteins regulate viral genome and antigenome synthesis to dictate the negative genome polarity. *J Virol*, 88(1), 690–698.

Isaacs, A. and Edney, M. (1950). Interference between inactive and active influenza viruses in the chick embryo. II. The site of interference. *Aust J Exp Biol Med Sci*, 28(2), 231–238.

Iwai, A., Marusawa, H., Takada, Y., Egawa, H., Ikeda, K., Nabeshima, M., Uemoto, S., and Chiba, T. (2006). Identification of novel defective HCV clones in liver transplant recipients with recurrent HCV infection. *J Viral Hepat*, 13(8), 523–531.

Jacobs, N.T., Onuoha, N.O., Antia, A., Steel, J., Antia, R., and Lowen, A.C. (2019). Incomplete influenza A virus genomes occur frequently but are readily complemented during localized viral spread. *Nat Commun*, 10(1), 3526.

Jaworski, E. and Routh, A. (2017). Parallel ClickSeq and Nanopore sequencing elucidates the rapid evolution of defective-interfering RNAs in Flock House virus. *PLoS Pathog*, 13(5), e1006365.

Jennings, P.A., Finch, J.T., Winter, G., and Robertson, J.S. (1983). Does the higher order structure of the influenza virus ribonucleoprotein guide sequence rearrangements in influenza viral RNA? *Cell*, 34(2), 619–627.

Johnston, M.D. (1981). The characteristics required for a Sendai virus preparation to induce high levels of interferon in human lymphoblastoid cells. *J Gen Virol*, 56(Pt 1), 175–184.

Kawai, A., Matsumoto, S., and Tanabe, K. (1975). Characterization of rabies viruses recovered from persistently infected BHK cells. *Virology*, 67(2), 520–533.

Kennedy, J.C. and Macdonald, R.D. (1982). Persistent infection with infectious pancreatic necrosis virus mediated by defective-interfering (DI) virus particles in a cell line showing strong interference but little DI replication. *J Gen Virol*, 58(Pt 2), 361–371.

Kennedy, S.I., Bruton, C.J., Weiss, B., and Schlesinger, S. (1976). Defective interfering passages of Sindbis virus: nature of the defective virion RNA. *J Virol*, 19(3), 1034–1043.

Killip, M.J., Young, D.F., Gatherer, D., Ross, C.S., Short, J.A., Davison, A.J., Goodbourn, S., and Randall, R.E. (2013). Deep sequencing analysis of defective genomes of parainfluenza virus 5 and their role in interferon induction. *J Virol*, 87(9), 4798–4807.

Kim, Y.N., Lai, M.M., and Makino, S. (1993). Generation and selection of coronavirus defective interfering RNA with large open reading frame by RNA recombination and possible editing. *Virology*, 194(1), 244–253.

Kingsbury, D.W. and Portner, A. (1970). On the genesis of incomplete Sendai virions. *Virology*, 42(4), 872–879.

Kingsbury, D.W., Portner, A., and Darlington, R.W. (1970). Properties of incomplete Sendai virions and subgenomic viral RNAs. *Virology*, 42(4), 857–871.

Kolakofsky, D. (1976). Isolation and characterization of Sendai virus DI-RNAs. *Cell*, 8(4), 547–555.

Kuge, S., Saito, I., and Nomoto, A. (1986). Primary structure of poliovirus defective-interfering particle genomes and possible generation mechanisms of the particles. *J Mol Biol*, 192(3), 473–487.

Kupke, S.Y., Riedel, D., Frensing, T., Zmora, P., and Reichl, U. (2019). A novel type of influenza A virus-derived defective interfering particle with nucleotide substitutions in its genome. *J Virol*, 93(4).

Lancaster, M.U., Hodgetts, S.I., Mackenzie, J.S., and Urosevic, N. (1998). Characterization of defective viral RNA produced during persistent infection of Vero cells with Murray Valley encephalitis virus. *J Virol*, 72(3), 2474–2482.

Lazzarini, R.A., Keene, J.D., and Schubert, M. (1981). The origins of defective interfering particles of the negative-strand RNA viruses. *Cell*, 26(2 Pt 2), 145–154.

Lehmann-Grube, F., Slenczka, W., and Tees, R. (1969). A persistent and inapparent infection of L cells with the virus of lymphocytic choriomeningitis. *J Gen Virol*, 5(1), 63–81.

Li, D. and Aaskov, J. (2014). Sub-genomic RNA of defective interfering (D.I.) dengue viral particles is replicated in the same manner as full length genomes. *Virology*, 468–470, 248–255.

Li, X.H., Heaton, L.A., Morris, T.J., and Simon, A.E. (1989). Turnip crinkle virus defective interfering RNAs intensify viral symptoms and are generated de novo. *Proc Natl Acad Sci USA*, 86(23), 9173–9177.

Li, D., Lott, W.B., Lowry, K., Jones, A., Thu, H.M., and Aaskov, J. (2011). Defective interfering viral particles in acute dengue infections. *PLoS One*, 6(4), e19447.

Li, Q., Tong, Y., Xu, Y., Niu, J., and Zhong, J. (2018). Genetic analysis of serum-derived defective hepatitis C virus genomes revealed novel viral cis elements for virus replication and assembly. *J Virol*, 92(7).

Liao, W.Y., Ke, T.Y., and Wu, H.Y. (2014). The 3'-terminal 55 nucleotides of bovine coronavirus defective interfering RNA harbor cis-acting elements required for both negative- and positive-strand RNA synthesis. *PLoS One*, 9(5), e98422.

Lopez, C.B., Moltedo, B., Alexopoulou, L., Bonifaz, L., Flavell, R.A., and Moran, T.M. (2004). TLR-independent induction of dendritic cell maturation and adaptive immunity by negative-strand RNA viruses. *J Immunol*, 173(11), 6882–6889.

Macdonald, R.D. and Yamamoto, T. (1978). Quantitative analysis of defective interfering particles in infectious pancreatic necrosis virus preparations. *Arch Virol*, 57(1), 77–89.

Magnus, P. v. (1947). Studies on interference in experimental influenza. *Ark Kemi Mineral Geol*, 24(1).

Makino, S., Taguchi, F., and Fujiwara, K. (1984). Defective interfering particles of mouse hepatitis virus. *Virology*, 133(1), 9–17.

Makino, S., Shieh, C.K., Keck, J.G., and Lai, M.M. (1988). Defective-interfering particles of murine coronavirus: mechanism of synthesis of defective viral RNAs. *Virology*, 163(1), 104–111.

Mandl, C.W., Holzmann, H., Meixner, T., Rauscher, S., Stadler, P.F., Allison, S.L., and Heinz, F.X. (1998). Spontaneous and engineered deletions in the 3' noncoding region of tick-borne encephalitis virus: construction of highly attenuated mutants of a flavivirus. *J Virol*, 72(3), 2132–2140.

Marchi, A., Nicoletti, L., Accardi, L., Di Bonito, P., and Giorgi, C. (1998). Characterization of Toscana virus-defective interfering particles generated *in vivo*. *Virology*, 246(1), 125–133.

Marcus, P.I. and Gaccione, C. (1989). Interferon induction by viruses. XIX. Vesicular stomatitis virus – New Jersey: high multiplicity passages generate interferon-inducing, defective-interfering particles. *Virology*, 171(2), 630–633.

Marcus, P.I. and Sekellick, M.J. (1977). Defective interfering particles with covalently linked [+/-]RNA induce interferon. *Nature*, 266(5605), 815–819.

Marriott, A.C. and Dimmock, N.J. (2010). Defective interfering viruses and their potential as antiviral agents. *Rev Med Virol*, 20(1), 51–62.

Martinez-Gil, L., Goff, P.H., Hai, R., Garcia-Sastre, A., Shaw, M.L., and Palese, P. (2013). A Sendai virus-derived RNA agonist of RIG-I as a virus vaccine adjuvant. *J Virol*, 87(3), 1290–1300.

Mawassi, M., Karasev, A.V., Mietkiewska, E., Gafny, R., Lee, R.F., Dawson, W.O., and Bar-Joseph, M. (1995a). Defective RNA molecules associated with citrus tristeza virus. *Virology*, 208(1), 383–387.

Mawassi, M., Mietkiewska, E., Hilf, M.E., Ashoulin, L., Karasev, A.V., Gafny, R., Lee, R.F., Garnsey, S.M., Dawson, W.O., and Bar-Joseph, M. (1995b). Multiple species of defective RNAs in plants infected with citrus tristeza virus. *Virology*, 214(1), 264–268.

McClure, M.A., Holland, J.J., and Perrault, J. (1980). Generation of defective interfering particles in picornaviruses. *Virology*, 100(2), 408–418.

McLaren, L.C. and Holland, J.J. (1974). Defective interfering particles from poliovirus vaccine and vaccine reference strains. *Virology*, 60(2), 579–583.

Meier, E., Harmison, G.G., Keene, J.D., and Schubert, M. (1984). Sites of copy choice replication involved in generation of vesicular stomatitis virus defective-interfering particle RNAs. *J Virol*, 51(2), 515–521.

Mendez, A., Smerdou, C., Gebauer, F., Izeta, A., and Enjuanes, L. (1995). Structure and encapsidation of transmissible gastroenteritis coronavirus (TGEV) defective interfering genomes. *Adv Exp Med Biol*, 380, 583–589.

Mercado-Lopez, X., Cotter, C.R., Kim, W.K., Sun, Y., Munoz, L., Tapia, K., and Lopez, C.B. (2013). Highly immunostimulatory RNA derived from a Sendai virus defective viral genome. *Vaccine*, 31(48), 5713–5721.

ter Meulen, V. and Martin, S.J. (1976). Genesis and maintenance of a persistent infection by canine distemper virus. *J Gen Virol*, 32(3), 431–440.

Meyer, B.J. and Southern, P.J. (1997). A novel type of defective viral genome suggests a unique strategy to establish and maintain persistent lymphocytic choriomeningitis virus infections. *J Virol*, 71(9), 6757–6764.

Mims, C.A. (1956). Rift Valley Fever virus in mice. IV. Incomplete virus; its production and properties. *Br J Exp Pathol*, 37(2), 129–143.

Molenkamp, R., Rozier, B.C., Greve, S., Spaan, W.J., and Snijder, E.J. (2000). Isolation and characterization of an arterivirus defective interfering RNA genome. *J Virol*, 74(7), 3156–3165.

Morgan, D.J. and Dimmock, N.J. (1992). Defective interfering influenza virus inhibits immunopathological effects of infectious virus in the mouse. *J Virol*, 66(2), 1188–1192.

Moscona, A. (1991). Defective interfering particles of human parainfluenza virus type 3 are associated with persistent infection in cell culture. *Virology*, 183(2), 821–824.

Moscona, A. and Peluso, R.W. (1993). Persistent infection with human parainfluenza virus 3 in CV-1 cells: analysis of the role of defective interfering particles. *Virology*, 194(1), 399–402.

van der Most, R.G., Bredenbeek, P.J., and Spaan, W.J. (1991). A domain at the 3' end of the polymerase gene is essential for encapsidation of coronavirus defective interfering RNAs. *J Virol*, 65(6), 3219–3226.

Muller, H., Lange, H., and Becht, H. (1986). Formation, characterization and interfering capacity of a small plaque mutant and of incomplete virus particles of infectious bursal disease virus. *Virus Res*, 4(3), 297–309.

Mura, M., Combredet, C., Najburg, V., Sanchez, David, R.Y., Tangy, F., and Komarova, A.V. (2017). Nonencapsidated 5' copy-back defective interfering genomes produced by recombinant measles viruses are recognized by RIG-I and LGP2 but not MDA5. *J Virol*, 91(20).

Murphy, S.K. and Parks, G.D. (1997). Genome nucleotide lengths that are divisible by six are not essential but enhance replication of defective interfering RNAs of the paramyxovirus simian virus 5. *Virology*, 232(1), 145–157.

Murphy, D.G., Dimock, K., and Kang, C.Y. (1987). Defective interfering particles of human parainfluenza virus 3. *Virology*, 158(2), 439–443.

Murray, B.K., Biswal, N., Bookout, J.B., Lanford, R.E., Courtney, R.J., and Melnick, J.L. (1975). Cyclic appearance of defective interfering particles of herpes simplex virus and the concomitant accumulation of early polypeptide VP175. *Intervirology*, 5(3–4), 173–184.

Ngunjiri, J.M., Lee, C.W., Ali, A., and Marcus, P.I. (2012). Influenza virus interferon-inducing particle efficiency is reversed in avian and mammalian cells, and enhanced in cells co-infected with defective-interfering particles. *J Interferon Cytokine Res*, 32(6), 280–285.

Ngunjiri, J.M., Buchek, G.M., Mohni, K.N., Sekellick, M.J., and Marcus, P.I. (2013). Influenza virus subpopulations: exchange of lethal H5N1 virus NS for H1N1 virus NS triggers de novo generation of defective-interfering particles and enhances interferon-inducing particle efficiency. *J Interferon Cytokine Res*, 33(3), 99–107.

Noppornpanth, S., Smits, S.L., Lien, T.X., Poovorawan, Y., Osterhaus, A.D.M.E., and Haagmans, B.L. (2007). Characterization of hepatitis C virus deletion mutants circulating in chronically infected patients. *J Virol*, 81(22), 7.

Notton, T., Sardanyes, J., Weinberger, A.D., and Weinberger, L.S. (2014). The case for transmissible antivirals to control population-wide infectious disease. *Trends Biotechnol*, 32(8), 400–405.

Odagiri, T. and Tobita, K. (1990). Mutation in NS2, a nonstructural protein of influenza A virus, extragenically causes aberrant replication and expression of the PA gene and leads to generation of defective interfering particles. *Proc Natl Acad Sci USA*, 87(15), 5988–5992.

Odagiri, T., Tominaga, K., Tobita, K., and Ohta, S. (1994). An amino acid change in the non-structural NS2 protein of an influenza A virus mutant is responsible for the generation of defective interfering (DI) particles by amplifying DI RNAs and suppressing complementary RNA synthesis. *J Gen Virol*, 75(Pt 1), 43–53.

O'Hara, P.J., Nichol, S.T., Horodyski, F.M., and Holland, J.J. (1984). Vesicular stomatitis virus defective interfering particles can contain extensive genomic sequence rearrangements and base substitutions. *Cell*, 36(4), 915–924.

Palma, E.L. and Huang, A. (1974). Cyclic production of vesicular stomatitis virus caused by defective interfering particles. *J Infect Dis*, 129(4), 402–410.

Patel, A.H. and Elliott, R.M. (1992). Characterization of Bunyamwera virus defective interfering particles. *J Gen Virol*, 73(Pt 2), 389–396.

Patel, J.R., Jain, A., Chou, Y.Y., Baum, A., Ha, T., and Garcia-Sastre, A. (2013). ATPase-driven oligomerization of RIG-I on RNA allows optimal activation of type-I interferon. *EMBO Reports*, 14(9), 780–787.

Paucker, K., Cantell, K., and Henle, W. (1962). Quantitative studies on viral interference in suspended L cells. III. Effect of interfering viruses and interferon on the growth rate of cells. *Virology*, 17, 324–334.

Penzes, Z., Tibbles, K.W., Shaw, K., Britton, P., Brown, T.D., and Cavanagh, D. (1995). Generation of a defective RNA of avian coronavirus infectious bronchitis virus (IBV). Defective RNA of coronavirus IBV. *Adv Exp Med Biol*, 380, 563–569.

Perales, C., Mateo, R., Mateu, M.G., and Domingo, E. (2007). Insights into RNA virus mutant spectrum and lethal mutagenesis events: replicative interference and complementation by multiple point mutants. *J Mol Biol*, 369(4), 985–1000.

Perrault, J. (1981). Origin and replication of defective interfering particles. *Curr Top Microbiol Immunol*, 93, 151–207.

Perrault, J. and Semler, B.L. (1979). Internal genome deletions in two distinct classes of defective interfering particles of vesicular stomatitis virus. *Proc Natl Acad Sci USA*, 76(12), 6191–6195.

Pfaller, C.K., Mastorakos, G.M., Matchett, W.E., Ma, X., Samuel, C.E., and Cattaneo, R. (2015). Measles virus defective interfering RNAs are generated frequently and early in the absence of C protein and can be destabilized by adenosine deaminase acting on RNA-1-like hypermutations. *J Virol*, 89(15), 7735–7747.

Poirier, E.Z., Mounce, B.C., Rozen-Gagnon, K., Hooikaas, P.J., Stapleford, K.A., Moratorio, G., and Vignuzzi, M. (2015). Low-fidelity polymerases of alphaviruses recombine at higher rates to overproduce defective interfering particles. *J Virol*, 90(5), 2446–2454.

Poirier, E.Z., Goic, B., Tome-Poderti, L., Frangeul, L., Boussier, J., Gausson, V., Blanc, H., Vallet, T., Loyd, H., Levi, L.I., Lanciano, S., Baron, C., Merkling, S.H., Lambrechts, L., Mirouze, M., Carpenter, S., Vignuzzi, M., and Saleh, M.C. (2018). Dicer-2-dependent generation of viral DNA from defective genomes of RNA viruses modulates antiviral immunity in insects. *Cell Host Microbe*, 23(3), 353–365.e358.

Pons, M. and Hirst, G.K. (1969). The single- and double-stranded RNA's and the proteins of incomplete influenza virus. *Virology*, 38(1), 4.

Popescu, M. and Lehmann-Grube, F. (1977). Defective interfering particles in mice infected with lymphocytic choriomeningitis virus. *Virology*, 77(1), 78–83.

Portner, A. and Kingsbury, D.W. (1971). Homologous interference by incomplete Sendai virus particles: changes in virus-specific ribonucleic acid synthesis. *J Virol*, 8(4), 388–394.

Portner, A. and Kingsbury, D.W. (1972). Identification of transcriptive and replicative intermediates in Sendai virus-infected cells. *Virology*, 47(3), 711–725.

Prince, A.M., Huima-Byron, T., Parker, T.S., and Levine, D.M. (1996). Visualization of hepatitis C virions and putative defective interfering particles isolated from low-density lipoproteins. *J Viral Hepat*, 3(1), 11–17.

Rabinowitz, S.G. and Huprikar, J. (1979). The influence of defective-interfering particles of the PR-8 strain of influenza A virus on the pathogenesis of pulmonary infection in mice. *J Infect Dis*, 140(3), 305–315.

Radloff, R.J. and Young, S.A. (1983). Defective interfering particles of encephalomyocarditis virus. *J Gen Virol*, 64(Pt 7), 1637–1641.

Ramseur, J.M. and Friedman, R.M. (1978). Prolonged infection of L cells with vesicular stomatitis virus. Defective interfering forms and temperature-sensitive mutants as factors in the infection. *Virology*, 85(1), 253–261.

Rao, D.D. and Huang, A.S. (1982). Interference among defective interfering particles of vesicular stomatitis virus. *J Virol*, 41(1), 210–221.

Rast, L.I., Rouzine, I.M., Rozhnova, G., Bishop, L., Weinberger, A.D., and Weinberger, L.S. (2016). Conflicting selection pressures will constrain viral escape from interfering particles: principles for designing resistance-proof antivirals. *PLoS Comput Biol*, 12(5), e1004799.

Re, G.G. and Kingsbury, D.W. (1988). Paradoxical effects of Sendai virus DI RNA size on survival: inefficient envelopment of small nucleocapsids. *Virology*, 165(2), 331–337.

Re, G.G., Gupta, K.C., and Kingsbury, D.W. (1983). Genomic and copy-back 3' termini in Sendai virus defective interfering RNA species. *J Virol*, 45(2), 659–664.

Resende Rde, O., de Haan, P., de Avila, A.C., Kitajima, E.W., Kormelink, R., Goldbach, R., and Peters, D. (1991). Generation of envelope and defective interfering RNA mutants of tomato spotted wilt virus by mechanical passage. *J Gen Virol*, 72(Pt 10), 2375–2383.

Rochon, D.M. (1991). Rapid de novo generation of defective interfering RNA by cucumber necrosis virus mutants that do not express the 20-kDa nonstructural protein. *Proc Natl Acad Sci USA*, 88(24), 11153–11157.

Rott, R. and Scholtissek, C. (1963). Investigations about the formation of incomplete forms of fowl plague virus. *J Gen Microbiol*, 33, 303–312.

Routh, A., Ordoukhanian, P., and Johnson, J.E. (2012). Nucleotide-resolution profiling of RNA recombination in the encapsidated genome of a eukaryotic RNA virus by next-generation sequencing. *J Mol Biol*, 424(5), 257–269.

Roux, L. and Holland, J.J. (1979). Role of defective interfering particles of Sendai virus in persistent infections. *Virology*, 93(1), 91–103.

Roux, L. and Waldvogel, F.A. (1981). Establishment of Sendai virus persistent infection: biochemical analysis of the early phase of a standard plus defective interfering virus infection of BHK cells. *Virology*, 112(2), 400–410.

Runge, S., Sparrer, K.M., Lassig, C., Hembach, K., Baum, A., Garcia-Sastre, A., Soding, J., Conzelmann, K.K., and Hopfner, K.P. (2014). *In vivo* ligands of MDA5 and RIG-I in measles virus-infected cells. *PLoS Pathog*, 10(4), e1004081.

Saira, K., Lin, X., DePasse, J.V., Halpin, R., Twaddle, A., Stockwell, T., Angus, B., Cozzi-Lepri, A., Delfino, M., Dugan, V., Dwyer, D.E., Freiberg, M., Horban, A., Losso, M., Lynfield, R., Wentworth, D.N., Holmes, E.C., Davey, R., Wentworth, D.E., and Ghedin, E. (2013). Sequence analysis of *in vivo* defective interfering-like RNA of influenza A H1N1 pandemic virus. *J Virol*, 87(14), 8064–8074.

Salinas, Y. and Roux, L. (2005). Replication and packaging properties of short Paramyxovirus defective RNAs. *Virus Res*, 109(2), 125–132.

Sanchez, G., Xu, X., Chermann, J.C., and Hirsch, I. (1997). Accumulation of defective viral genomes in peripheral blood mononuclear cells of human immunodeficiency virus type 1-infected individuals. *J Virol*, 71(3), 2233–2240.

Sanchez-Aparicio, M.T., Garcin, D., Rice, C.M., Kolakofsky, D., Garcia-Sastre, A., and Baum, A. (2017). Loss of Sendai virus C protein leads to accumulation of RIG-I immunostimulatory defective interfering RNA. *J Gen Virol*, 98(6), 1282–1293.

Sanjuan, R., Moya, A., and Elena, S.F. (2004). The distribution of fitness effects caused by single-nucleotide substitutions in an RNA virus. *Proc Natl Acad Sci USA*, 101(22), 8396–8401.

Santak, M., Markusic, M., Balija, M.L., Kopac, S.K., Jug, R., Orvell, C., Tomac, J., and Forcic, D. (2015). Accumulation of defective interfering viral particles in only a few passages in Vero cells attenuates mumps virus neurovirulence. *Microbes Infect*, 17(3), 228–236.

Sarmiento, R.E., Tirado, R., and Gomez, B. (2002). Characteristics of a respiratory syncytial virus persistently infected macrophage-like culture. *Virus Res*, 84(1–2), 45–58.

Schmaljohn, C. and Blair, C.D. (1977). Persistent infection of cultured mammalian cells by Japanese encephalitis virus. *J Virol*, 24(2), 580–589.

Schroder, C.H., Furst, B., Weise, K., and Gray, C.P. (1984). A study of interfering herpes simplex virus DNA preparations containing defective genomes of either class I or II and the identification of minimal requirements for interference. *J Gen Virol*, 65(Pt 3), 493–506.

Schubert, M., Keene, J.D., Lazzarini, R.A., and Emerson, S.U. (1978). The complete sequence of a unique RNA species synthesized by a DI particle of VSV. *Cell*, 15(1), 103–112.

Scott, P.D., Meng, B., Marriott, A.C., Easton, A.J., and Dimmock, N.J. (2011). Defective interfering influenza virus confers only short-lived protection against influenza virus disease: evidence for a role for adaptive immunity in DI virus-mediated protection *in vivo*. *Vaccine*, 29(38), 6584–6591.

Sekellick, M.J. and Marcus, P.I. (1978). Persistent infection. I Interferon-inducing defective-interfering particles as mediators of cell sparing: possible role in persistent infection by vesicular stomatitis virus. *Virology*, 85(1), 175–186.

Sekellick, M.J. and Marcus, P.I. (1980). Viral interference by defective particles of vesicular stomatitis virus measured in individual cells. *Virology*, 104(1), 247–252.

Shivakoti, R., Siwek, M., Hauer, D., Schultz, K.L., and Griffin, D.E. (2013). Induction of dendritic cell production of type I and type III interferons by wild-type and vaccine strains of measles virus: role of defective interfering RNAs. *J Virol*, 87(14), 7816–7827.

Sidhu, M.S., Crowley, J., Lowenthal, A., Karcher, D., Menonna, J., Cook, S., Udem, S., and Dowling, P. (1994). Defective measles virus in human subacute sclerosing panencephalitis brain. *Virology*, 202(2), 10.

Sinarachatanant, P. and Huang, A.S. (1977). Effects of temperature and antibody on the cyclid growth of vesicular stomatitis virus. *J Virol*, 21(1), 161–167.

Singanayagam, A., Zambon, M., Lalvani, A., and Barclay, W. (2017). Can defective interfering RNAs affect the live attenuated influenza vaccine? – Authors' reply. *Lancet Infect Dis*, 17(12), 1235–1236.

Snijder, E.J., den Boon, J.A., Horzinek, M.C., and Spaan, W.J. (1991). Characterization of defective interfering RNAs of Berne virus. *J Gen Virol*, 72(Pt 7), 1635–1643.

Sokol, F., Neurath, A.R., and Vilcek, J. (1964). Formation of incomplete Sendai virus in embryonated eggs. *Acta Virol*, 8, 59–67.

Stampfer, M., Baltimore, D., and Huang, A.S. (1969). Ribonucleic acid synthesis of vesicular stomatitis virus. I. Species of ribonucleic acid found in Chinese hamster ovary cells infected with plaque-forming and defective particles. *J Virol*, 4(2), 154–161.

Strahle, L., Garcin, D., and Kolakofsky, D. (2006). Sendai virus defective-interfering genomes and the activation of interferon-beta. *Virology*, 351(1), 101–111.

Strahle, L., Marq, J.B., Brini, A., Hausmann, S., Kolakofsky, D., and Garcin, D. (2007). Activation of the beta interferon promoter by unnatural Sendai virus infection requires RIG-I and is inhibited by viral C proteins. *J Virol*, 81(22), 12227–12237.

Sugiyama, K., Suzuki, K., Nakazawa, T., Funami, K., Hishiki, T., Ogawa, K., Saito, S., Shimotohno, K.W., Suzuki, T., Shimizu, Y., Tobita, R., Hijikata, M., Takaku, H., and Shimotohno, K. (2009). Genetic analysis of hepatitis C virus with defective genome and its infectivity *in vitro*. *J Virol*, 83(13), 6922–6928.

Sun, Y., Jain, D., Koziol-White, C.J., Genoyer, E., Gilbert, M., Tapia, K., Panettieri, R.A., Jr., Hodinka, R.L., and Lopez, C.B. (2015). Immunostimulatory defective viral genomes from respiratory syncytial virus promote a strong innate antiviral response during infection in mice and humans. *PLoS Pathog*, 11(9), e1005122.

Sun, Y., Kim, E.J., Felt, S.A., Taylor, L.J., Agarwal, D., Grant, G.R., and Lopez, C.B. (2019). A specific sequence in the genome of respiratory syncytial virus regulates the generation of copy-back defective viral genomes. *PLoS Pathog*, 15(4), e1007707.

Tapia, K., Kim, W.K., Sun, Y., Mercado-Lopez, X., Dunay, E., Wise, M., Adu, M., and Lopez, C.B. (2013). Defective viral genomes arising *in vivo* provide critical danger signals for the triggering of lung antiviral immunity. *PLoS Pathog*, 9(10), e1003703.

Thyagarajan, B. and Bloom, J.D. (2014). The inherent mutational tolerance and antigenic evolvability of influenza hemagglutinin. *Elife*, 3.

Tominaga, K., Kato, S., Negishi, M., and Takano, T. (1996). A high frequency of defective vif genes in peripheral blood mononuclear cells from HIV type 1-infected individuals. *AIDS Res Hum Retroviruses*, 12(16), 1543–1549.

Treuhaft, M.W. (1983). A colorimetric assay for quantification of defective interfering particles of respiratory syncytial virus. *J Gen Virol*, 64(Pt 6), 1301–1309.

Treuhaft, M.W. and Beem, M.O. (1982). Defective interfering particles of respiratory syncytial virus. *Infect Immun*, 37(2), 439–444.

Uchida, S., Watanabe, S., and Kato, M. (1966). Incomplete growth of simian virus 40 in African green monkey kidney culture induced by serial undiluted passages. *Virology*, 28(1), 135–141.

Valdovinos, M.R. and Gomez, B. (2003). Establishment of respiratory syncytial virus persistence in cell lines: association with defective interfering particles. *Intervirology*, 46(3), 190–198.

Vasilijevic, J., Zamarreno, N., Oliveros, J.C., Rodriguez-Frandsen, A., Gomez, G., Rodriguez, G., Perez-Ruiz, M., Rey, S., Barba, I., Pozo, F., Casas, I., Nieto, A., and Falcon, A. (2017). Reduced accumulation of defective viral genomes contributes to severe outcome in influenza virus infected patients. *PLoS Pathog*, 13(10), e1006650.

Vasou, A., Sultanoglu, N., Goodbourn, S., Randall, R.E., and Kostrikis, L.G. (2017). Targeting pattern recognition receptors (PRR) for vaccine adjuvantation: from synthetic PRR agonists to the potential of defective interfering particles of viruses. *Viruses*, 9(7).

Vignuzzi, M. and Lopez, C.B. (2019). Defective viral genomes are key drivers of the virus–host interaction. *Nat Microbiol*, 4(7), 1075–1087.

Viola, M.V., Scott, C., and Duffy, P.D. (1978). Persistent measles virus infection *in vitro* and in man. *Arthritis Rheum*, 21(5 Suppl), S47–51.

Wagner, R.R., Levey, A.H., Snyder, R.M., Ratcliff, G.A., Jr., and Hyatt, D.F. (1963). Biologic properties of two plaque variants of vesicular stomatitis virus (Indiana Serotype) *J Immunol*, 91, 112–122.

Whistler, T., Bellini, W.J., and Rota, P.A. (1996). Generation of defective interfering particles by two vaccine strains of measles virus. *Virology*, 220(2), 480–484.

White, K.A. and Morris, T.J. (1994). Nonhomologous RNA recombination in tombusviruses: generation and evolution of defective interfering RNAs by stepwise deletions. *J Virol*, 68(1), 14–24.

White, K.A. and Morris, T.J. (1999). Defective and defective interfering RNAs of monopartite plus-strand RNA plant viruses. *Curr Top Microbiol Immunol*, 239, 1–17.

White, C.L., Thomson, M., and Dimmock, N.J. (1998). Deletion analysis of a defective interfering Semliki Forest virus RNA genome defines a region in the nsP2 sequence that is required for efficient packaging of the genome into virus particles. *J Virol*, 72(5), 4320–4326.

Wichgers Schreur, P.J. and Kortekaas, J. (2016). Single-molecule FISH reveals non-selective packaging of Rift Valley fever virus genome segments. *PLoS Pathog*, 12(8), e1005800.

Wiktor, T.J., Dietzschold, B., Leamnson, R.N., and Koprowski, H. (1977). Induction and biological properties of defective interfering particles of rabies virus. *J Virol*, 21(2), 626–635.

Xiao, C.T., Liu, Z.H., Yu, X.L., Ge, M., Li, R.C., Xiao, B.R., and Zhou, H.R. (2011). Identification of new defective interfering RNA species associated with porcine reproductive and respiratory syndrome virus infection. *Virus Res*, 158(1–2), 33–36.

Xu, J., Mercado-Lopez, X., Grier, J.T., Kim, W.K., Chun, L.F., Irvine, E.B., Del Toro Duany, Y., Kell, A., Hur, S., Gale, M., Jr., Raj, A., and Lopez, C.B. (2015). Identification of a natural viral RNA motif that optimizes sensing of viral RNA by RIG-I. *MBio*, 6(5).

Xu, J., Sun, Y., Li, Y., Ruthel, G., Weiss, S.R., Raj, A., Beiting, D., and Lopez, C.B. (2017). Replication defective viral genomes exploit a cellular pro-survival mechanism to establish paramyxovirus persistence. *Nat Commun*, 8(1), 799.

Xue, J., Chambers, B.S., Hensley, S.E., and Lopez, C.B. (2016). Propagation and characterization of influenza virus stocks that lack high levels of defective viral genomes and hemagglutinin mutations. *Front Microbiol*, 7, 326.

Yagi, S., Mori, K., Tanaka, E., Matsumoto, A., Sunaga, F., Kiyosawa, K., and Yamaguchi, K. (2005). Identification of novel HCV subgenome replicating persistently in chronic active hepatitis C patients. *J Med Virol*, 77(3), 399–413.

Yoon, S.W., Lee, S.Y., Won, S.Y., Park, S.H., Park, S.Y., and Jeong, Y.S. (2006). Characterization of homologous defective interfering RNA during persistent infection of Vero cells with Japanese encephalitis virus. *Mol Cells*, 21(1), 112–120.

Yoshida, A., Kawabata, R., Honda, T., Sakai, K., Ami, Y., Sakaguchi, T., and Irie, T. (2018). A single amino acid substitution within the paramyxovirus Sendai virus nucleoprotein is a critical determinant for production of interferon-beta-inducing copyback-type defective interfering genomes. *J Virol*, 92(5).

Yount, J.S., Kraus, T.A., Horvath, C.M., Moran, T.M., and Lopez, C.B. (2006). A novel role for viral-defective interfering particles in enhancing dendritic cell maturation. *J Immunol*, 177(7), 4503–4513.

Yount, J.S., Gitlin, L., Moran, T.M., and Lopez, C.B. (2008). MDA5 participates in the detection of paramyxovirus infection and is essential for the early activation of dendritic cells in response to Sendai virus defective interfering particles. *J Immunol*, 180(7), 4910–4918.

Ziegler, C.M., Eisenhauer, P., Bruce, E.A., Weir, M.E., King, B.R., Klaus, J.P., Krementsov, D.N., Shirley, D.J., Ballif, B.A., and Botten, J. (2016). The lymphocytic choriomeningitis virus matrix protein PPXY late domain drives the production of defective interfering particles. *PLoS Pathog*, 12(3), e1005501.

6

Enteric Viruses and the Intestinal Microbiota

Matthew PHILLIPS[1], Bria F. DUNLAP[2], Megan T. BALDRIDGE[2], Stephanie M. KARST[1]

[1] *University of Florida College of Medicine, Gainesville, USA*

[2] *Washington University School of Medicine, St. Louis, USA*

6.1. Introduction

The human intestinal tract is home to a highly diverse group of commensal microorganisms that play significant roles in gut homeostasis and health. This gut microbiota is critical for proper development of the host immune system, normal gut metabolism and nutrient uptake, protection from invasive bacterial species and maintenance of normal gut physiology. During development, a lack of gut commensals results in immature gut associated lymphoid tissues, diminished secretory IgA levels, and fewer tissue resident lymphocytes (Abt and Artis 2009; Kabat *et al.* 2014). Immune homeostasis is also dependent on the sampling of the gut lumen by intestinal epithelial cells and dendritic cells, with dysbiosis, or an imbalance in the gut microbes, contributing to the loss of tolerance and pathologic levels of inflammation (Wang *et al.* 2019). The normal gut microbial population also prevents colonization by pathogenic bacteria through the production of bacteriocins or competition for nutrients (Pickard *et al.* 2017). Finally, the gut microbiota aid in

---

For a color version of all figures in this book, see www.iste.co.uk/saleh/virology.zip.

Virology,
coordinated by María-Carla SALEH and Félix AUGUSTO REY. © ISTE Ltd 2021

digestion and metabolism by providing enzymes that are not present in the human genome which allow the breakdown of complex polysaccharides and polyphenols, as well as vitamin synthesis (Rowland *et al.* 2018). Enteric viruses encounter this environment during fecal–oral transmission; however, its impact on viral pathogenesis was largely a mystery until recent years. Intriguingly, a multitude of diverse enteric viruses can bind to the surface of bacteria commonly found in the human gut (Miura *et al.* 2013; Almand *et al.* 2017; Berger *et al.* 2017; Erickson *et al.* 2018). The focus of this chapter is the consequence of interactions, both direct and indirect, between the intestinal microbiota and viruses. In early studies of the effects of the intestinal microbiota on enteric virus infections, these interactions were shown to generally benefit enteric viruses. Yet, evidence of a more complex and nuanced relationship between enteric viruses and the gut microbiota is rapidly emerging.

6.2. Enteric picornaviruses

The picornavirus family is a group of non-enveloped, positive-sense RNA viruses and is one of the largest virus families, containing many human pathogens that can cause a wide range of diseases. These diseases include gastroenteritis (e.g. enterovirus 71), encephalitis and myocarditis (e.g. encephalomyocarditis virus), acute flaccid myelitis (e.g. enterovirus D68 and poliovirus), hand-foot-and-mouth disease (e.g. coxsackievirus) and hepatitis (e.g. hepatitis A virus). While a few of these viruses transmit via the respiratory route, most human viruses in the picornavirus family transmit via fecal–oral transmission. Although several of these viruses cause disease in various tissues, they must be able to productively infect gastrointestinal tissue in order to replicate and spread. One of the most well-studied picornaviruses, poliovirus, causes neurovirulence but transmits fecal-orally. Poliovirus continues to be an effective tool in investigating picornavirus–host–microbiota interactions. Because the primary site of poliovirus infection is the gastrointestinal tract, researchers questioned whether intestinal microbiota influenced the ability of the virus to infect this host tissue (Kuss *et al.* 2011). Indeed, poliovirus-susceptible mice pretreated with a cocktail of oral antibiotics (Abx) display reduced disease and reduced viral replication following oral poliovirus inoculation. Supporting a stimulatory role for commensal bacteria, poliovirus disease severity is unaffected by Abx treatment when the virus is inoculated intraperitoneally (i.e. bypassing the intestinal lumen); the reintroduction of commensal bacteria to mice enhances poliovirus disease severity; and the presence of Abx-resistant bacteria under Abx treatment conditions is sufficient to promote poliovirus infection. Similar results were obtained for another enteric picornavirus, coxsackievirus B3 (CVB3), although CVB3 is more sensitive to Abx treatment than poliovirus (Robinson *et al.* 2019). Bacterial enhancement of poliovirus infectivity is multifactorial, as described in the next sections.

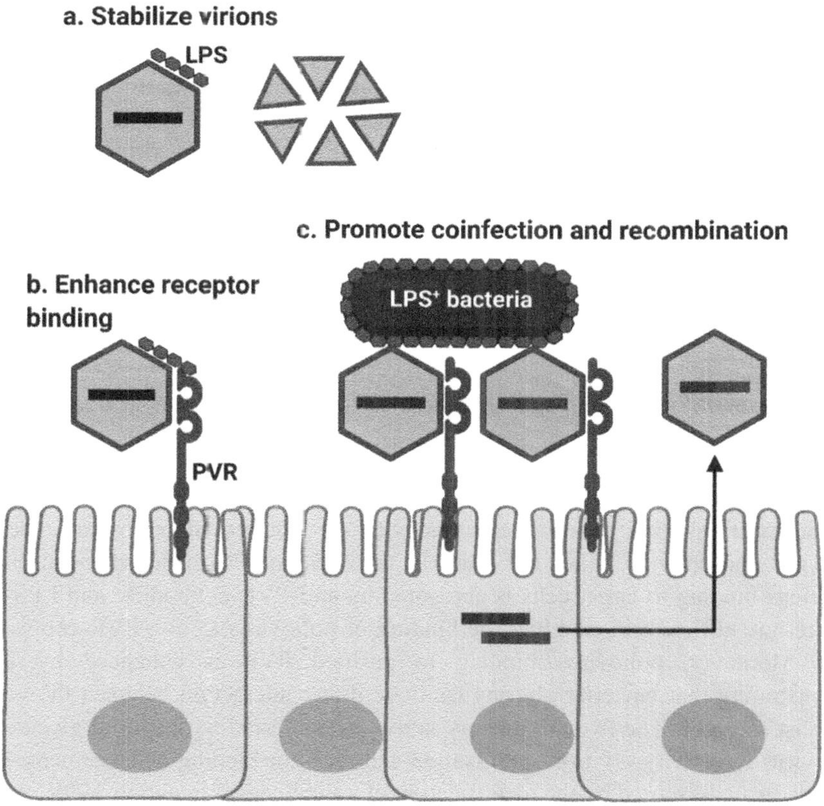

Figure 6.1. *Poliovirus interactions with gut microbiota. Poliovirus infection in the gut is enhanced by the microbiota through at least three distinct mechanisms. (a) The poliovirus virion binds to bacterial components, such as lipopolysaccharide (LPS) that stabilizes the capsid. (b) Viral binding to microbial glycans such as LPS or PG enhance poliovirus interaction with its cell surface receptor. (c) Microbial enhancement of poliovirus increases coinfection probability, thereby leading to increased exchange of genetic material between polioviruses via recombination*

6.2.1. *Intestinal microbiota enhance poliovirus stability*

Both Gram-negative and Gram-positive bacteria can enhance poliovirus thermostability in a manner that involves direct interaction between the virus and bacterial surface, but does not require live bacteria (Kuss *et al.* 2011). Bacterial surface polysaccharides such as lipopolysaccharide (LPS) and peptidoglycan (PG) are sufficient in enhancing poliovirus thermostability (Figure 6.1a), demonstrating a key role for bacterial surface glycans (Kuss *et al.* 2011). Robinson *et al.* determined that N-acetylglucosamine (GlcNAc)-containing polysaccharides longer than six residues are sufficient in providing enhanced thermostability, and even bleach resistance, to poliovirus particles (Robinson *et al.* 2014). Picornaviruses from three different genera display increased bleach resistance when bound to bacteria, although only a subset of these viruses display increased thermostability (Aguilera *et al.* 2019). For poliovirus, increased particle stability in the presence of bacterial glycans limits the premature release of viral genomes and promotes inter-host transmission by increasing environmental stability (Robinson *et al.* 2014).

6.2.2. *Bacterial glycans facilitate virion attachment to target cells*

In addition to stabilizing virions, bacterial LPS and PG enhance poliovirus binding to permissive cells (Figure 6.1b). Multiple lines of evidence support that this enhancement involves virion interaction with the established poliovirus receptor (PVR) (Robinson *et al.* 2014). For example, the LPS- and PG-mediated enhancement of virions binding to target cells is abrogated by anti-PVR antibodies; and LPS and PG are not sufficient in enabling the binding of poliovirus to non-PVR-expressing cells. Moreover, poliovirus binding to purified PVR is enhanced by LPS, demonstrating that bacterial glycans facilitate direct interaction between the virion and host receptor. The two stimulatory activities conferred by bacterial glycans on poliovirus infectivity – virion stabilization and receptor binding enhancement – are distinct, as underscored by the identification of a viral capsid mutation T99K, which reduces LPS binding in a manner affecting the bacterial enhancement of virion stability, but not LPS-mediated enhancement of receptor binding (Robinson *et al.* 2014).

6.2.3. *Intestinal microbiota promote poliovirus recombination*

As described in the Introduction, a variety of enteric viruses, including poliovirus, bind directly to the surface of bacteria (Miura *et al.* 2013; Almand *et al.* 2017; Berger *et al.* 2017; Erickson *et al.* 2018). As multiple virions can be visualized bound to a single bacterium, it seemed plausible that bacterial binding could serve to concentrate virions during a natural infection and promote coinfections of target cells. In an elegant series of experiments, Erickson *et al.* demonstrated that interactions with bacteria do indeed promote poliovirus coinfection and viral recombination (Figure 6.1c) (Erickson *et al.* 2018). Genetically engineered poliovirus strains expressing different fluorescent markers display an increased frequency of viral coinfections when incubated with bacteria. In fact, at least six genetically distinct polioviruses could infect one host cell under very low MOI conditions, in the presence of bacteria. Most importantly, these coinfections result in enhanced viral recombination and consequently improved viral adaptability. As proof-of-principle, it was shown that bacteria facilitate the viral coinfection of two mutant polioviruses (e.g. a heat-sensitive strain and a drug-sensitive strain), in order to produce a recombinant virus with increased fitness (e.g. a heat-resistant strain and a drug-resistant strain). Although a similar phenomenon has not been observed for other types of viruses, it is entirely plausible that commensal bacteria serve to enhance the viral coinfections and recombinations for other viruses that directly bind bacteria, including reovirus and norovirus (Miura *et al.* 2013; Almand *et al.* 2017; Berger *et al.* 2017).

Studies of poliovirus interactions with the gut microbiota outline several themes. First, the gut microbiota is an essential component in understanding viral phenotypes and pathogenesis in the gut. Using poliovirus as a model, it was discovered that the microbiota has the ability to support effective viral infection, evidenced by decreased poliovirus titers in mice lacking microbiota. Second, it is possible for non-bacteriophage viruses to interact directly with bacteria and bacterial components. Finally, the outcomes of these interactions are likely to be varied and diverse, as evidenced by the ability of the gut microbiota to increase poliovirus capsid stability, enhance poliovirus receptor binding and modulate poliovirus recombination. The following sections will explore even more mechanisms of virus–microbiota interactions that can enhance, and sometimes inhibit, the infection and pathology of several different groups of enteric viruses.

6.3. Mouse mammary tumor virus

Mouse mammary tumor virus (MMTV) is a retrovirus that has been well studied as a causative agent of mammary tumors (Dudley *et al.* 2016). It can be passed from a persistently infected mother to her offspring through the ingestion of maternal milk, where it replicates in the immune cells of the Peyer's patches after being transcytosed across M cells. During maturation of the mouse, the virus is trafficked in lymphocytes and infects the mammary gland epithelial cells, the main reservoir of the virus in the host. This completes a transmission cycle wherein enteric routes are required to establish infection in a new host and the transmission of the virus from infected mice requires the infection of the mammary gland.

6.3.1. *MMTV binds LPS, which in turn promotes a tolerogenic immune environment conducive to viral persistence*

Early pathogenesis studies of MMTV revealed that mice deficient in the LPS sensor toll-like receptor 4 (TLR4), display decreased milk-borne MMTV transmission, and that MMTV triggers the production of the immunosuppressive cytokine IL-10 via TLR4 (Outzen *et al.* 1985; Rassa *et al.* 2002; Jude *et al.* 2003). Subsequent studies demonstrated that, like *Tlr4*$^{-/-}$ mice, germ-free and Abx-treated mice are unable to transmit MMTV through successive generations, but transmission can be rescued in germ-free mice by reconstitution with a defined bacterial community (Kane *et al.* 2011). These findings raised the possibility that commensal bacterial LPS is responsible for activating TLR4 during MMTV oral transmission, resulting in IL-10 production. Indeed, MMTV binds LPS in the gut lumen of pups and MMTV-LPS complexes potently induce IL-10 production in a TLR4-dependent manner, when applied to splenocytes *in vitro* (Kane *et al.* 2011). In a series of elegant studies, it was ultimately shown that LPS bound to MMTV engages TLR4 on dendritic cells and macrophages, resulting in the induction of IL-6, which in turn drives the B cell expression of IL-10 (Figure 6.2) (Jude *et al.* 2003; Kane *et al.* 2011; Wilks *et al.* 2015). The expression of IL-10 during the primary MMTV infection of Peyer's patches creates a tolerogenic environment in which infected pups not only become persistently MMTV-infected, but they also become immune-tolerant, as indicated by their failure to produce virus-specific antibodies when immunized with MMTV antigens later in life (Kane *et al.* 2011).

6.3.2. *MMTV incorporates host LPS-binding proteins into its envelope*

While researchers showed that MMTV binds LPS and triggers an immunomodulatory host response, the nature of the interaction between MMTV and LPS remained undefined until 2015. Remarkably, it was then discovered that MMTV acquires mammalian LPS-binding proteins (e.g. CD14, MD2 and TLR4) in its envelope during budding from an infected cell (Wilks *et al.* 2015). Several lines of evidence supported this conclusion. First, MMTV produced from *Tlr4*$^{-/-}$ mice or *Md2*$^{-/-}$ mice was associated with reduced amounts of LPS. Second, the virus from *Md2*$^{-/-}$ mice was compromised in its ability to establish infection. Third, MMTV produced from mice deficient in all LPS-binding proteins was unable to activate TLR4. Thus, MMTV is well adapted to take advantage of the commensal bacterial products present in the gut.

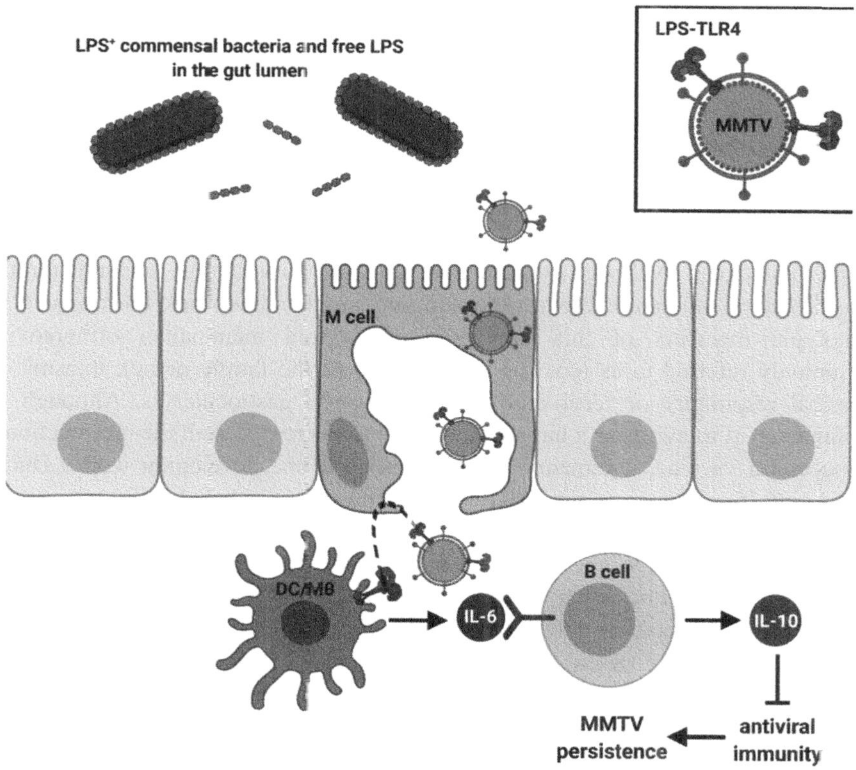

Figure 6.2. *MMTV interactions with gut microbiota. Mouse mammary tumor virus (MMTV) is transmitted to the offspring of infected female mice in the mother's milk. After entering the gut of suckling mouse pups, MMTV can pick up LPS in the lumen of the gut, potentially though hijacked cell-surface host proteins that can bind LPS such as TLR4, MD2 and CD14 that are on the viral envelope. MMTV is transcytosed across microfold cells (M cells) into the Peyer's patch. LPS bound to MMTV signals through TLR4 found on macrophages or dendritic cells to drive production of IL-6. IL-6 then induces B cells to produce IL-10 – a classical inhibitory cytokine that dampens antiviral immune responses and allows MMTV to replicate effectively in the Peyer's patch*

6.4. Reoviruses

The name reovirus was derived as an acronym from the **R**espiratory **E**nteric **O**rphan viruses, since the first viruses discovered in this family could be isolated from the respiratory or enteric tracts, but were not associated with a human disease. The reovirus family is a group of segmented, double-stranded RNA viruses that can infect a broad range of hosts in nature, including protists, fungi, fish and mammals.

Despite the origin of the name, the *Reoviridae* family includes a number of human pathogens. Although some members of this virus family transmit through insect vectors and cause febrile illness in humans (Colorado tick fever virus), the prototypic members of this family, rotavirus and mammalian orthoreovirus (commonly referred to as reovirus as the origin of the family name), transmit via classical respiratory or fecal–oral routes and cause gastroenteritis. Although the prominence of rotavirus as a human pathogen has decreased with the introduction of a successful vaccine, it remains a health burden in the developing world. During enteric infection, rotavirus and reovirus infect intestinal epithelial cells and cross the epithelial barrier to infect cells in the Peyer's patches. Similar to findings for poliovirus and MMTV, the intestinal microbiota enhance reovirus and rotavirus infections in mice. For example, Abx treatment results in reduced susceptibility of *Ifnar1*$^{-/-}$ mice to reovirus infection, as indicated by reduced intestinal titers and decreased pathology (Kuss *et al.* 2011). Likewise, Abx-treated and germ-free wild-type mice display delayed and reduced rotavirus shedding compared to conventionally colonized mice (Uchiyama *et al.* 2014). Finally, neonatal wild-type mice develop diarrhea when infected with rotavirus; however, disease severity and duration are both reduced in Abx-treated pups (Uchiyama *et al.* 2014).

6.4.1. *Intestinal microbiota enhance reovirus stability*

In terms of underlying stimulatory mechanisms, reovirus binds Gram-positive and Gram-negative bacteria and displays increased thermostability when incubated with either type of bacteria (Berger *et al.* 2017). Incubation of a virus with LPS or PG is sufficient to enhance thermostability of both the mature reovirus particles and infectious subvirion particles (ISVPs), which are generated through proteolysis in the gut lumen of infected hosts. However, bacterial interactions do not appear to affect reovirus binding to its host receptor JAM-A, nor do they prevent antibody-mediated neutralization, suggesting that bacterial glycans bind to a site on the reovirus capsid, distinct from the receptor binding site and neutralizing epitopes (Berger *et al* 2017).

In apparent contradiction to these studies demonstrating the bacterial enhancement of reovirus and rotavirus infection, multiple lines of investigation reveal that specific bacteria or bacterial constituents can suppress rotavirus infection through a variety of distinct mechanisms.

For example, it was reported in 2008 that the administration of bacterial flagellin reduces the susceptibility of mice to rotavirus infection (Vijay-Kumar *et al.* 2008; Zhang *et al.* 2014). Considering the potent immunostimulatory activity of flagellin (Vijay-Kumar and Gewirtz 2009), researchers speculated that flagellin confers protection to rotavirus, by activating antiviral immune responses. While flagellin-mediated protection occurs independent of lymphocytes or IFN, two host sensors of flagellin – surface-expressed TLR5 and intracellular NOD-like receptor C4 (NLRC4) – contribute to the protective effect of flagellin against rotavirus via distinct mechanisms (Zhang *et al.* 2014). Using bone marrow chimeric mice and dendritic cell depletions, it was determined that the activation of TLR5 on dendritic cells is necessary and sufficient for its protection (Figure 6.3a): TLR5 engagement of flagellin on dendritic cells results in IL-12/IL-23-dependent activation of innate lymphoid cells (ILCs), which in turn produces IL-17 and IL-22. Accordingly, researchers then used a combination of immune-knockout mice and cytokine neutralizations to demonstrate that ILCs and IL-22, but not IL-17, are required for flagellin's TLR5-dependent antiviral activity (Zhang *et al.* 2014).

More recent work reveals that IL-22 drives IEC proliferation and migration towards the villous tips, resulting in the luminal shedding of the highly differentiated villous tip epithelial cells that serve as the primary rotavirus target (Zhang *et al.* 2019). In terms of NLRC4-dependent protection against rotavirus, when NLRC4 recognizes flagellin, it triggers the assembly of inflammasomes and pyroptosis involving the activation of IL-18, resulting in the death of RV-infected IECs (Figure 6.3b) (Zhang *et al.* 2019).

6.4.2. *Immunostimulatory properties of bacterial flagellin inhibit rotavirus infection*

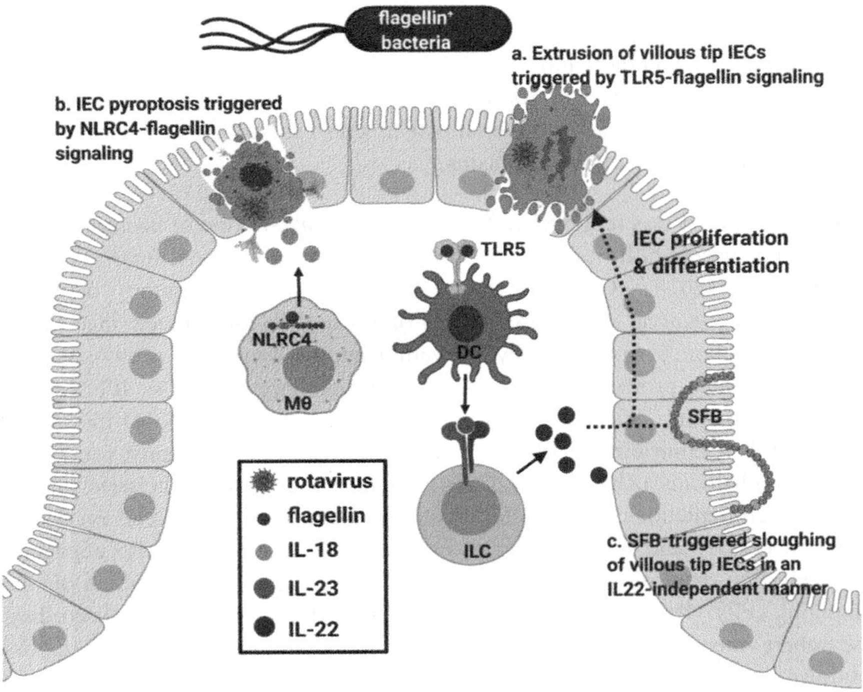

Figure 6.3. *Rotavirus interactions with gut microbiota. Although the gut microbiota enhance rotavirus infection on a tissue level through mechanisms that are not yet clear, specific bacteria and bacterial components are able to control rotavirus infection. (a, b) Bacterial flagellin can act in an immunostimulatory manner to induce antiviral responses to rotavirus. This occurs through the receptors TLR5 and NLRC4, both of which directly bind flagellin. (a) Signaling through TLR5 on dendritic cells (DC) causes an immune signaling cascade beginning with production of IL-23 that induces innate lymphoid cells (ILC) to produce IL-22. In the gut, IL-22 drives IEC proliferation and differentiation which leads to IECs in the villous tips containing rotavirus to be sloughed off. (b) Flagellin signaling through NLRC4 on macrophages induces expression of the pro-inflammatory IL-18. IL-18 then causes infected IECs to undergo pyroptosis. (c) Segmented filamentous bacteria (SFB) can also contribute to control of rotavirus infection. Similar to the mechanism of action of IL-22, SFBs trigger IEC proliferation, differentiation and migration leading to the rotavirus-infected IECs at the villous tips to be sloughed off into the intestinal lumen. However, the mechanism by which SFBs cause this is IL-22-independent*

6.4.3. *Segmented filamentous bacteria have direct and indirect antiviral activity against rotavirus*

Another recent report describes a distinct bacterial inhibitory process targeting rotavirus infection in mice. Researchers at Georgia State University (GSU) discovered that *Rag1*$^{-/-}$ mice purchased from Jackson Laboratories (JAX) can be chronically infected with rotavirus; however, they became resistant after being bred at GSU (Shi *et al.* 2019). Cohousing or fecal transplants from GSU-bred mice to JAX-purchased mice confers the resistant phenotype, implicating the intestinal microbiota. The feces from GSU-bred *Rag1*$^{-/-}$ mice can protect neonatal mice from rotavirus diarrhea and are sufficient in curing chronic rotavirus infection. These feces block rotavirus, vesicular stomatitis virus (VSV), and influenza A virus infectivity *in vitro*, demonstrating that a microbiota constituent has broad and direct antiviral activity. This activity does not appear to be mediated by the activation of host immunity, since GSU feces does not induce cytokines known to block rotavirus replication, including IL-22, IL-18, IFN-λ or IFN-γ, nor does the neutralization of IL-22, IFN-λ or IL-17 reduce the resistance of GSU-bred *Rag1*$^{-/-}$ mice. The protective agent was demonstrated to be a new strain of segmented filamentous bacteria (SFB-G): (1) it passed through a 5 μm filter, but not a 0.22 μm, filter, withstood heating at 60°C, but not 75°C, and was sensitive to antibacterial, but not antifungal, agents; (2) it was enriched in the feces of GSU-bred mice compared to the feces of JAX-bred mice; (3) SFB-G was sufficient in reducing rotavirus infection of the epithelial cells *in vitro*; and (4) SFB-G was sufficient in preventing rotavirus infection in *Rag1*$^{-/-}$ mice and suppressing rotavirus-induced diarrhea in wild-type neonatal mice. SFB are known to promote enterocyte proliferation and migration along the crypt–villous axis. Considering that rotavirus primarily infects villous tip epithelial cells, increased epithelial cell turnover could contribute to the resistance phenotype (Figure 6.3c), similar to how bacterial flagellin-induced IL-22 protects against rotavirus infection (Zhang *et al.* 2019). Indeed, feces from GSU-bred mice promote enterocyte proliferation, migration and subsequent luminal shedding. This protective mechanism is distinct from flagellin-mediated protection since it occurs in an IL-22-independent manner (Shi *et al.* 2019).

6.4.4. *How to reconcile the seemingly contradictory observations of bacterial enhancement and bacterial suppression of rotavirus infection*

Although several specific mechanisms of microbiota-induced control of rotavirus have been uncovered, including flagellin- and SFB-mediated triggering of villous tip IEC sloughing and flagellin-activated IEC pyroptosis, the overall gut microbiota enhances reovirus and rotavirus infection and pathology (Kuss *et al.* 2011; Uchiyama *et al.* 2014). Because little is known about the mechanisms of the microbial enhancement of rotavirus, it is currently difficult to reconcile these conflicting observations. One possibility is that the role of the gut microbiota in rotavirus infection is dependent on the composition of the microbiota. The precise species of bacteria that are present, and in what proportions, could dictate whether the microbiota functions in a stimulatory or inhibitory manner. This is supported by the GSU studies in which the enteric microbiota from mice bred in their facility inhibited rotavirus, in contrast to other studies. The caveat to these experiments is that this inhibition occurred in stool from *Rag1*$^{-/-}$ mice, so the microbial makeup and its effect on rotavirus infection of the mouse gut may also be influenced by the host genetic context of the mouse line being used. The flagellin studies were performed by exogenous administration of flagellin to infected mice, which may not be reflective of the environment during gut homeostasis. Instead these studies may reflect specific gut environments, perhaps when bacterial death results in a large increase in free flagellin, or the invasion of flagellated bacterial species leads to an increase in flagellin being sampled from the intestinal lumen, or crossing the intestinal barrier to induce immune signaling. Further studies of the mechanisms of bacteria–rotavirus interactions are needed in order to reconcile the seemingly contradictory roles of gut bacteria in the modulation of rotavirus infection.

6.5. Noroviruses

Noroviruses (NoVs) are non-enveloped positive-sense RNA viruses that comprise a genus of the family *Caliciviridae*. Based on their genomic sequence, they are classified into multiple genogroups, with a recent classification system reporting 10 genogroups (GI through GX), which are then further subdivided into genotypes (Zheng *et al.* 2006; Chhabra *et al.* 2019). NoVs infect a wide range of mammals including humans, mice, dogs, cats, pigs, sheep and cattle (Liu *et al.* 1999; Karst *et al.* 2003; Vinjé *et al.* 2019), and more recently, NoVs have been found in bats, porpoises, and sea lions (de Graaf *et al.* 2017; Kocher *et al.* 2018; Teng *et al.* 2018). GI, GII and GIV genogroups contain viruses that infect humans, with GII.4 strains representing the predominant cause of epidemic gastroenteritis outbreaks (Pringle *et al.* 2015). These human noroviruses (HNoVs) cause nearly one-fifth of the cases of acute gastroenteritis globally and are now recognized as the most common cause of diarrhea in all cohorts, regardless of age (Wobus *et al.* 2006; Patel *et al.* 2009; Payne *et al.* 2013; Ahmed *et al.* 2014). Infection is generally characterized by one to two days of vomiting, diarrhea, abdominal pain and malaise. Healthy individuals typically resolve symptoms after this short duration whereas children, the elderly and immunocompromised may experience a more severe and protracted illness.

HNoV has been cultivated in human B cell lines (Jones *et al.* 2014, 2015) and enterocytes in stem cell-derived human intestinal enteroid cultures (Ettayebi *et al.* 2016). Consistent with a dual tropism for epithelial and immune cells *in vivo*, the HNoV protein has been detected in intestinal samples from immunocompromised patients in enterocytes, macrophages, T cells and dendritic cells (Karandikar *et al.* 2016), although further studies in immunocompetent individuals are needed. Murine NoVs (MNoVs) belong to GV and include acute virulent strains that are cleared by the immune system (e.g. MNV-1), as well as persistent avirulent strains that infect mice indefinitely (e.g. MNV-3, MNV-CR6) (Hsu *et al.* 2006; Thackray *et al.* 2007; Kahan *et al.* 2011; Strong *et al.* 2012). All MNoV strains isolated from laboratory mice replicate robustly *in vitro* in macrophages and dendritic cell lines, and certain strains have been demonstrated to replicate in B cells and T cells (Wobus *et al.* 2004; Jones *et al.* 2014; Grau *et al.* 2017). Acute MNoV bypasses the epithelial barrier *in vivo* by passing through the microfold (M) cells (Gonzalez-Hernandez *et al.* 2013, 2014; Kolawole *et al.* 2016) and then infects macrophages, dendritic cells, B cells and T cells in gut-associated lymphoid tissues (Grau *et al.* 2017). In contrast, persistent MNoV chronically infects tuft cells, specialized chemosensory intestinal epithelial cells (Lee *et al.* 2017; Wilen *et al.* 2018). Over the past several years, there have been new insights into important regulatory roles for the microbiota in the regulation of HNoV *in vitro* and MNoV *in vivo*. In the following sections, we describe the ways in which the microbiota support or inhibit NoV infection and how both NoV and the microbiota interact with host systems.

6.5.1. *Intestinal microbiota can promote norovirus infection*

Similar to poliovirus, MMTV, reovirus and rotavirus, initial studies revealed that commensal bacteria promote acute and persistent MNoV infections. Specifically, virus titers are reduced in Abx-treated mice, while fecal transplants from untreated mice to Abx-treated mice restore infection (Jones *et al.* 2014; Baldridge *et al.* 2015). Multiple stimulatory mechanisms have been described, including: (a) enhancement of virion binding to host cells; (b) promotion of secretory IgA, which reduces antiviral immunity; and (c) stimulation of the number of target cells.

a) There are several examples of commensal bacteria and the metabolites produced by bacterial enzymes enhancing NoV binding to target cells. First, although the receptor for HNoV has not been determined, histo-blood group antigens (HBGAs) are glycans that serve as attachment factors on host cells (Harrington *et al.* 2002; Lindesmith *et al.* 2003). HBGAs produced by commensal bacteria can bind HNoV virus-like particles (Miura *et al.* 2013) and enhance the infection of human B cells by HNoV strain GII.4 (Figure 6.4a) (Jones *et al.* 2014).

Indeed, HNoV capsid binding to exogenous HBGA alone (Huang *et al.* 2003) is sufficient to increase infection levels in B cell cultures (Jones *et al.* 2014). Second, bacterially bio-transformed bile acids enhance the binding of MNoV to its proteinaceous receptor CD300lf, which is expressed on immune cells and tuft cells, with the presence of a bile acid binding pocket demonstrated in the co-crystal structure for MNoV VP1 with CD300lf (Figure 6.4b) (Nelson *et al.* 2018). The addition of human bile to organoid cultures increases HNoV replication in a dose-dependent manner NoV infection (Ettayebi *et al.* 2016), and bile acids can enhance the binding of certain HNoV strains to HBGAs (Kilic *et al.* 2019), thus implicating bile acids as being proviral for HNoV as well (Ettayebi *et al.* 2016).

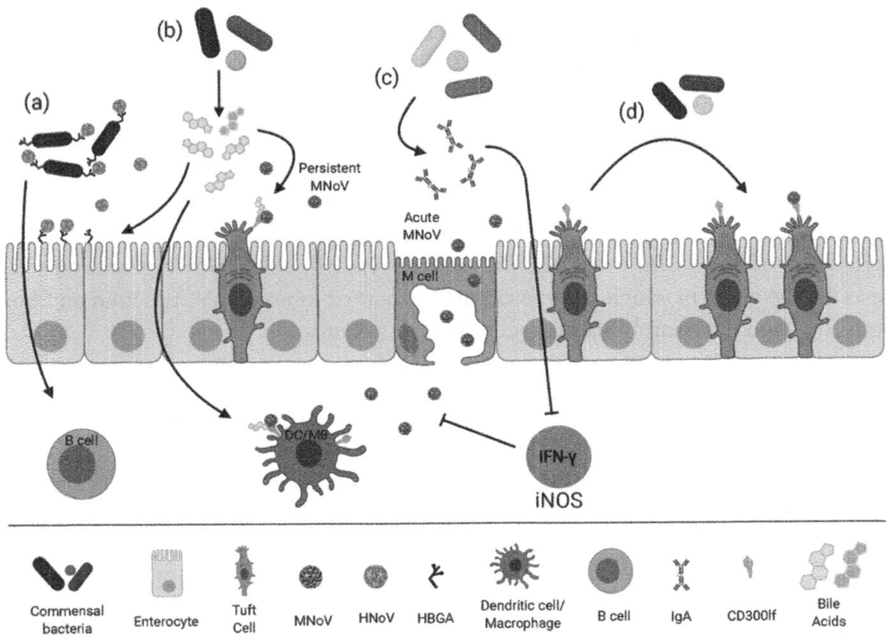

Figure 6.4. *Proviral effects of the microbiota on norovirus infection. (a) HBGA-producing bacteria bind HNoV and can facilitate HNoV infection of B cells* in vitro. *(b) Bile acids improve HNoV infection of enterocytes* in vitro, *and also improve MNoV binding to target cells by enabling viral capsid binding to its receptor CD300lf. (c) Commensal bacteria sustain IgA levels in the gut, decreasing type II interferon which can limit acute MNoV infection. (d) Microbiota-mediated maintenance of tuft cell numbers provides permissive cells for persistent MNoV infection*

b) Acute MNV-1 infection is reduced in *Pigr*$^{-/-}$ mice that lack intestinal IgA due to increased IFN-γ and iNOS levels (Figure 6.4c) (Turula *et al.* 2018). Without IgA, the microbiota contributes to altered host responses (Turula *et al.* 2018), supporting an important role for IgA in limiting the potential inflammatory effects of the microbiota which normally helps to promote MNoV infection. These studies suggest that the normal intestinal microbiota maintains IgA levels that are sufficient in enhancing MNoV infection.

c) Tuft cells are the reservoir for persistent MNoV (Wilen *et al.* 2018). Abx treatment of mice reduces the number of tuft cells in the epithelium, thus limiting the number of permissive cells for MNoV infection (Wilen *et al.* 2018). Cytokines IL-4 and IL-25 can induce tuft cell numbers in Abx-treated mice, thereby restoring infection (Wilen *et al.* 2018). The mechanisms that the microbiota use to regulate tuft cell numbers have yet to be uncovered.

6.5.2. *Intestinal microbiota can trigger antiviral immune responses during norovirus infection*

Few studies have explored the effects of intestinal microbiota composition on HNoV infection, although one study has linked an increase in gut bacterial families Ruminococcaceae and Faecalibacterium to decreased susceptibility to HNoV and rotavirus infection (Rodríguez-Díaz *et al.* 2017). There is, however, an abundance of emerging research exploring the antiviral effects of the microbiota on MNoV infection. Under the conditions of Abx treatment, the microbiota has recently been shown to mediate antiviral effects against MNV-1 in the proximal, but not the distal, intestine via bile acid priming of IFN-λ (Figure 6.5a) (Grau *et al.* 2020). In Abx-treated mice depleted of their intestinal microbiota, the MNV-1 infection shifts from the proximal to the distal gut in wild-type mice. The increased virus titers observed in the proximal gut of Abx-treated wild-type mice are phenocopied in colonized *Ifnlr1-/-* mice, suggesting that bacteria inhibit MNV-1 proximal gut infection via priming a type III IFN response. Evidence supports the concept that bacterially biotransformed bile acids enhance the type III IFN response in the proximal gut (Grau *et al.* 2020): (a) treatment of mice with the single antibiotic clindamycin that reduces bile acid metabolism in the gut (Sun *et al.* 2018) is sufficient to block the restriction of the MNV-1 infection in the proximal gut; (b) commensal bacteria, such as *Clostridium scindens* that biotransform bile acids (Wells and Hylemon 2000; Buffie *et al.* 2015), can rescue this inhibition; and

(c) bacteria-regulated bile acids can restore viral inhibition in Abx-treated mice (Grau *et al.* 2020). Furthermore, *in vitro* bile acid priming enhances the viral induction of IFN-λ (Grau *et al.* 2020). Additional studies are needed to fully understand why this antiviral mechanism operates in the proximal, but not the distal, gut but one possible explanation is that key bile acid receptors involved in immune regulation are expressed in regionally distinct manners (Zhang *et al.* 2003; Bookout *et al.* 2006; Inagaki *et al.* 2006; Higashiyama *et al.* 2008; Grau *et al.* 2020). As described for rotavirus, different bacterial species or products have also been found to restrict MNoV replication. A dietary metabolite of vitamin A and retinoic acid enriches *Lactobacillus* species in the gut, resulting in significant decreases in MNoV replication, possibly via a *Lactobacillus*-induced induction of type I IFN signaling (Figure 6.5b) (Lee and Ko 2016). More studies are needed to distinguish between direct antiviral effects of retinoic acid versus those of *Lactobacillus*; however, the upregulation of *Lactobacillus* has also been associated with antiviral effects against rotavirus and influenza infections (Kikuchi *et al.* 2014; Sindhu *et al.* 2014). Another study found that oral administration of a poly-γ glutamic acid (γ-PGA), an extracellular biopolymer produced by the *Bacillus* species, limits MNV-1 replication by increasing IFN-β levels in the small intestine (Figure 6.5c) (Lee *et al.* 2018). Interestingly, the viral constituents of the microbiota can also mediate antiviral effects during MNoV infection. Chronic murine astrovirus present in some immunocompromised mice can stimulate high IFN-λ levels and limit the host's susceptibility to persistent MNoV infection (Figure 6.5d) – a phenotype that can be transferred to other immunocompromised mice by fecal transplant (Ingle *et al.* 2019). Thus far, all instances of MNoV restriction by the microbiota are mediated through the modulation of IFNs, highlighting the importance of interactions between the microbiota and the host immune system.

In summary, there are numerous complex interactions between NoVs, the host immune system and the commensal microbiota. Studies in mice strongly support context-dependent proviral and antiviral roles for the microbiota in controlling infection, which suggests that variations in the microbiota may be an important factor underlying the differential responses to infection in humans. Future studies using small animal models and new *in vitro* systems for HNoV cultivation will continue to clarify these interactions and give us further insights into NoV pathogenesis and suggest new innovative approaches for counteracting infection.

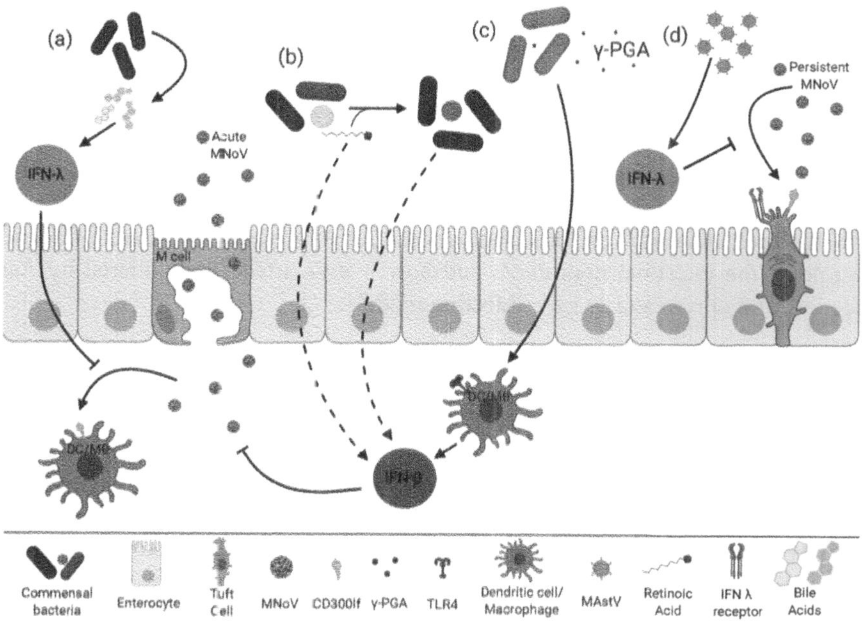

Figure 6.5. *Antiviral effects of the microbiota on norovirus infection. (a) Commensal bacteria facilitate antiviral effects against MNV-1 in the duodenum by biotransforming bile acids that prime an IFN-λ response. (b) The addition of retinoic acid, which itself has antiviral effects* in vitro*, enriches Lactobacillus in the gut which is associated with induction of type I IFN to block MNoV replication. (c) Bacillus species produce poly-γ glutamic acid (γ-PGA), a non-LPS TLR4 agonist that upregulates IFN-β, limiting acute MNoV infection of its myeloid and lymphoid target cells. (d) IFN-λ stimulation by chronic murine astrovirus infection in immunocompromised mice results in resistance to persistent MNoV infection which infects tuft cells*

6.6. Astroviruses

Astroviruses (AstV) are non-enveloped, positive-sense enteric RNA viruses with a wide genetic diversity among both mammals and birds (Cortez *et al.* 2017; Moser and Schultz-Cherry 2005). They are transmitted via the fecal–oral route and infect intestinal epithelial cells (Ingle *et al.* 2019; Kolawole *et al.* 2019) after entry via endosomes (Donelli *et al.* 1992; Marvin *et al.* 2016). Human astrovirus (HAstV) infection can cause gastroenteritis in children, as well as in the elderly and in immunocompromised groups. HAstV has been shown to increase the permeability of intestinal epithelial cells, leading to a loss of water and nutrients, and ultimately intestinal dysregulation and diarrhea (Moser *et al.* 2007; Nighot *et al.* 2010;

Meliopoulos *et al.* 2016). Asymptomatic infection can occur in humans and other animal species, with the endemic murine astrovirus (MuAstV) infection of many laboratory mouse facilities serving as a prominent example (Compton *et al.* 2017). Despite AstV being one of the leading causes of pediatric diarrhea, there are still a number of gaps in our understanding of both viral pathogenesis and host immune responses, due to limitations in culturing methods and a lack of small animal models, although recent advances are starting to overcome these challenges. Current AstV research is focused on the examination of host immune responses to infection, the role of the intestinal microbiota and how it may affect or be altered by AstV infection, and *in vivo* and *in vitro* infection models.

6.6.1. *Host interferon responses reduce astrovirus replication and infection*

The modulation of host immune factors, specifically IFN, is often necessary for successful replication of enteric pathogens, and AstVs are no different (Marvin *et al.* 2016). Human astrovirus type 1 (HAstV-1), a "classical" HAstV strain, induces type I IFNs and is sensitive to both endogenous and exogenous IFN-β, which limit viral replication in Caco-2 cells, a human intestinal adenocarcinoma line commonly used for HAstV studies (Marvin *et al.* 2016; Cortez *et al.* 2019; Kolawole *et al.* 2019). In the presence of exogenous IFN-β, Caco-2 cell permeability is also greatly reduced during HAstV-1 infection, suggesting that type I IFNs can protect the intestinal barrier during AstV infection (Marvin *et al.* 2016). Immunodeficient mice, including CD1 nude, $Ifnar1^{-/-}$, NOD SCID $Il2rg^{-/-}$, $Rag2^{-/-}Il2rg^{-/-}$ and $Rag1^{-/-}$ exhibit higher viral loads and a persistent shedding of murine AstVs, compared to wild-type mice, consistent with important roles for innate and adaptive immunity in AstV clearance (Yokoyama *et al.* 2012; Marvin *et al.* 2016; Ingle *et al.* 2019). A recent study also found that some strains of murine AstV can robustly stimulate type III IFN or IFN-λ, at least in highly immunocompromised $Rag2^{-/-}Il2rg^{-/-}$ mice, thereby conferring resistance to other enteric viruses including persistent murine norovirus and rotavirus (Ingle *et al.* 2019). It remains unclear how murine AstV may be able to evade the effects of high IFN-λ levels *in vivo*.

6.6.2. *Dysbiosis can occur after AstV infection*

Early data suggests that AstV can alter the intestinal microbiota of both infected human patients and Jamaican fruit bats (Ma *et al.* 2011; Wasimuddin *et al.* 2018). In a study examining 52 patients with virus-induced diarrhea, 13 of whom were infected with AstV, the abundance of traditional commensal bacteria, such as Bacteroidetes, decreased to be replaced with more opportunistic pathogens, including

Enterococcus, Streptococcus and *Shigella* (Ma *et al.* 2011). In Jamaican fruit bats, both alpha and beta diversity metrics were altered after AstV infection, with changes depending on host age: the younger the host, the more dramatic the microbiome changes (Wasimuddin *et al.* 2018). Similar to the report in humans, levels of pathogenic microbes rise in bats after AstV infection, regardless of age (Wasimuddin *et al.* 2018). Disruption of the gut microbiota by AstV infection may predispose hosts to be more susceptibile to coinfections (Ma *et al.* 2011; Wasimuddin *et al.* 2018). There is currently very limited data regarding whether microbiota influences AstV infection; however, it has been recently demonstrated that murine AstV is capable of infecting germ-free mice (Ingle *et al.* 2019), suggesting that there is not a strict requirement for commensal bacteria for AstV infection.

6.6.3. In vivo *and* in vitro *culture systems for AstV pathogenesis studies*

A major challenge for AstV studies has been a lack of tractable model systems beyond HAstV studies in differentiated Caco-2 cells. Turkey astrovirus type 2 has been recently developed as a well-defined model system to study AstV replication and cell tropism, although there are limitations, including access to reagents, proper facilities and turkeys themselves (Marvin *et al.* 2016). In addition, various mouse models are being developed that can mimic the characteristics present in asymptomatic human patients. As in asymptomatic human infection, murine AstV replication appears to be generally restricted to the intestinal tract alone (Cortez *et al.* 2019). Different murine AstV strains, characterized by whole genome sequencing, are associated with different durations of persistent shedding in wild-type mice; however, to date, no distinct pathology has been found to be associated with murine AstV infection (Cortez *et al.* 2019). Further explorations in these murine models may continue to offer insights into AstV pathogenesis in humans.

Recent advances in culturing HAstVs in human intestinal enteroids (HIE) also hold great promise for our understanding of AstV (Kolawole *et al.* 2019). HIEs from different segments of the intestinal tract (duodenum, jejunum, ileum and colon) of different human patients all showed similar levels of viral infection across segments, but differential levels of infection between patients. This system has permitted the characterization of the cellular tropism of VA1, a "non-classical" HAstV, revealing viral replication in progenitor cells, enterocytes and goblet cells. Other HAstV strains were also shown to robustly replicate in HIE cultures (Kolawole *et al.* 2019). In summary, the advent of a variety of new systems to explore AstV pathogenesis will likely lead to a number of upcoming break-through AstV studies.

6.7. Overall conclusion

The intestinal lumen is home to trillions of commensal microorganisms which have profound impacts on the host immune system, metabolism and physiology. Viruses can adapt rapidly to their host environment and develop ways to co-opt host systems for their own ends. A wealth of evidence has emerged in recent years revealing that enteric viruses interact with these commensal bacteria in both direct and indirect manners, and that these interactions have significant effects on viral pathogenesis. Considering the varied types of viruses that transmit via the intestine, it is not surprising that enteric viruses have evolved many different strategies to use bacteria to their advantage. Picornaviruses and reovirus can bind bacterial components, such as LPS and PG, to stabilize their virion structures and survive harsh extracellular environments (Kuss *et al.* 2011; Robinson *et al.* 2014; Berger *et al.* 2017). Bacterial components or metabolites can also facilitate virus–receptor interactions, as is the case with noroviruses and picornaviruses, thereby allowing viruses to efficiently enter target cells (Miura *et al.* 2013; Jones *et al.* 2014; Robinson *et al.* 2014; Nelson *et al.* 2018). Other viruses, such as MMTV, take advantage of the tolerogenic environments induced by normal commensal bacteria and even incorporate LPS into its envelope to induce these environments (Kane *et al.* 2011; Wilks *et al.* 2015). While numerous examples of bacterial enhancement have been described for enteric viruses, it is also becoming appreciated that the intestinal microbiota can be inhibitory to enteric viruses in certain contexts. For example, bacterially biotransformed bile acids can prime a key host antiviral response, type III IFN, to restrict norovirus infection (Grau *et al.* 2020).

With the well-established examples of bacterial enhancement and the bacterial inhibition of enteric viruses in mind, it is important to recognize that these influences have been largely defined in hosts depleted of a majority of their microbiota, where more nuanced and complex outcomes would not be apparent. This point is underscored by studies of norovirus and rotavirus in which both proviral and antiviral mechanisms have been uncovered in the gut–bacteria–virus axis. For example, while certain bacterial components can induce a tolerogenic environment, too much signaling can boost antiviral signaling, as evidenced by the effect of flagellin treatment on rotavirus infection (Zhang *et al.* 2014). Specific types of bacteria may change how host tissues respond to virus challenges as well, such as the segmented filamentous bacteria that promote tissue resistance to rotavirus (Shi *et al.* 2019). Finally, the human gut is functionally compartmentalized, with profound differences in host immunity, metabolic state and physiology in the various regions of the gut (Mowat and Agace 2014). Norovirus–bacteria interactions are highly region-specific, being antiviral in the proximal gut and proviral in the distal gut (Grau *et al.* 2020). It will be critical moving forward, that the field considers the effects of individualized host microbiomes and distinct bacterial communities when building a biologically relevant and complete picture of the influence of the commensal bacteria on enteric viruses. They should also consider discrete intestinal environments in order to determine how the bacteria and regional physiology shape virus pathogenesis.

6.8. References

Abt, M.C., and Artis, D. (2009). The intestinal microbiota in health and disease: The influence of microbial products on immune cell homeostasis. *Curr. Opin. Gastroenterol.*, 25, 496–502. https://doi.org/10.1097/MOG.0b013e328331b6b4.

Aguilera, E.R., Nguyen, Y., Sasaki, J., and Pfeiffer, J.K. (2019). Bacterial stabilization of a panel of picornaviruses. *mSphere 4.* https://doi.org/10.1128/mSphere.00183-19.

Ahmed, S.M., Hall, A.J., Robinson, A.E., Verhoef, L., Premkumar, P., Parashar, U.D., Koopmans, M., and Lopman, B.A. (2014). Global prevalence of norovirus in cases of gastroenteritis: A systematic review and meta-analysis *Lancet Infect. Dis.*, 14, 725–730. https://doi.org/10.1016/S1473-3099(14)70767-4.

Almand, E.A., Moore, M.D., Outlaw, J., and Jaykus, L.-A. (2017). Human norovirus binding to select bacteria representative of the human gut microbiota. *PLOS ONE*, 12, e0173124. https://doi.org/10.1371/journal.pone.0173124.

Baldridge, M.T., Nice, T.J., McCune, B.T., Yokoyama, C.C., Kambal, A., Wheadon, M., Diamond, M.S., Ivanova, Y., Artyomov, M., and Virgin, H.W. (2015). Commensal microbes and interferon-λ determine persistence of enteric murine norovirus infection. *Science*, 347, 266–269. https://doi.org/10.1126/science.1258025.

Berger, A.K., Yi, H., Kearns, D.B., and Mainou, B.A., (2017). Bacteria and bacterial envelope components enhance mammalian reovirus thermostability. *PLOS Pathog.*, 13, e1006768. https://doi.org/10.1371/journal.ppat.1006768.

Bookout, A.L., Jeong, Y., Downes, M., Yu, R.T., Evans, R.M., and Mangelsdorf, D.J. (2006). Anatomical profiling of nuclear receptor expression reveals a hierarchical transcriptional network. *Cell*, 126, 789–799. https://doi.org/10.1016/j.cell.2006.06.049.

Buffie, C.G., Bucci, V., Stein, R.R., McKenney, P.T., Ling, L., Gobourne, A., No, D., Liu, H., Kinnebrew, M., Viale, A., Littmann, E., van den Brink, M.R.M., Jenq, R.R., Taur, Y., Sander, C., Cross, J.R., Toussaint, N.C., Xavier, J.B., and Pamer, E.G. (2015). Precision microbiome reconstitution restores bile acid mediated resistance to clostridium difficile. *Nature*, 517, 205–208. https://doi.org/10.1038/nature13828.

Chhabra, P., de Graaf, M., Parra, G.I., Chan, M.C.-W., Green, K., Martella, V., Wang, Q., White, P.A., Katayama, K., Vennema, H., Koopmans, M.P.G., and Vinjé, J. (2019). Updated classification of norovirus genogroups and genotypes. *J. Gen. Virol.*, 100, 1393–1406. https://doi.org/10.1099/jgv.0.001318.

Compton, S.R., Booth, C.J., and Macy, J.D. (2017). Murine astrovirus infection and transmission in neonatal CD1 mice. *J. Am. Assoc. Lab. Anim. Sci. JAALAS*, 56, 402–411.

Cortez, V., Meliopoulos, V.A., Karlsson, E.A., Hargest, V., Johnson, C., and Schultz-Cherry, S. (2017). Astrovirus biology and pathogenesis. *Annu. Rev. Virol.*, 4, 327–348. https://doi.org/10.1146/annurev-virology-101416-041742.

Cortez, V., Sharp, B., Yao, J., Livingston, B., Vogel, P., and Schultz-Cherry, S. (2019). Characterizing a murine model for astrovirus using viral isolates from persistently infected immunocompromised mice. *J. Virol.*, 93. https://doi.org/10.1128/JVI.00223-19.

Donelli, G., Superti, F., Tinari, A., and Marziano, M.L. (1992). Mechanism of astrovirus entry into Graham 293 cells. *J. Med. Virol.*, 38, 271–277. https://doi.org/10.1002/jmv.1890380408.

Dudley, J.P., Golovkina, T.V., and Ross, S.R. (2016). Lessons learned from mouse mammary tumor virus in animal models. ILAR J., 57, 12–23. https://doi.org/10.1093/ilar/ilv044.

Erickson, A.K., Jesudhasan, P.R., Mayer, M.J., Narbad, A., Winter, S.E., and Pfeiffer, J.K. (2018). Bacteria facilitate enteric virus co-infection of mammalian cells and promote genetic recombination. *Cell Host Microbe*, 23, 77–88.e5. https://doi.org/10.1016/j.chom.2017.11.007.

Ettayebi, K., Crawford, S.E., Murakami, K., Broughman, J.R., Karandikar, U., Tenge, V.R., Neill, F.H., Blutt, S.E., Zeng, X.-L., Qu, L., Kou, B., Opekun, A.R., Burrin, D., Graham, D.Y., Ramani, S., Atmar, R.L., and Estes, M.K. (2016). Replication of human noroviruses in stem cell–derived human enteroids. *Science*, aaf5211. https://doi.org/10.1126/science.aaf5211.

Gonzalez-Hernandez, M.B., Liu, T., Blanco, L.P., Auble, H., Payne, H.C., and Wobus, C.E. (2013). Murine norovirus transcytosis across an *in vitro* polarized murine intestinal epithelial monolayer is mediated by M-like cells. *J. Virol.*, 87, 12685–12693. https://doi.org/10.1128/JVI.02378-13.

Gonzalez-Hernandez, M.B., Liu, T., Payne, H.C., Stencel-Baerenwald, J., Ikizler, M., Yagita, H., Dermody, T.S., Williams, I.R., and Wobus, C.E. (2014). Efficient norovirus and reovirus replication in the mouse intestine requires microfold (M) cells. *J. Virol.*, 88, 6934–6943. https://doi.org/10.1128/JVI.00204-14.

de Graaf, M., Bodewes, R., van Elk, C.E., van de Bildt, M., Getu, S., Aron, G.I., Verjans, G.M.G.M., Osterhaus, A.D.M.E., van den Brand, J.M.A., Kuiken, T., and Koopmans, M.P.G. (2017). Norovirus infection in harbor porpoises. *Emerg. Infect.* Dis., 23, 87–91. https://doi.org/10.3201/eid2301.161081.

Grau, K.R., Roth, A.N., Zhu, S., Hernandez, A., Colliou, N., DiVita, B.B., Philip, D.T., Riffe, C., Giasson, B., Wallet, S.M., Mohamadzadeh, M., and Karst, S.M. (2017). The major targets of acute norovirus infection are immune cells in the gut-associated lymphoid tissue. *Nat. Microbiol.*, 2, 1586. https://doi.org/10.1038/s41564-017-0057-7.

Grau, K.R., Zhu, S., Peterson, S.T., Helm, E.W., Philip, D., Phillips, M., Hernandez, A., Turula, H., Frasse, P., Graziano, V.R., Wilen, C.B., Wobus, C.E., Baldridge, M.T., and Karst, S.M. (2020). The intestinal regionalization of acute norovirus infection is regulated by the microbiota via bile acid-mediated priming of type III interferon. *Nat. Microbiol.*, 5, 84–92. https://doi.org/10.1038/s41564-019-0602-7.

Harrington, P.R., Lindesmith, L., Yount, B., Moe, C.L., and Baric, R.S. (2002). Binding of Norwalk virus-like particles to ABH histo-blood group antigens is blocked by antisera from infected human volunteers or experimentally vaccinated mice. *J. Virol.*, 76, 12335–12343. https://doi.org/10.1128/JVI.76.23.12335-12343.2002.

Higashiyama, H., Kinoshita, M., and Asano, S. (2008). Immunolocalization of farnesoid X receptor (FXR) in mouse tissues using tissue microarray. *Acta Histochem.*, 110, 86–93. https://doi.org/10.1016/j.acthis.2007.08.001.

Hsu, C.C., Riley, L.K., Wills, H.M., and Livingston, R.S. (2006). Persistent infection with and serologic crossreactivity of three novel murine noroviruses. *Comp. Med.*, 56, 247–251.

Huang, P., Farkas, T., Marionneau, S., Zhong, W., Ruvoën-Clouet, N., Morrow, A.L., Altaye, M., Pickering, L.K., Newburg, D.S., LePendu, J., and Jiang, X. (2003). Noroviruses bind to human ABO, Lewis, and secretor histo–blood group antigens: Identification of 4 distinct strain – specific patterns. *J. Infect. Dis.*, 188, 19–31. https://doi.org/10.1086/375742.

Inagaki, T., Moschetta, A., Lee, Y.-K., Peng, L., Zhao, G., Downes, M., Yu, R.T., Shelton, J.M., Richardson, J.A., Repa, J.J., Mangelsdorf, D.J., and Kliewer, S.A. (2006). Regulation of antibacterial defense in the small intestine by the nuclear bile acid receptor. *Proc. Natl. Acad. Sci. USA*, 103, 3920–3925. https://doi.org/10.1073/pnas.0509592103.

Ingle, H., Lee, S., Ai, T., Orvedahl, A., Rodgers, R., Zhao, G., Sullender, M., Peterson, S.T., Locke, M., Liu, T.-C., Yokoyama, C.C., Sharp, B., Schultz-Cherry, S., Miner, J.J., and Baldridge, M.T. (2019). Viral complementation of immunodeficiency confers protection against enteric pathogens via interferon-λ. *Nat. Microbiol.*, 4, 1120. https://doi.org/10.1038/s41564-019-0416-7.

Jones, M.K., Watanabe, M., Zhu, S., Graves, C.L., Keyes, L.R., Grau, K.R., Gonzalez-Hernandez, M.B., Iovine, N.M., Wobus, C.E., Vinjé, J., Tibbetts, S.A., Wallet, S.M., and Karst, S.M. (2014). Enteric bacteria promote human and murine norovirus infection of B cells. *Science* 346, 755–759.

Jones, M.K., Grau, K.R., Costantini, V., Kolawole, A.O., de Graaf, M., Freiden, P., Graves, C.L., Koopmans, M., Wallet, S.M., Tibbetts, S.A., Schultz-Cherry, S., Wobus, C.E., Vinjé, J., and Karst, S.M. (2015). Human norovirus culture in B cells. *Nat. Protoc.*, 10, 1939–1947. https://doi.org/10.1038/nprot.2015.121.

Jude, B.A., Pobezinskaya, Y., Bishop, J., Parke, S., Medzhitov, R.M., Chervonsky, A.V., and Golovkina, T.V. (2003). Subversion of the innate immune system by a retrovirus. *Nat. Immunol.* 4, 573–578. https://doi.org/10.1038/ni926.

Kabat, A.M., Srinivasan, N., and Maloy, K.J. (2014). Modulation of immune development and function by intestinal microbiota. *Trends Immunol.* 35, 507–517. https://doi.org/10.1016/j.it.2014.07.010.

Kahan, S.M., Liu, G., Reinhard, M.K., Hsu, C.C., Livingston, R.S., and Karst, S.M. (2011). Comparative murine norovirus studies reveal a lack of correlation between intestinal virus titers and enteric pathology. *Virology* 421, 202–210. https://doi.org/10.1016/j.virol.2011.09.030.

Kane, M., Case, L.K., Kopaskie, K., Kozlova, A., MacDearmid, C., Chervonsky, A.V., and Golovkina, T.V. (2011). Successful transmission of a retrovirus depends on the commensal microbiota. *Science*, 334, 245–249. https://doi.org/10.1126/science.1210718.

Karandikar, U.C., Crawford, S.E., Ajami, N.J., Murakami, K., Kou, B., Ettayebi, K., Papanicolaou, G.A., Jongwutiwes, U., Perales, M.-A., Shia, J., Mercer, D., Finegold, M.J., Vinjé, J., Atmar, R.L., and Estes, M.K. (2016). Detection of human norovirus in intestinal biopsies from immunocompromised transplant patients. *J. Gen. Virol.*, https://doi.org/10.1099/jgv.0.000545.

Karst, S.M., Wobus, C.E., Lay, M., Davidson, J., and Virgin, H.W. (2003). STAT1-dependent innate immunity to a Norwalk-like virus. *Science*, 299, 1575–1578. https://doi.org/10.1126/science.1077905.

Kikuchi, Y., Kunitoh-Asari, A., Hayakawa, K., Imai, S., Kasuya, K., Abe, K., Adachi, Y., Fukudome, S.-I., Takahashi, Y., and Hachimura, S. (2014). Oral administration of Lactobacillus plantarum strain AYA enhances IgA secretion and provides survival protection against influenza virus infection in mice. *PloS One*, 9, e86416. https://doi org/10.1371/journal.pone.0086416.

Kilic, T., Koromyslova, A., and Hansman, G.S. (2019). Structural basis for human norovirus capsid binding to bile acids. *J. Virol.*, 93. https://doi.org/10.1128/JVI.01581-18.

Kocher, J.F., Lindesmith, L.C., Debbink, K., Beall, A., Mallory, M.L., Yount, B.L., Graham, R.L., Huynh, J., Gates, J.E., Donaldson, E.F., and Baric, R.S. (2018). Bat caliciviruses and human noroviruses are antigenically similar and have overlapping histo-blood group antigen binding profiles. *mBio*, 9. https://doi.org/10.1128/mBio.00869-18.

Kolawole, A.O., Gonzalez-Hernandez, M.B., Turula, H., Yu, C., Elftman, M.D., and Wobus, C.E. (2016). Oral norovirus infection is blocked in mice lacking Peyer's patches and mature M cells. *J. Virol.*, 90, 1499–1506. https://doi.org/10.1128/JVI.02872-15.

Kolawole, A.O., Mirabelli, C., Hill, D.R., Svoboda, S.A., Janowski, A.B., Passalacqua, K.D., Rodriguez, B.N., Dame, M.K., Freiden, P., Berger, R.P., Vu, D.-L., Hosmillo, M., O'Riordan, M.X.D., Schultz-Cherry, S., Guix, S., Spence, J.R., Wang, D., and Wobus, C.E. (2019). Astrovirus replication in human intestinal enteroids reveals multi-cellular tropism and an intricate host innate immune landscape. *PLoS Pathog.*, 15, e1008057. https://doi.org/10.1371/journal.ppat.1008057.

Kuss, S.K., Best, G.T., Etheredge, C.A., Pruijssers, A.J., Frierson, J.M., Hooper, L.V., Dermody, T.S., and Pfeiffer, J.K. (2011). Intestinal microbiota promote enteric virus replication and systemic pathogenesis. *Science*, 334, 249–252. https://doi.org/10.1126/science.1211057.

Lee, H., and Ko, G. (2016). Antiviral effect of vitamin A on norovirus infection via modulation of the gut microbiome. *Sci. Rep.*, 6, 25835. https://doi.org/10.1038/srep25835.

Lee, S., Wilen, C.B., Orvedahl, A., McCune, B.T., Kim, K.-W., Orchard, R.C., Peterson, S.T., Nice, T.J., Baldridge, M.T., and Virgin, H.W. (2017). Norovirus cell tropism is determined by combinatorial action of a viral non-structural protein and host cytokine. *Cell Host Microbe*, 22, 449–459.e4. https://doi.org/10.1016/j.chom.2017.08.021.

Lee, W., Kim, M., Lee, S.-H., Jung, H.-G., and Oh, J.-W. (2018). Prophylactic efficacy of orally administered Bacillus poly-γ-glutamic acid, a non-LPS TLR4 ligand, against norovirus infection in mice. *Sci. Rep.*, 8, 8667. https://doi.org/10.1038/s41598-018-26935-y.

Lindesmith, L., Moe, C., Marionneau, S., Ruvoen, N., Jiang, X., Lindblad, L., Stewart, P., LePendu, J., and Baric, R. (2003). Human susceptibility and resistance to Norwalk virus infection. *Nat Med*, 9, 548–553. https://doi.org/10.1038/nm860.

Liu, B.L., Lambden, P.R., Gunther, H., Otto, P., Elschner, M., and Clarke, I.N. (1999). Molecular characterization of a bovine enteric calicivirus: Relationship to the Norwalk-like viruses. *J Virol*, 73, 819–825.

Ma, C., Wu, X., Nawaz, M., Li, J., Yu, P., Moore, J.E., and Xu, J., (2011). Molecular characterization of fecal microbiota in patients with viral diarrhea. *Curr. Microbiol.*, 63, 259–266. https://doi.org/10.1007/s00284-011-9972-7.

Marvin, S.A., Huerta, C.T., Sharp, B., Freiden, P., Cline, T.D., and Schultz-Cherry, S. (2016). Type I interferon response limits astrovirus replication and protects against increased barrier permeability *in vitro* and *in vivo*. *J. Virol.*, 90, 1988–1996. https://doi.org/10.1128/JVI.02367-15.

Meliopoulos, V.A., Marvin, S.A., Freiden, P., Moser, L.A., Nighot, P., Ali, R., Blikslager, A., Reddivari, M., Heath, R.J., Koci, M.D., and Schultz-Cherry, S. (2016). Oral administration of astrovirus capsid protein is sufficient to induce acute diarrhea *in vivo*. *mBio*, 7. https://doi.org/10.1128/mBio.01494-16.

Miura, T., Sano, D., Suenaga, A., Yoshimura, T., Fuzawa, M., Nakagomi, T., Nakagomi, O., and Okabe, S. (2013). Histo-blood group antigen-like substances of human enteric bacteria as specific adsorbents for human noroviruses. *J. Virol.*, 87, 9441–9451. https://doi.org/10.1128/JVI.01060-13.

Moser, L.A., and Schultz-Cherry, S. (2005). Pathogenesis of astrovirus infection. *Viral Immunol.*, 18, 4–10. https://doi.org/10.1089/vim.2005.18.4.

Moser, L.A., Carter, M., and Schultz-Cherry, S. (2007). Astrovirus increases epithelial barrier permeability independently of viral replication. *J. Virol.*, 81, 11937–11945. https://doi.org/10.1128/JVI.00942-07.

Mowat, A.M., and Agace, W.W. (2014). Regional specialization within the intestinal immune system. *Nat. Rev. Immunol.*, 14, 667–685. https://doi.org/10.1038/nri3738.

Nelson, C.A., Wilen, C.B., Dai, Y.-N., Orchard, R.C., Kim, A.S., Stegeman, R.A., Hsieh, L.L., Smith, T.J., Virgin, H.W., and Fremont, D.H. (2018). Structural basis for murine norovirus engagement of bile acids and the CD300lf receptor. *Proc. Natl. Acad. Sci. USA*, 115, E9201–E9210. https://doi.org/10.1073/pnas.1805797115.

Nighot, P.K., Moeser, A., Ali, R.A., Blikslager, A.T., and Koci, M.D. (2010). Astrovirus infection induces sodium malabsorption and redistributes sodium hydrogen exchanger expression. *Virology*, 401, 146–154. https://doi.org/10.1016/j.virol.2010.02.004.

Outzen, H.C., Corrow, D., and Shultz, L.D. (1985). Attenuation of exogenous murine mammary tumor virus virulence in the C3H/HeJ mouse substrain bearing the Lps mutation. *J. Natl. Cancer Inst.*, 75, 917–923. https://doi.org/10.1093/jnci/75.5.917.

Patel, M.M., Hall, A.J., Vinjé, J., and Parashar, U.D. (2009). Noroviruses: A comprehensive review. *J. Clin. Virol.*, 44, 1–8. https://doi.org/10.1016/j.jcv.2008.10.009.

Payne, D.C., Vinjé, J., Szilagyi, P.G., Edwards, K.M., Staat, M.A., Weinberg, G.A., Hall, C.B., Chappell, J., Bernstein, D.I., Curns, A.T., Wikswo, M., Shirley, S.H., Hall, A.J., Lopman, B., and Parashar, U.D. (2013). Norovirus and medically attended gastroenteritis in U.S. children. *N. Engl. J. Med.*, 368, 1121–1130. https://doi.org/10.1056/NEJMsa1206589.

Pickard, J.M., Zeng, M.Y., Caruso, R., and Núñez, G. (2017). Gut microbiota: Role in pathogen colonization, immune responses, and inflammatory disease. *Immunol. Rev.*, 279, 70–89. https://doi.org/10.1111/imr.12567.

Pringle, K., Lopman, B., Vega, E., Vinje, J., Parashar, U.D., and Hall, A.J. (2015). Noroviruses: Epidemiology, immunity and prospects for prevention. *Future Microbiol.*, 10, 53–67. https://doi.org/10.2217/fmb.14.102.

Rassa, J.C., Meyers, J.L., Zhang, Y., Kudaravalli, R., and Ross, S.R. (2002). Murine retroviruses activate B cells via interaction with toll-like receptor 4. *Proc. Natl. Acad. Sci.*, 99, 2281–2286. https://doi.org/10.1073/pnas.042355399.

Robinson, C.M., Jesudhasan, P.R., and Pfeiffer, J.K. (2014). Bacterial lipopolysaccharide binding enhances virion stability and promotes environmental fitness of an enteric virus. *Cell Host Microbe*, 15, 36–46. https://doi.org/10.1016/j.chom.2013.12.004.

Robinson, C.M., Acevedo, M.A.W., McCune, B.T., and Pfeiffer, J.K. (2019). Related enteric viruses have different requirements for host microbiota in mice. *J. Virol.* https://doi org/10.1128/JVI.01339-19.

Rodríguez-Díaz, J., García-Mantrana, I., Vila-Vicent, S., Gozalbo-Rovira, R., Buesa, J., Monedero, V., and Collado, M.C. (2017). Relevance of secretor status genotype and microbiota composition in susceptibility to rotavirus and norovirus infections in humans. *Sci. Rep.*, 7, 45559. https://doi.org/10.1038/srep45559.

Rowland, I., Gibson, G., Heinken, A., Scott, K., Swann, J., Thiele, I., and Tuohy, K. (2018). Gut microbiota functions: Metabolism of nutrients and other food components. *Eur. J. Nutr.*, 57, 1–24. https://doi.org/10.1007/s00394-017-1445-8.

Shi, Z., Zou, J., Zhang, Z., Zhao, X., Noriega, J., Zhang, B., Zhao, C., Ingle, H., Bittinger, K., Mattei, L.M., Pruijssers, A.J., Plemper, R.K., Nice, T.J., Baldridge, M.T., Dermody, T.S., Chassaing, B., and Gewirtz, A.T. (2019). Segmented filamentous bacteria prevent and cure rotavirus infection. *Cell*, 179, 644–658.e13. https://doi.org/10.1016/j.cell.2019.09.028.

Sindhu, K.N.C., Sowmyanarayanan, T.V., Paul, A., Babji, S., Ajjampur, S.S.R., Priyadarshini, S., Sarkar, R., Balasubramanian, K.A., Wanke, C.A., Ward, H.D., and Kang, G., (2014). Immune response and intestinal permeability in children with acute gastroenteritis treated with Lactobacillus rhamnosus GG: A randomized, double-blind, placebo-controlled trial. *Clin. Infect. Dis. Off. Publ. Infect. Dis. Soc. Am.*, 58, 1107–1115. https://doi.org/10.1093/cid/ciu065.

Strong, D.W., Thackray, L.B., Smith, T.J., and Virgin, H.W. (2012). Protruding domain of capsid protein is necessary and sufficient to determine murine norovirus replication and pathogenesis *in vivo*. *J. Virol.*, 86, 2950–2958. https://doi.org/10.1128/JVI.07038-11.

Sun, X., Winglee, K., Gharaibeh, R.Z., Gauthier, J., He, Z., Tripathi, P., Avram, D., Bruner, S., Fodor, A., and Jobin, C. (2018). Microbiota-derived metabolic factors reduce campylobacteriosis in mice. *Gastroenterology*, 154, 1751–1763.e2. https://doi.org/10.1053/j.gastro.2018.01.042.

Teng, J.L.L., Martelli, P., Chan, W.-M., Lee, H.H., Hui, S.-W., Lau, C.C.Y., Tse, H., Yuen, K.-Y., Lau, S.K.P., and Woo, P.C.Y. (2018). Two novel noroviruses and a novel norovirus genogroup in California sea lions. *J. Gen. Virol.*, 99, 777–782. https://doi.org/10.1099/jgv.0.001071.

Thackray, L.B., Wobus, C.E., Chachu, K.A., Liu, B., Alegre, E.R., Henderson, K.S., Kelley, S.T., and Virgin, H.W. (2007). Murine noroviruses comprising a single genogroup exhibit biological diversity despite limited sequence divergence. *J Virol*, 81, 10460–10473. https://doi.org/10.1128/JVI.00783-07.

Turula, H., Bragazzi Cunha, J., Mainou, B.A., Ramakrishnan, S.K., Wilke, C.A., Gonzalez-Hernandez, M.B., Pry, A., Fava, J., Bassis, C.M., Edelman, J., Shah, Y.M., Corthesy, B., Moore, B.B., and Wobus, C.E. (2018). Natural secretory immunoglobulins promote enteric viral infections. *J. Virol.*, https://doi.org/10.1128/JVI.00826-18.

Uchiyama, R., Chassaing, B., Zhang, B., and Gewirtz, A.T. (2014). Antibiotic treatment suppresses rotavirus infection and enhances specific humoral immunity. *J. Infect. Dis.*, 210, 171–182. https://doi.org/10.1093/infdis/jiu037.

Vijay-Kumar, M., and Gewirtz, A.T. (2009). Flagellin: Key target of mucosal innate immunity. *Mucosal Immunol.* 2, 197–205. https://doi.org/10.1038/mi.2009.9.

Vijay-Kumar, M., Aitken, J.D., Sanders, C.J., Frias, A., Sloane, V.M., Xu, J., Neish, A.S., Rojas, M., and Gewirtz, A.T. (2008). Flagellin treatment protects against chemicals, bacteria, viruses, and radiation. *J. Immunol.*, 180, 8280–8285. https://doi.org/10.4049/jimmunol.180.12.8280.

Vinjé, J., Estes, M.K., Esteves, P., Green, K.Y., Katayama, K., Knowles, N.J., L'Homme, Y., Martella, V., Vennema, H., and White, P.A., Ictv Report Consortium, null (2019). ICTV virus taxonomy profile: Caliciviridae. *J. Gen. Virol.*, 100, 1469–1470. https://doi.org/10.1099/jgv.0.001332.

Wang, L., Zhu, L., and Qin, S. (2019). Gut microbiota modulation on intestinal mucosal adaptive immunity. *J. Immunol. Res.*, 2019, 4735040. https://doi.org/10.1155/2019/4735040.

Wasimuddin, null, Brändel, S.D., Tschapka, M., Page, R., Rasche, A., Corman, V.M., Drosten, C., and Sommer, S. (2018). Astrovirus infections induce age-dependent dysbiosis in gut microbiomes of bats. *ISME J.*, 12, 2883–2893. https://doi.org/10.1038/s41396-018-0239-1.

Wells, J.E., and Hylemon, P.B. (2000). Identification and characterization of a bile acid 7α-dehydroxylation operon in *Clostridium* sp. strain TO-931, a highly active 7α-dehydroxylating strain isolated from human feces. *Appl. Environ. Microbiol.*, 66, 1107–1113. https://doi.org/10.1128/AEM.66.3.1107-1113.2000.

Wilen, C.B., Lee, S., Hsieh, L.L., Orchard, R.C., Desai, C., Hykes, B.L., McAllaster, M.R., Balce, D.R., Feehley, T., Brestoff, J.R., Hickey, C.A., Yokoyama, C.C., Wang, Y.-T., MacDuff, D.A., Kreamalmayer, D., Howitt, M.R., Neil, J.A., Cadwell, K., Allen, P.M., Handley, S.A., van Lookeren Campagne, M., Baldridge, M.T., and Virgin, H.W. (2018). Tropism for tuft cells determines immune promotion of norovirus pathogenesis. *Science*, 360, 204–208. https://doi.org/10.1126/science.aar3799.

Wilks, J., Lien, E., Jacobson, A.N., Fischbach, M.A., Qureshi, N., Chervonsky, A.V., and Golovkina, T.V. (2015). Mammalian lipopolysaccharide receptors incorporated into the retroviral envelope augment virus transmission. *Cell Host Microbe*, 18, 456–462. https://doi.org/10.1016/j.chom.2015.09.005.

Wobus, C.E., Karst, S.M., Thackray, L.B., Chang, K.-O., Sosnovtsev, S.V., Belliot, G., Krug, A., Mackenzie, J.M., Green, K.Y., and Virgin, H.W. (2004). Replication of norovirus in cell culture reveals a tropism for dendritic cells and macrophages. *PLoS Biol.*, 2, e432. https://doi.org/10.1371/journal.pbio.0020432.

Wobus, C.E., Thackray, L.B., and Virgin, H.W. (2006). Murine norovirus: A model system to study norovirus biology and pathogenesis. *J Virol*, 80, 5104–5112. https://doi.org/10.1128/JVI.02346-05.

Yokoyama, C.C., Loh, J., Zhao, G., Stappenbeck, T.S., Wang, D., Huang, H.V., Virgin, H.W., and Thackray, L.B. (2012). Adaptive immunity restricts replication of novel murine astroviruses. *J. Virol.*, 86, 12262–12270. https://doi.org/10.1128/JVI.02018-12.

Zhang, Y., Kast-Woelbern, H.R., and Edwards, P.A. (2003). Natural structural variants of the nuclear receptor farnesoid X receptor affect transcriptional activation. *J. Biol. Chem.*, 278, 104–110. https://doi.org/10.1074/jbc.M209505200.

Zhang, B., Chassaing, B., Shi, Z., Uchiyama, R., Zhang, Z., Denning, T.L., Crawford, S.E., Pruijssers, A.J., Iskarpatyoti, J.A., Estes, M.K., Dermody, T.S., Ouyang, W., Williams, I.R., Vijay-Kumar, M., and Gewirtz, A.T. (2014). Prevention and cure of rotavirus infection via TLR5/NLRC4–mediated production of IL-22 and IL-18. *Science*, 346, 861–865. https://doi.org/10.1126/science.1256999.

Zhang, Z., Zou, J., Shi, Z., Zhang, B., Etienne-Mesmin, L., Chassaing, B., and Gewirtz, A.T. (2019). IL-22 induced cell extrusion and IL-18-induced pyroptosis prevent and cure rotavirus infection. *bioRxiv*, 774984. https://doi.org/10.1101/774984.

Zheng, D.-P., Ando, T., Fankhauser, R.L., Beard, R.S., Glass, R.I., and Monroe, S.S. (2006). Norovirus classification and proposed strain nomenclature. *Virology*, 346, 312–323. https://doi.org/10.1016/j.virol.2005.11.015.

7

Plant–Virus–Vector Interactions

Swapna Priya Rajarapu[1], Diane E. Ullman[2], Marilyne Uzest[3], Dorith Rotenberg[1], Norma A. Ordaz[4] and Anna E. Whitfield[1]

[1] *Department of Entomology and Plant Pathology, North Carolina State University, Raleigh, USA*

[2] *Department of Entomology and Nematology, University of California, Davis, USA*

[3] *BGPI, University of Montpellier, INRA, CIRAD, Montpellier SupAgro, France*

[4] *Department of Plant Pathology, University of California, Davis, USA*

7.1. Introduction

Transmission of plant viruses by their arthropod vectors is mediated by specific interactions between the plant, virus and vector. The determinants underlying evolution of the known vector transmission strategies include: location of transmissible virus within the vector arthropod (non-circulative vs. circulative); duration of the vector–virus relationship (non-persistent, semi-persistent and persistent); and replication status of the virus in the insect (non-propagative vs. propagative). Most plant viruses are transmitted horizontally by their arthropod vectors, although reoviruses, rhabdoviruses and tenuiviruses may also be vertically transmitted. This chapter highlights recent progress and longstanding questions in vector biology. The viral and arthropod determinants of vector competence for viruses, transmitted in non-circulative and circulative transmission modes, have revealed that even viruses transmitted via diverse transmission modes may interact with similar insect proteins during vector transmission. For example, insect cuticular proteins have been shown to be receptors for a non-circulative virus, as well as interacting with circulative viruses. Bacterial endosymbionts of vector insects play a role in vertical

For a color version of all figures in this book, see www.iste.co.uk/saleh/virology.zip.

Virology,
coordinated by María-Carla SALEH and Félix AUGUSTO REY. © ISTE Ltd 2021

transmission of plant reoviruses, presenting the possibility that this may be a general method of virus transport into eggs for this group of viruses. During the last decade, development and applications of -omics technologies advanced our understanding of direct and indirect effects of viruses on their vectors; providing new insights into plant virus transmission, and opening up an area of study previously reserved only for model organisms. Other pivotal work in the field has focused on understanding the role of vector feeding on virus transmission and the multitrophic interactions between viruses, vectors and plants, and the indirect effects of pathogens on vector behavior and population dynamics. The basic research discussed in this chapter has practical implications for developing new control strategies for these complex and emerging vector-borne pathogens.

The field of plant vector biology has a long history of impactful research by women scientists. From Marion Watson's seminal research describing the transmission parameters that are still in use today, to the recent work highlighted in this chapter, the interdisciplinary nature of vector biology has provided an environment that promotes the integration of diverse ideas and lateral thinking (Watson *et al.* 1939). Plant vector biologists will undoubtedly continue to embrace new ideas and technologies, and diverse perspectives, to fully address these complex pathosystems that threaten global food security.

7.2. Non-circulative virus transmission

Non-circulative virus transmission refers to viruses that are transiently and reversibly retained on the cuticle which lines the inner surface of the vector mouthparts (Figure 7.1) (Kennedy *et al.* 1962; Harris 1977). Retention within the vector is generally brief, and viruses are lost upon molting. The time periods required for virus acquisition and inoculation are short, with no latent period prior to virus inoculation. Vectors that have just acquired viruses are immediately capable of inoculating them in a new host plant. Historically, these viruses were classified according to the duration of acquisition and retention within the vector. Two categories were distinguished: non-persistent viruses acquired efficiently within seconds; semi-persistent viruses acquired within minutes to hours (Watson *et al.* 1939; Day *et al.* 1954; Sylvester 1956); and viruses being inoculated all within seconds or minutes. Semi-persistent viruses are mostly phloem-limited, while non-persistent viruses are not and are also present in epidermal, mesophyll or parenchyma cells. Semi-persistent viruses were initially assumed to be retained in the foregut of their vectors, and non-persistent viruses in the stylets. Although this is generally the case, an exception to this rule is the cauliflower mosaic virus (CaMV), classified as semi-persistent (Markham *et al.* 1987) and shown to bind receptors at the tip of aphid stylets (Uzest *et al.* 2007). The two terminologies based on the type of virus–vector interaction and on the duration of the transmission phases co-exist in the literature (Harris 1977; Kennedy *et al.* 1962).

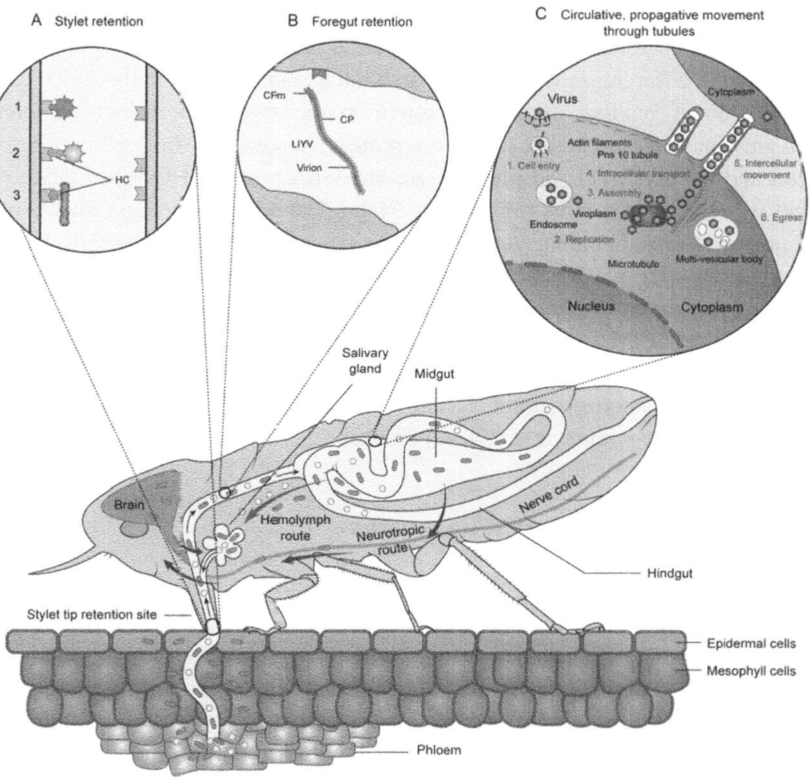

Figure 7.1. *Virus localization sites in insect vectors. Reprinted from Virology, 479–480, Whitfield, A.E., Rotenberg, D., and Falk, B.W.,* Insect vector-mediated transmission of plant viruses, *278–89, 2015, with permission from Elsevier*

Note on Figure 7.1.– Non-circulative viruses are retained in the insect stylet (A) or foregut (B). Non-propagative circulative (yellow circles) viruses are generally phloem limited and penetrate the insect body via the midgut or hindgut. Circulative viruses use a hemolymph route to reach the salivary glands. In contrast, circulative, propagative viruses (red ovals) enter the insect at the anterior region of the midgut and/or filter chamber region. Propagative viruses replicate in the midgut cells and other insect tissues. Some propagative viruses are phloem-limited while others are widely distributed in plant tissues. The salivary glands are the final destination for circulative transmission, and viruses reach the salivary glands via the hemolymph or other routes, such as the nervous tissue (neurotropic route) or through connective tissues. Reoviruses use tubules to move cell-to-cell in the midgut and to traverse the basal lamina (C). Inset A: Magnification of an insect stylet showing the proposed site of virion attachment at the tip of the stylet in the common duct region. Numbers designate the different strategies for virion binding and retention in the stylet: capsid strategy, direct binding of capsid protein to the stylet (1), helper component strategies for caulimoviruses, two virus proteins serve as a "bridge" between the virion and the stylet (2) and potyviruses, one virus protein (HC-Pro) binds to the aphid stylet and to the virus (3). Inset B: Magnification of the foregut retention site and proposed capsid binding strategy used by criniviruses. The minor capsid protein (CPm) is the viral attachment protein. Inset C: The steps in the reovirus infection cycle and the spread to adjacent cells modeled on rice dwarf virus. Rice dwarf virus enters cells using the endocytic pathway and after virion release from the vesicle, the replication cycle begins. Progeny virions move cell-to-cell via tubule structures composed of virus non-structural protein. This enables virions to move directly from one cell to another without an extracellular phase.

7.2.1. *Vectors of non-circulative viruses*

Viruses infecting all tissues can be acquired in the early phases of vector feeding, during short probing events, sometimes within a single puncture. This mode of transmission is particularly fast and efficient, and it is widespread among phytoviruses. Non-circulative viruses are generally transmitted by insects with piercing-sucking mouthparts, mostly by aphids, and to a lesser extent by whiteflies, mealybugs and leafhoppers (Gray and Banerjee 1999; Hogenhout *et al.* 2008; Stafford *et al.* 2012). Aphid mouthparts and feeding habits seem to fit perfectly with optimized uptake and transport of viruses. Their stylets form long, thin needles. They are steerable, highly flexible, but strong, and can bend in between adjacent cells to navigate intercellularly towards the phloem while causing little damage to the plant (Pollard 1969, 1973; Gray and Banerjee 1999; Cerkvenik *et al.* 2018). In addition, aphids rapidly evaluate whether the plant is suitable or not when they land on it, determining if they can remain and feed for longer periods. During the first contact, they will insert their stylets within the superficial tissues and will probe the plant for suitability by puncturing and partly ingesting cell contents via their stylets (Powell 2005; Fereres and Moreno 2009). If the plant is accepted, aphids continue inserting their stylets intercellularly, while making small intracellular punctures along the path to the sieve elements. If the plant is not acceptable, then aphids leave very quickly, sometimes after a single puncture of 5–10 seconds, which is enough to efficiently acquire or efficiently inoculate non-circulative viruses (Martin *et al.* 1997; Powell 2005;

Fereres and Moreno 2009). Aphids make these short probes irrespective of their host, thereby contributing to efficient viral spread (McLean and Kinsey 1968; Powell *et al.* 1999). Moreover, short probes lasting less than one minute are frequent for aphids and are ideal feeding behaviors for transmission of non-circulative, stylet-borne viruses (Calatayud *et al.* 1994; Johnson and Walker 1999).

Intercellular probing with intracellular puncturing is not restricted to aphids and has been reported in whiteflies and mealybugs (Calatayud *et al.* 1994; Johnson and Walker 1999). The rarity of non-persistent transmission by these insects is thought to be related to their propensity to produce generally fewer intracellular punctures than aphids and not in the early phases of the feeding process, as well as their lack of brief short probes that are common to aphids.

7.2.2. *Virus–vector interactions are highly specific*

Interactions between non-circulative viruses and their vectors are highly specific, and the mechanisms underlying virus transmission are remarkably sophisticated. For example, different aphid species transmit the same virus species with the same or different efficiencies, and a given vector species can transmit some non-circulative virus species and not others. Moreover, on the virus side, minor changes in the viral determinant involved in transmission can lead to altered vector transmission phenotypes (Perry *et al.* 1994, 1998; Blanc *et al.* 1998; Moreno *et al.* 2005). This is perfectly illustrated with the cauliflower mosaic virus (CaMV). A single amino acid substitution at position 6 of the viral protein P2 affects CaMV transmission of aphid vector species differently. Transmission may be unchanged, reduced or abolished, depending on the substitution residue (Moreno *et al.* 2005). The most stringent phenotype was obtained when the glutamine residue was substituted by a Tyrosine (Q6Y mutant). This mutant lost its capacity to bind aphid stylets, which explains why it was no longer aphid-transmissible and is therefore a useful negative control in the quest for identifying the virus receptor (Moreno *et al.* 2005; Uzest *et al.* 2007).

The use of viral strains with altered transmission properties, the development of genetic approaches to generate mutants or reassortant viruses, and the establishment of insect feeding on artificial diets were instrumental in characterizing the viral determinants involved in transmission (Megahed and Pirone 1966; Pirone and Megahed 1966; Kaper and Geelen 1971; Habili and Francki 1974). All of these studies demonstrated the existence of two distinct strategies adopted by viruses to interact with their vectors, the capsid and the helper strategy, depending on the molecule used to mediate virus retention in the vector, either the coat protein (CP), or a non-structural viral protein called a "helper", used as a "bridge" between insect receptors and the capsid (Pirone and Blanc 1996).

7.2.3. *Capsid strategy*

The best-characterized virus transmitted using the capsid strategy is the type member of the genus *Cucumovirus*, *Cucumber mosaic virus* (CMV), also renowned for its wide host range. It has been described as having infected more than 1,200 plant species and is transmitted by more than 80 aphid species (Scholthof *et al.* 2011). Many molecular studies have been developed to determine the CP motifs directly involved in vector interaction, and several residues – in particular 25, 129, 162, 168 and 214 – were initially suspected to play a direct role in virus retention (Perry *et al.* 1994; Perry *et al.* 1998; Ng and Perry 1999; Ng *et al.* 2000;). However, the three-dimensional structure of CMV particles revealed that these residues were poorly accessible at the surface of the capsid (Smith *et al.* 2000), and rather involved in virion stability (Ng *et al.* 2000; Perry 2001). From the 3D structure, two other domains were investigated by mutational analyses and shown to be involved in virus transmission: the metal-ion binding ßH-ßI loop (amino acids 191-198) conserved in cucumoviruses and exposed at the surface of the capsid; and internal residues involved in stabilizing salt bridges at the quasi-threefold axis of symmetry (Liu *et al.* 2002; Bricault and Perry 2013). Mutations of these domains impacted transmission. Changing the charges at the surface of the capsid in the ßH-ßI loop resulted in disruption of CMV–aphid interaction. Bricault and Perry also showed that amino acids, distal to the surface and present in a pore at the quasi-threefold axis involved in acid–base pair interactions, were crucial to maintain CMV aphid-transmissibility (Bricault and Perry 2013).

7.2.4. *Helper strategy*

The need for a helper protein mandatory for vector transmission has been demonstrated for several viruses. Purified particles acquired through membrane lost their ability to be transmitted by insect vectors (Pirone and Megahed 1966; Kassanis and Govier 1971a; Lung and Pirone 1973; Govier and Kassanis 1974; Lung and Pirone 1974). This transmission strategy has been intensively characterized for potyviruses and caulimoviruses (Figure 7.1).

The helper of potyviruses is HC-Pro, a multi-functional protein of 50 kDa (recently reviewed in Valli *et al.* 2017). Two conserved motifs, each one involved in CP or vector interactions, were identified in helix-rich domains at the N- and C-terminus of the protein. In the N-terminal domain, the motif KITC likely interacts with receptors in aphid stylets. Hence, it was shown that mutations in this motif disrupted aphid transmission without affecting virus–helper interactions (Blanc *et al.* 1997). In the C-terminal domain, the amino acid triad PTK was shown to interact with the consensus DAG motif of the CP (Huet *et al.* 1994; Peng *et al.* 1998). The active form of HC-Pro within the transmissible complex involves the multimerization of the protein through disulfide bridges (Ruiz-Ferrer *et al.* 2005; Berthelot *et al.* 2019a). The 3D structures of several potyviruses have recently been released in the literature (Zamora *et al.* 2017; Cuesta *et al.* 2019; Kežar *et al.* 2019), as has the protease domain of HC-pro (Guo *et al.* 2011). Unfortunately, the high-resolution structure of the entire protein, which would make it possible to fully characterize the formation of the transmissible complex and decipher virus–vector interactions, is still lacking.

CaMV has been extensively studied, from the early sophisticated events occurring at the plant–cell level for optimized virus acquisition by aphids, to virus–vector interactions per se and the characterization of CaMV receptors in aphid stylets. The CaMV transmissible complex involves three protein partners: the coat protein and the viral proteins P2 and P3 (Leh 1999). P2 is the helper protein and interacts with receptors in aphid stylets and with the virus particle (Blanc *et al.* 1993; Hebrard *et al.* 2001). Its sole known function is to ensure viral transmission by aphids (Howarth *et al.* 1981). Another distinctive feature when compared to HC-Pro is that P2 does not directly bind to the CP but instead interacts via its C-terminal alpha-helix domain with the viral protein P3 that decorates the surface of the capsid (Leh 1999; Leh *et al.* 2001; Plisson *et al.* 2005). Structural analyses indicated that P3 proteins are tightly associated with the icosahedral virion. The P3 C-terminal domains are anchored in the viral capsid, whereas their N-terminal ectodomains form a regular network of anti-parallel coiled-coils at the surface (Plisson *et al.* 2005). In addition to its interaction with the helper protein, P2, that mediates retention of the whole complex in aphid stylets and the long distance inter-host spread, P3 also interacts with the N-terminal coiled-coil domain of the movement protein P1, thereby promoting intra-plant movement of viruses and systemic infection (Stavolone *et al.* 2005). The protein P3 exists in different conformations, depending on its cellular localization and its partners (Leclerc *et al.* 1998; Plisson *et al.* 2005; Hoh *et al.* 2010). The 3D structure of the N-terminal domain of unliganded P3 revealed a unique tetramer organization with two subdomains, separated by the amino acid triad Cys-Pro-Cys and forming a CPC interchain disulfide ring. This CPC motif is thought to interact as a redox sensor that could trigger the different conformational shifts of P3 in local environmental changes, thereby regulating interactions with various partners during the virus lifecycle (Hoh *et al.* 2010). This versatility may be particularly important during the transmission process for reconstituting whole transmissible complexes, the three viral partners mostly separated in different compartments until the arrival of the aphid vector (Drucker *et al.* 2002). Connected to the regulation of dynamic formation of transmissible complexes, a novel concept emerged for non-circulative plant viruses, the possibility that different virus morphs exist, some dedicated to replication or storage, others to aphid transmission. This concept was proposed by Martinière and colleagues who have shown that viruses could "perceive" the feeding activities of the vector, and immediately produce active transmissible morphs leading to optimal virus uptake and efficient aphid transmission (Bak *et al.* 2010; Martinière *et al.* 2013). This sophisticated phenomenon, known as Transmission Activation or TA, is discussed below (Drucker and Then 2015).

7.3. Circulative virus transmission

Circulative virus transmission is characterized by viral entry into the insect body at a specific region of the alimentary canal, following acquisition during insect feeding (Figure 7.1). After this initial entry step, the virus is able to traverse membrane and basal lamina barriers to circulate through the insect body and eventually reach salivary glands. The site of virus binding and entry into gut and salivary glands is specific to the virus–vector combination. Viruses have been shown to bind to the filter chamber, midgut and hindgut of the insect and associate with specialized salivary tissues of their respective vectors. Because these viruses circulate through the insect body, they have a longer interaction duration with the vector and have also been described as persistent viruses. In general, the acquisition access and inoculation access periods for these viruses are longer than those for the non-circulative viruses. Viruses are passaged between insect life stages, termed trans-stadial passages, and circulative viruses are retained in the vector for weeks, up to the lifetime of the insect. Circulative virus transmission also has a latent period, which is the length of time required for the virus to be acquired and to reach the salivary gland tissues for delivery into a new plant host. The latent period is analogous to the animal arbovirus term, "extrinsic incubation period". The latent period for circulative viruses varies greatly and one contributing factor to this variation is that some viruses replicate in their insect vectors. Thus, the circulative-persistent viruses are further subdivided into propagative vs. non-propagative viruses. Interestingly, all the phytoviruses that replicate in their arthropod vectors are classified in virus orders and families that also include animal-infecting viruses, and it appears that these plant-infecting viruses evolved from arthropod viruses (Junglen 2016; Whitfield *et al.* 2018; Q. Chen *et al.* 2019). For the propagative viruses, there does appear to be a minimum dose of virus required for virus infection of the vector and greater acquisition efficiency is correlated with high virus titer in the plant and longer duration of acquisition feeding (Rotenberg *et al.* 2009).

7.3.1. *Vectors of circulative viruses*

There is significant overlap between the insect vectors that transmit both circulative and non-circulative viruses. The majority of vectors are insects in the order Hemiptera, and other vectors include Thysanopterans and Arachnids (mites). In contrast to transmission parameters that facilitate non-circulative transmission, extended feeding periods for acquisition and inoculation generally improve the efficiency of circulative virus transmission (reviewed in Stafford *et al.* 2012) . Many non-propagative, circulative viruses are phloem-limited viruses and thus the minimum time for efficient acquisition and inoculation of the virus varies from minutes to hours, in conjunction with the time required for the vector to reach the phloem. The luteoviruses, nanoviruses and geminiviruses are transmitted in the circulative, non-propagative manner by aphids, whiteflies, leafhoppers and treehoppers. The propagative viruses, rhabdoviruses, bunyaviruses (tospoviruses, figmoviruses and phenuiviruses) and reoviruses are transmitted by aphids, leafhoppers, planthoppers, thrips, whiteflies and mites. Insect feeding behaviors play a critical role in virus transmission. As described earlier, whiteflies and aphids have short intracellular test feeding probes and if they find that a plant is acceptable, they initiate long feeding probes following an intercellular route, until they reach the phloem tissues. In contrast, leafhoppers and planthoppers will probe the plants intracellularly and then continue this more destructive pathway to the vasculature, i.e. phloem and xylem tissues. Thrips vectors are also piercing-sucking insects and they ingest primarily from epidermal and mesophyll cells (reviewed in Montero-Astua *et al.* 2016a). Thrips display three different feeding probes:

a) non-ingestion probes that include salivation into plant cells and these probes are hypothesized to be the primary feeding behavior for virus inoculation because the majority of cells remain intact;

b) short ingestion probes that will empty a cell of its contents;

c) long ingestion typified by long bouts of feeding and visible cibarial pumping (Stafford *et al.* 2011).

The eriophyid mites and *Breviplapus* “flat” mites also transmit circulative viruses and are extremely small (200 – 400 μM in length). Because of their small size, these arthropods feed from epidermal and mesophyll cells. In general, circulative viruses have a more specific relationship with their vectors because plants that are effective acquisition and inoculation hosts for the virus are also hosts of arthropod vectors.

7.4. Receptors in vectors of non-circulative viruses

When acquired by vectors, non-circulative viruses will bind to receptors present at the surface of the cuticle of the mouthparts (Figure 7.1(A) and 7.1(B)). These receptors have long been of interest because they represent targets for alternative control strategies to limit viral spread.

7.4.1. *Receptors in aphid stylets*

The aphid stylet bundle is composed of two mandibular stylets and two inner maxillary stylets containing the food and the salivary canals (Forbes 1969). At the distal tip, the two canals merge into a common one that was long believed to host viral receptors of potyviruses and cucumoviruses (Martin *et al.* 1997; Powell 2005). The development of a novel approach based on *in vitro* detection of CaMV helper protein at the surface of dissected stylets definitely evidenced the presence of such receptors in the common canal. The viral protein P2 fused to green fluorescent protein (GFP) and produced in *sf9* cells bound exclusively the surface of the cuticle in a tiny area at the tip of maxillary stylets of vector species, whereas the Q6Y mutant deficient for virus transmission did not (Uzest *et al.* 2007). A significant step towards the identification of CaMV receptors has been the discovery of an organ emerging at the surface of the cuticle of the common canal, named the acrostyle (Uzest *et al.* 2010). This organ has unique surface properties, distinct from the rest of the cuticle of aphid stylets. It displays cuticular proteins all over its surface, some of which interact directly with P2 (Uzest *et al.* 2010). An immunolabeling systematic approach was developed to identify proteins in the acrostyle, and a series of antibodies specific to annotated cuticular proteins from the CPR family were produced and incubated on dissected stylets. The CPR family is the largest cuticular protein family in aphids, containing the R&R consensus sequence described by Rebers and Riddiford (Willis 2010; Mathers *et al.* 2017). Several peptides from the RR-1 and RR-2 subgroups were detected in the acrostyle, and two RR-1 proteins were formally identified (Uzest *et al.* 2010; Webster *et al.* 2017, 2018). These two proteins, named Stylin-01 and Stylin-02, have a common C-terminal domain shown to be highly accessible at the stylet–solvent interface. Complementary experiments designated Stylin-01 as the best receptor candidate of CaMV. First, P2 and the antibody targeting the peptide at the surface compete for binding to the acrostyle. Secondly, silencing the *stylin-01* gene expression impacted the ability of aphid vector to transmit CaMV (Webster *et al.* 2018).

The acrostyle is a unique distinctive physical structure in the common duct. It may also bear the receptors of other non-circulative viruses. For example, Stylin-01, previously named MpCp4 in the aphid species *Myzus persicae* (Dombrovsky *et al.* 2007), interacts with the CP of CMV in a yeast two-hybrid system (Liang and Gao 2017). Binding of CMV particles on the acrostyle, or involvement of Stylin-01 in CMV transmission, has not yet been confirmed and will require further investigations. Recently, CRISPR-Cas9 technology has been successfully developed in aphids, and the first *Stylin-01* knock out mutant lineages have been generated (Le Trionnaire *et al.* 2019). This promising tool will be useful to fully understand the role of Stylin-01 in non-circulative virus transmission and to characterize its intrinsic function in aphids.

7.4.2. *Receptors in vector foreguts*

The mechanisms of transmission of foregut-borne viruses are not well described and no receptor candidates have been identified so far. Most of the knowledge of phloem-restricted plant viruses that are semi-persistently transmitted came from studies of lettuce yellows virus (LIYV) (Figure 7.1(B)). The LIYV is the type member of *Crinivirus* and is transmitted by whiteflies. It has a long flexuous rod shape, composed of four viral proteins and does not require any helper for transmission (Tian *et al.* 1999; Ng 2004). Using artificial diet feeding, and an immunofluorescence localization system, Chen *et al.* have shown that LIYV particles were specifically retained in the anterior foregut or cibarium of their whitefly vector, *Bemisia tabaci* biotype A. This retention site was directly correlated with effective transmission. Furthermore, using biochemical and genetic approaches, the authors provided evidence that the minor coat protein CPm, shown to be essential for virus transmission (Tian *et al.* 1999; Stewart *et al.* 2010), mediates LIYV binding to its vector (Chen *et al.* 2011). Since then, this approach has allowed the detection of other criniviruses in the foregut or cibarium of their insect vector, the cucurbit chlorotic yellows virus in the whitefly *B. tabaci* Mediterranean (MED) biotype, and the Citrus tristeza virus in the cibarium of the brown citrus aphid *Toxoptera citricida* (Killiny *et al.* 2016; Li *et al.* 2016).

7.5. Receptors in vectors of circulative viruses

7.5.1. *Circulative virus binding and transcytosis*

The non-propagative circulative viruses have small icosahedral particles composed of a single capsid protein, or in the case of the luteoviruses, also incorporate the capsid + readthrough domain (RTD) in some of the structural proteins (reviewed in Wilson *et al.* 2019). The capsid proteins are the major determinants of vector transmission specificity and this has been determined with capsid swapping experiments and deletion/mutation analysis (Briddon *et al.* 1990; Chay *et al.* 1996; Brault *et al.* 2000, 2003, 2005; Kaplan *et al.* 2007; Peter *et al.* 2008). Virus particles are ingested during feeding on plant tissue, and for the geminiviruses the viruses enter the vector midgut at the anterior region of the whitefly and leafhopper midgut and the filter chamber region (Medina *et al.* 2006; Ammar *et al.* 2009; Wang *et al.* 2014).

The luteoviruses have a virus–vector species-specific entry site at either the midgut or hindgut (Gildow 1985; Reinbold *et al.* 2003; reviewed in Gray and Gildow 2003) and the RTD protein is thought to mediate vector specificity in the salivary glands. The nanoviruses may have a helper protein (N) that is not a structural protein but it is required for virus transmission by the aphid vector (Franz *et al.* 1999; Grigoras *et al.* 2018). Nanoviruses cross into the aphid body through the anterior midgut (Bressan and Wantanabe 2011; Wantanabe and Bressan 2013). After transcytosis through the vector gut epithelial cells, viruses encounter the basal lamina and the mechanism of transport through this structure is not well described. The circulative, non-propagative viruses can be injected into the hemolymph, and transmitted efficiently, suggesting this is a site that they naturally encounter in their dissemination pathway. The virus particles eventually accumulate in salivary gland tissues and from there they are transmitted with saliva into a new plant host.

7.5.2. *Circulative virus receptors*

The small size of insect vectors and lack of sequence resources hampered identification of receptors in plant virus vectors until recently. The first aphid receptor for a luteovirid was identified to be alanyl aminopeptidase N (APN), a protein that is present in the insect gut (Linz *et al.* 2015). The APN protein was identified to bind pea enation mosaic virus (PEMV) particles in far-Western (overlay) assays and the interaction between the viral structural protein and APN was confirmed using multiple methodologies (Linz *et al.* 2015). In gain-of-function experiments, expression of the putative receptor, APN, in *Sf9* cells enabled PEMV entry into this refractory cell line, thus demonstrating the functional role of APN as a virus receptor. This work was conducted with the model hemipteran, the pea aphid, which was notably the first hemipteran genome sequenced. Using another approach with a related luteovirid-aphid system, Mulot *et al.* (2018) identified that the *Myzus persicae* ephrin receptor interacted with the turnip yellows virus coat protein in yeast two-hybrid assays. Using RNAi to knock-down ephrin receptor expression, they observed a reduction in virus acquisition and transmission efficiency, providing compelling evidence of a role for ephrin receptor as a virus receptor. Additional vector proteins have been identified as interactors with viral capsid proteins and, notably, work with aphid and whitefly vectors identified a role for insect cyclophilin in the transmission process of viruses from two different non-circulative virus families, the *Geminiviridae* and *Luteoviridae* (Tamborindeguy *et al.* 2013; Kanakala *et al.* 2016, 2019). For additional detail on the circulative virus interactions with vector proteins, please refer to the recent review by Wilson *et al.* (2019).

7.6. Circulative, propagative virus binding and entry

7.6.1. *Circulative, propagative viruses binding and entry*

For circulative, propagative viruses, viral surface proteins serve as attachment proteins that enable virus binding and subsequent fusion with vector host cells (reviewed in Whitfield *et al.* 2015; Chen *et al.* 2019). In contrast to the non-propagative, circulative viruses that transcytose through gut epithelial cells, the propagative viruses uncoat in the gut epithelial cells and replicate. Progeny virus particles move cell-to-cell in the gut and are then released from the cell basal membrane. Among the propagative viruses, the dissemination pathway in the vector varies, depending on the virus family and specific arthropod vector. Some propagative viruses, i.e. reoviruses, have specialized tubule structures that facilitate release from the gut cells and may be the mechanism for movement through the basal lamina (Figure 7.1) (Katayama *et al.* 2007; Jia *et al.* 2014; Wei and Li 2016).

Orthotospovirus have been observed in tubular salivary glands, specialized structures in thrips that are bound to the midgut, thus providing a connection between the gut and the principal salivary glands (Ullman *et al.* 1989; Montero-Astua *et al.* 2016). Rhabdoviruses infect many different tissues in the vector, including the hemocytes and neural system, suggesting these viruses may use a hemolymph or neurotropic route for dissemination in vectors (Ammar and Hogenhout 2008). For successful inoculation of a new plant, the viruses must reach the salivary glands; in the salivary tissues the virus must again traverse the basal lamina and basal membrane to reach the salivary gland cells for another round of replication and spread within this tissue. Rhabdoviruses, orthotospoviruses and reoviruses have been observed accumulating at the luminal surface of salivary gland cells, the site for deposition into salivary ducts and delivery to plant cells during salivation (Ammar and Hogenhout 2008, Montero-Astua *et al.* 2016b, Wei and Li 2016).

To understand the viral determinants of entry into gut cells of insect vectors, we will focus on a few of the well-characterized virus–vector systems that are of economic and food security importance. The model tospovirus, tomato spotted wilt virus (TSWV), is a typical bunyavirus with two surface glycoproteins (Gn, Gc). Experiments examining Gn and Gc function have revealed that Gn is the viral attachment protein that binds to the thrips vector midgut (Whitfield *et al.* 2004, 2008 Montero-Astúa *et al.* 2014). Comparisons with animal-infecting bunyaviruses suggest that Gc is a fusion protein and in accordance with that hypothesis, Gc does undergo a conformational change a low pH (Garry and Garry 2004; Whitfield *et al.* 2005). Because Gn is involved in the initial binding step during entry to thrips guts, excess Gn, in a purified form and when expressed in transgenic plants, can disrupt virus entry and results in a significant decrease in thrips transmission efficiency (Whitfield *et al.* 2008; Montero-Astúa *et al.* 2014). The role of the TSWV glycoproteins in virus transmission by the thrips vector is undisputed. However, the related tenuiviruses that are fellow members of the order *Bunyavirales* have an intriguing mechanism of transmission, recently suggested to include a role for the viral-encoded glycoproteins. The tenuiviruses encode two glycoproteins like the tospoviruses, but they do not form enveloped virus particles. The four genome segments are encapsidated in nucleoprotein (N) and are thought to constitute the viral particle, and several nucleocapsid protein (N)-vector proteins have been defined (reviewed in Liu *et al.* 2018). Surprisingly, Lu *et al.* (2019) found that the two GPs of the tenuivirus, rice stripe virus (RSV), serve analogous functions in the insect vector as other bunyavirus GPs, and interact with the RNPs to promote virus entry into the vector. The Gn homolog, NSvc2-N, binds to planthopper vector midguts and serves as a viral attachment protein (Lu *et al.* 2019). The Gc homolog, NSvc2-C, mediates membrane fusion and both GPs interact with the N protein. Based on this evidence, Lu *et al.* (2019) hypothesized that the RSV GPs interact with the nucleocapsid/viral particles and serve as the molecular bridge/helper component for virus acquisition by the planthopper vector (Lu *et al.* 2019). Despite dramatic differences in particle structure for orthotospoviruses and tenuiviruses, these findings demonstrate functional conservation of the viral GPs.

Plant viruses in the family *Reovirus* have been grouped into three genera: *Phytoreovirus*, *Fijivirus* and *Oryzavirus*. These viruses have icosahedral particles with a complex structure that encapsidates 9-12 dsRNA genome segments. The availability of cell lines derived from insect vectors has facilitated the study of virus entry for this group of plant viruses (reviewed in Wei and Li 2016). Using insect cell lines, the detailed steps in virus entry and escape from cells have been described and, most notably, the observation of "tubule-like" structures that serve as a conduit for virus intercellular spread (Figure 7.1(C)). Rice dwarf virus (RDV) interactions with leafhopper vectors is one of the best characterized reoviruses. The minor outer capsid protein, P2, forms a spike structure at the 5-fold axis of symmetry on the outer shell of the particle (Nakagawa *et al.* 2003; Miyazaki *et al.* 2015). P2 mediates virus entry into insect cells by serving as a viral attachment and fusion protein and virions enter cells by receptor-mediated, clathrin-dependent endocytosis (Wei *et al.* 2007). After entry and replication in insect cells, RDV particles can be seen tightly packed in rows within tubule structures. The tubule structure is made of Pns10, another viral protein, and tubules form via an interaction with actin and other host actin-binding proteins (Wei *et al.* 2006, 2008). The tubule structures were first observed in cell cultures infected with reoviruses and then their presence and relevance to dissemination in insect vectors was confirmed. Tubules appear to serve as an effective mechanism of transport for virus particles between midgut epithelial cells and as an escape route through the gut basal lamina (Chen *et al.* 2012; Jia *et al.* 2014). The tubule structures have also been documented in vector salivary glands and likely play a role in virus entry and escape in these tissues as well (Wei and Li 2016).

7.6.2. *Receptors in vectors of circulative, propagative viruses*

For the circulative, propagative viruses, numerous insect vector proteins have been identified which interact with viral structural proteins. The virus–vector protein interactions for tenuiviruses and orthotospoviruses are some of the best characterized. Experiments with the RSV vector, the small brown planthopper (SBPH) *Laodelphax striatellus* Fallén (Hemiptera: Delphacidae), have revealed multiple protein interactors with the viral nucleocapsid protein (PC3). An *L. striatellus* cuticular protein, CPR1, was identified as interacting with RSV PC3 and the viral and vector protein colocalized in hemocytes (Liu *et al.* 2015). Intriguingly, Badillo-Vargas *et al.* (2019) characterized the first thrips proteins that directly interact with TSWV Gn protein, and among those interactors, they found three cuticle proteins in larval thrips that interacted with TSWV proteins in gel overlay assays. The TSWV Gn interaction with cuticle proteins provides additional insight into RNA-Seq experiments indicating that cuticle protein transcripts were dysregulated in virus-infected larval thrips (Schneweis *et al.* 2017). The endocucticle glycoprotein, endocuticle glycoprotein-Gn, which interacted with TSWV Gn, also co-localized with the viral protein and has a unique virus-interaction domain that is not found in other organisms. Cuticle proteins of diverse insect vectors have been found to bind directly to viral attachment proteins and are now being recognized as important players in vector competency (reviewed in Deshoux *et al.* 2018).

Other proteins identified in planthoppers and thrips that interact with circulative, propagative viruses may play a role in virus entry and intercellular transport. A candidate RSV receptor, sugar transporter 6, was identified in *L. striatellus* by Qin *et al.* (2018). Sugar transporter 6 is highly expressed in the midgut of *L. striatellus* and localized to the surface of insect cell membranes, the expected site of a receptor (Qin *et al.* 2018). Expression of the *L. striatellus* sugar transporter 6 in a heterologous cell system enabled RSV entry into this non-permissive cell type and knock-down of sugar transporter 6 transcripts in planthoppers reduced virus entry but did not affect virus replication. Intriguingly, this proposed receptor protein interacted with the nucleocapsid protein of RSV and was not tested for interaction with the glycoproteins that may also play a role in tenuivirus entry (Qin *et al.* 2018; Lu *et al.* 2019). Defining the entry mechanisms for tenuiviruses and the role of the "helper" glycoproteins and the nucleocapsid protein will be of significant interest in the near future. Cyclophilin is another protein identified to bind TSWV Gn, which continues to appear in diverse virus–vector protein interaction studies

(Badillo-Vargas *et al.* 2019). In experiments with circulative viruses, geminiviruses and luteovirids, cyclophilin was shown to be important in their respective vectors, whiteflies and aphids (Kanakala and Ghanim 2016; Tamborindeguy *et al.* 2013). Planthopper vectors of viruses appear to have robust RNAi responses that enable functional analysis of virus-interacting proteins (Yao *et al.* 2013). In contrast, functional analysis of proteins in thrips vectors has been limited because RNAi has only been successful in adult thrips, and larval thrips are the only stage at which the insect can acquire the virus (Badillo-Vargas *et al.* 2015, Han *et al.* 2019). Future work examining the conserved and unique virus–vector protein interactions will help us understand the molecular basis of vector competence.

7.6.3. *Vertical transmission of propagative, circulative viruses*

Some of the most exciting research with propagative, circulative viruses has revealed a role for specific vector proteins and symbionts in vertical transmission of plant viruses in their arthropod vectors. Generally, the tenuiviruses and reoviruses are efficiently transmitted from mother to offspring. This transmission strategy has an impact on virus epidemiology and persistence of viruses in nature with the infected eggs serving as an important source of virus inoculum. Research with RSV and *L. striatellus* identified a nucleocapsid (PC3)-vitellogenin interaction and a novel route for viral invasion of insect eggs (Huo *et al.* 2014). Localization experiments with the planthopper revealed that PC3 and vitellogenin colocalize in the ovarial germarium and the viral RNPs enter nurse cells of the germarium, and subsequently move into the oocytes via the nutritive cords (Huo *et al.* 2014; Liu *et al.* 2018). Vitellogenin processed in hemocytes interacts with RSV, leading the authors to hypothesize that this form of vitellogenin protects viral particles in the hemolymph and aids in invasion of the insect eggs (Huo *et al.* 2018). For the plant reovirus, RDV, insect symbionts have been implicated in virus transport into leafhopper eggs (Jia *et al.* 2018). RDV hijacks the movement of leafhopper symbiont, *Sulcia*, into eggs by binding to the envelope of the bacterium (Jia *et al.* 2018). In related experiments, RDV was shown to also be transported by another symbiont, *Nasuia* (Wu *et al.* 2019a). Notably, this mechanism of virus–microbe interaction differed, and it appears that RDV enters the outer envelope of *Nasuia* and is retained in the periplasmic space of the bacterium. It will be interesting to determine if symbiont-mediated transmission is a common vertical transmission pathway for other viruses that replicate in insects. The reovirus, rice gall dwarf virus (RGDV), is efficiently paternally transmitted by the green rice leafhopper, *Recilia dorsalis* (Mao *et al.* 2019). Protein interaction assays revealed that the viral protein, P8, binds to heparan sulfate proteoglycan on the surface of sperm heads. This paternal transmission route was very efficient, with up to 60% of offspring carrying the virus (Mao *et al.* 2019).

7.7. Virus transmission morphs for non-circulative viruses

An intriguing discovery is that plant viruses respond instantly to the presence of the vector and become "activated" for transmission. Within infected plant cells, the helper protein P2 of CaMV is entirely confined in a specific cytoplasmic inclusion: the transmission body (TB) (Martinière *et al.* 2013). This TB also contains P3 and very few virus particles scattered throughout the inclusion. Most of the virus particles are in fact stored in numerous electron-dense inclusion bodies, also known as viral factories (VFs) (Espinoza *et al.* 1991; Drucker *et al.* 2002). When aphids insert their stylets in plant tissues, they induce plant responses (reviewed in Jaouannet *et al.* 2014) that provoke TB disintegration in the cells adjacent to the stylet track. This phenomenon is extremely fast and totally reversible (Martinière *et al.* 2013). It starts with massive entry of tubulin in the TB, immediately followed by relocation of P2 on the microtubule network. Virions also associate with this P2-microtubule network, probably through VFs-TB-microtubule dynamic exchanges (Bak *et al.* 2013). As a consequence, transmissible complexes are distributed homogeneously on the microtubule network at the cell periphery, enhancing the chances of the virus being taken up by aphids during their intracellular random punctures.

First described for CaMV, TA has been recently evidenced for a second virus, the potyvirus turnip mosaic virus (TuMV) (Berthelot *et al.* 2019a). Both viruses seem to deviate early plant defense reactions, derived from rapid calcium signaling pathways or the production of reactive oxygen species (ROS), to their own profit to produce transmission morphs (Berthelot *et al.* 2019b). However, the mechanisms leading to TA are completely different. Hence, HC-Pro and virus particles are dispersed in the cytoplasm of infected cells, and do not relocalize upon stylet penetration (Berthelot *et al.* 2019a; Zheng *et al.* 2011). A change in the HC-Pro oligomerization state is instead observed, with the formation of intermolecular disulfide bridges probably in response to changes in the cellular redox potential. This oligomerization shift is correlated with an increase in virus transmission. Berthelot *et al.* (2019a) proposed that only HC-Pro oxidized forms would be able to associate with viral particles favoring the formation of transmissible complexes, therefore mediating vector-binding. These data illustrate how non-circulative viruses tightly regulate their transmission by producing virus transmission morphs, at the right time, when the vector enters into contact with the infected plant.

7.8. "Omics" tools for studying virus–arthropod interactions

"Omics" tools typically include approaches to studying the genome, transcriptome, proteome and metabolome of an organism and its respective tissues or cells. The study of genetic composition of an organism is termed genomics and is fundamental for studying the evolution of biological processes. On the other hand, functional genomics tools, such as transcriptomics, proteomics and metabolomics, shed light on the genes expressed by the organism in response to biotic or abiotic stress. Improved sensitivity, affordability and flexibility of these techniques have enabled a deeper understanding of ecological interactions of arthropod vectors with their biotic partners, including viruses. Here we focus broadly on outcomes of "omics" tools applied to decipher virus–arthropod interactions.

Genome sequences have been invaluable tools for understanding genetic mechanisms of fundamental processes in organismal biology. Knowledge of arthropod vector genomes paved the way for testing hypotheses centered on the genetics of vector competence (Beerntsen *et al.* 2000; Matthews *et al.* 2018) and provided scaffolds for transcriptome-wide analyses of expression in bodies and tissue systems, in response to infectious agents (Geng *et al.* 2018; Hasegawa *et al.* 2018). While the majority of vector genomic studies have been dominated by arthropod vectors of animal-infecting pathogens, there are well-used, published genomic resources of three hemipteran arthropod vectors of plant viruses, including whiteflies (Chen *et al.* 2016), planthoppers (Xue *et al.* 2014), and aphids (Wenger *et al.* 2017; International Aphid Genomics Consortium 2010). The genome of *F. occidentalis*, the western flower thrips, a vector of a persistently transmitted orthotospoviruses, was sequenced via the i5k genome sequencing project (NCBI Bioproject PRJNA203209, Rotenberg *et al.* 2020) and can be added to the growing list of agricultural insect vector genomes. With the use of contemporary genetic technologies, the genome can also provide a means for studying epigenetic mechanisms of vectoral capacity (Sharakhov and Sharakhova, 2015) and is required for developing genome-editing (CRISPR-Cas9) and transgenesis tools for studying gene function (Kistler *et al.* 2015; Xue *et al.* 2018) or disease vector population suppression (Hammond *et al.* 2016). Such tools have been successfully applied to generate knockout mutants in brown planthopper and pea aphid, a first step towards understanding the transmission mechanisms of plant viruses using genetically modified insects (Xue *et al.* 2018; Le Trionnaire *et al.* 2019).

Global and targeted functional genomics tools, such as RNA interference, transcriptomics and proteomics, are used to elucidate molecular responses of an organism to a given condition and have thus been attractive tools for investigating virus–arthropod interactions. Determining constitutive transcriptional patterns of insect vectors such as *G. nigrifrons*, *F. occidentalis* and *N. lugens*, at both tissue and organismal level, helped in understanding the role of genes involved in the fundamental physiology of the insect vector and also provided molecular resources for the functional characterization of candidate genes, potentially involved in vector competence (Y. Chen *et al.* 2012; Stafford-Banks *et al.* 2014; Zhao *et al.* 2016;

Guschinskaya *et al.* 2020). For example, Chen *et al.* (2015) used the transcriptome to determine the responses of genes involved in innate immunity in *G. nifrons* infected and uninfected with maize fine streak virus and further tested the role of these genes in virus transmission using RNAi. Comparative transcriptomics across several vectors can also shed light on the common molecular features shared by these insect vectors. A comparison of transcriptomes of three hopper species, *Peregrinus maidis*, *Sogatella furcifera* and *G. nigrifrons*, showed a 16% similarity in virus responsive transcripts which mostly encode protein-binding proteins (Martin *et al.* 2017). Along with this, understanding the global transcriptome shifts in infected insect vectors relative to uninfected insects is insightful in generating a hypothesis to test the biological relevance of the differentially expressed genes using functional tools such as RNAi (Whitfield *et al.* 2011; Lee *et al.* 2013; Cassone *et al.* 2014a; 2014b; Wang *et al.* 2016; Kaur *et al.* 2017; Martin *et al.* 2017; Schneweis *et al.* 2017; Shrestha *et al.* 2017; Gamage *et al.* 2018; Geng *et al.* 2018; Hasegawa *et al.* 2018). Trancriptome studies across different insect vectors have consistently shown an upregulation of innate immunity genes when infected with persistent viruses. The immune system of insect vectors, and the effects of virus infection on vector innate immune pathways, is discussed in section 7.9. Launching an immune response against the virus might be crucial for the insect vector to maintain non-pathogenic levels of virus and thus make them efficient vectors. Although immune response was commonly observed across all the transcriptomes, there were some specific responses observed for each vector–virus interaction during the course of infection. For example, during *B. tabaci* and tomato chlorosis virus interaction, there was an upregulation of glucose transporter, glucosidase activity and uric acid pathway genes. A comparison of transcriptomes of *F. occidentalis* to TSWV infection showed downregulation of the majority of transcripts in larvae, relative to propupae and adult stages. Most of the downregulated genes in larvae were annotated as cuticular proteins (Schneweis *et al.* 2017). Transcriptomic differences were also observed within the same stage across different time points in *B. tabaci* and tomato yellow leaf curl virus interactions, wherein only seven transcripts were downregulated 48 hours after infection of adults (Hasegawa *et al.* 2018).

Another advantage of deep sequencing of vector transcriptomes is capturing viral sequences in the process (Whitfield *et al.* 2011; Lee *et al.* 2013; Cassone *et al.* 2014b; Martin *et al.* 2017). Many persistent, circulative and/or propagative plant viruses, transmitted by arthropods, produce polyadenylated transcripts (mRNAs have polyA tails) and most RNA-sequencing strategies enrich for mRNAs. Since not all viral transcripts are polyadenylated, developing metatranscriptomic methods using the ribodepletion approach to degrade the ribosomal RNA and enrich coding RNA for library preparation, would provide a better coverage of the virome within infected hosts. Additionally, small RNA sequencing studies have provided information on the viral derived transcripts, which aided in understanding the vector anti-viral defense mechanisms against RNA viruses (Wu *et al.* 2010; Li *et al.* 2013; Fletcher *et al.* 2016). However, the application of these techniques to agricultural vectors is in its infancy and would benefit from hypothesis-driven studies to efficiently utilize these techniques in understanding insect vector–virus interactions. Finally, new developments in single cell transcriptomics, in combination with other molecular and cell biology techniques, will shed light on the vector–virus interactions at a cellular level.

In contrast to transcriptomics, proteomics determines the composition and abundance of proteins in tissues and cells and often provides more information about post-translation modifications, and, as such, contributes to understanding the primary metabolic pathway regulations. Proteomic comparisons of infected and uninfected arthropod vectors have been studied to identify viral responsive proteins in vectors such as *F. occidentalis*, *Rhopalosiphum padi* and *B. tabaci* using both gel-based and gel-free proteomic approaches (Badillo-Vargas *et al.* 2012; Wang *et al.* 2015; Zhao *et al.* 2019). Analysis of non-vector and vector populations of *Schizaphis graminum* using two-dimensional difference gel electrophoresis (DIGE) revealed aphid and bacterial endosymbiont proteins differentially expressed between transmission-competent and refractive aphids (Cilia *et al.* 2011). Due to minimal sample requirements, the gel-free quantitative proteomics approach is an efficient technique to study realtively small vectors and their corresponding tissues. Among the proteins upregulated in virus infection in *B. tabaci* and *R. padi*, using gel-free proteomics, cytoskeletal reorganization, primary energy metabolism and proteasome pathways were well represented in *R. padi* infected with barley yellow dwarf virus; whereas structural proteins such as laminin, dystroglycan, integrin and cuticle protein were upregulated in *B. tabaci* infected with tomato yellow leaf curl virus (Wang *et al.* 2015; Zhao *et al.* 2019). Together, these studies illuminate the proteins that could help in virus mobility within the insect vector. With increasing access to efficient stereomicroscope and dissecting tools, it has been possible to study tissue level proteomics of insect vectors. Proteome profiles of the infected and uninfected ova of *L. striatellus*, infected with rice stripe virus, showed an upregulation of ribosome, metabolism and signaling pathways (Liu *et al.* 2016). The more advanced label-free shotgun proteomics requires minimal sample processing steps and also provides maximum coverage, increasing the probability of identifying rare proteins, which is expected in specific interactions of insect vector with the respective virus (O'Neill 2019). More recently, gel-free and label-free proteomics of *A. pisum* stylets has shed light on the cuticular proteins accessible to non-circulative viruses (Deshoux *et al.* 2020). In contrast to global discovery proteomics, targeted proteomics was used to identify virus-associated plant proteins in insect vectors, to determine the plant components assisting in virus acquisition (Cilia *et al.* 2012).

The chemical environment of vital vector tissues, such as the stylet, gut or hemolymph in insect vectors, manifests the mechanisms supporting the attachment, retention, and/or replication of the virus within the vector (Antunes *et al.* 2013; Killiny *et al.* 2017). Metabolomics is a powerful tool for profiling the chemical composition of samples under a given condition. A study has been performed characterizing the metabolic changes in *S. furcifera* associated with virus, Southern rice black-streaked dwarf virus, and temperature stress. This study showed an increase in fatty acids and nucleic acids upon virus infection but a decrease in sugars and polyols under temperature stress (Zhang *et al.* 2018). Metabolite profiling under such biotic or abiotic stress is more informative in combination with proteomic or transcriptomic data. In general, the significance of functional genomics techniques increases when they are used in combination with each other, allowing for a systems biology approach to address complex interactions between two or more organisms, such as in insect vector and virus interactions. For a more detailed narrative on "omics" technologies used in plant–virus–arthropod interactions, refer to Wilson *et al.* (2019).

7.9. Vector innate immunity in response to viruses

Insects lack adaptive immune responses and thus evolved innate immune defenses as a primary defense against microbial pathogens. Innate immune responses can be loosely classified as cell-mediated, humoral responses that synergize to recognize and respond to pathogen attack. Although insect innate immunity to bacterial and fungal pathogens is well studied, here we focus on the response to viral infection of their vectors. Innate immune responses in insects are typically mounted by the entry of pathogen into the insect body. These responses are expected in insect host vectors of persistent and semi-persistent viruses. Innate immune pathways directly or indirectly mounted by viral infection in insect vectors include the RNAi, Toll, JAK/STAT and IMD pathway (Kingsolver *et al.* 2013). We discuss the modulation of these pathways in arthropod vectors of plant viruses and most of the observations are inferred from transcriptomic and a few proteomic studies, with little functional or biochemical evidence supporting the observations. In addition, perturbations of these pathways also depend on the virus–vector system.

The RNA interference (RNAi) pathway is the primary antiviral defense mechanism in insects. Core cellular machinery of the RNAi pathway includes Dicer protein, RNA induced silencing complex (RISC) and Argonaute-2. Dicer recognizes the viral double stranded RNA and cleaves into 20–25 nucleotide length duplex small interfering RNA (siRNA) molecules. Argonaute-2 processes siRNA into a single stranded "guide strand" which is incorporated into the RISC complex, facilitating the specific binding and degradation of the target RNA. The RNAi mechanism is broadly applicable to most of the virus and an induction of the RNAi pathway is observed in several vector–plant virus interactions studied so far. Interestingly, the intensity of the RNAi response to the virus depends on the virus species (Li *et al.* 2013). In addition, the RNAi pathway in insect vectors is crucial in controlling levels of viral load within the vector and thus affecting the persistent virus–arthropod relationship of rice gall dwarf virus, a reovirus, with its leafhopper *Recilia dorsalis* (Lan *et al.* 2016). Thus, RNAi would also serve as an applied tool in disrupting the dissemination of the plant viruses with their hosts.

The Janus kinase/signal transducer and activator of the transcription (JAK/STAT) pathway is well studied in insects and is highly conserved between mammals and insects. The JAK/STAT pathway is induced by the binding of cytokines to the cytokine receptors, triggering downstream tyrosine kinases which in turn activates the transcription factors and STAT proteins, which transcribe the downstream effector proteins (reviewed by Bang 2019). The JAK/STAT pathway is induced in the arthropod vectors of circulative, propagative or non-propagative viruses. However, similar to the Toll pathway, the JAK/STAT pathway response is specific to individual vector–virus systems such as it is induced in thrips-tospovirus interactions (Schneweis *et al.* 2017; Shrestha *et al.* 2017) and not affected in planthopper–tenuivirus interactions (Zhao *et al.* 2016).

The Toll signaling pathway is well studied in insects and is presumed to be their ancestral signaling pathway. Induction of the Toll pathway results in the production of antifungal and antimicrobial peptides. Toll signaling is triggered by Spätzle protein, activated by serine proteases upon binding to pathogen recognition molecules and inhibited by serine protease inhibitors. Positive or negative regulation of Toll signaling pathways is specific to the vector–virus system, development stage and infected tissue, whereas in *B. tabaci* and begomovirus interactions the Toll pathway is suppressed. During the development of infected vectors, the Toll pathway is downregulated in pupal stages of *F. occidentalis* infected with tomato spotted wilt virus (Schneweis *et al.* 2017). Toll pathways are involved in development in insects and a downregulation of this pathway during pupation could be due to developmental changes occurring during this stage (Imler and Zheng 2004). At a tissue level, the Toll pathway was induced in the salivary glands and inhibited in the alimentary canal tissue of *L. striatellus*, infected with rice stripe virus (Zhao *et al.* 2016). The ability of a virus to suppress host immune defenses might depend on the co-evolutionary relationship shared by the vector host and the respective virus, such as in the case of *B. tabaci*, whiteflies and begomoviruses (Luan *et al.* 2011).

Although insect vectors possess an immune deficiency (Imd) pathway, some of the canonical members observed in other model organisms are not found in the genomes of the insect vectors of plant viruses, such as aphid vectors (Gerardo *et al.* 2010). The Imd pathway is activated upon the binding of pathogen molecules to extracellular or transmembrane pattern recognition receptors. The downstream cascade is mediated by an IMD protein phosphorylate relish transcription factor via the IKK complex. Caspase cleaves the relish protein and facilitates its movement into the nucleus where respective effector genes are transcribed. Similar to the Toll and JAK/STAT pathways, the Imd pathway responses are specific to vector–virus systems (Wang *et al.* 2016; Zhao *et al.* 2016). Although most of the canonical members of this pathway are identified in the studied insect genomes and transcriptomes, IMD, the most essential player, was not identified in the insect vectors of plant viruses, which could probably have diverged in the sequence composition from the model organisms or is replaced by another non-canonical member of the pathway (Gerardo *et al.* 2010). Due to the lack of genomes for many insect vectors, an absence of typical IMD protein cannot be concluded without functional and comparative studies.

Interestingly, several of the transcriptomic studies on insect vectors infected with plant viruses demonstrated an upregulation of prophenoloxidase (PPO) pathway members (Xu *et al.* 2012; Cassone *et al.* 2014; Wang *et al.* 2016; Schneweis *et al.* 2017; Shrestha *et al.* 2017; Geng *et al.* 2018). The PPO activation cascade is the first line of defense against invading pathogens and is well characterized in response to bacterial and fungal pathogens The pathway is activated by pattern recognition receptors binding to specific pathogen molecules (patterns). Downstream serine proteases are triggered to convert inactive PPO enzyme to active phenol oxidase which triggers the melanization of the pathogen. Reactive oxygen species produced during melanization could also degrade the invading pathogens. The PPO pathway coordinates with other mechanisms such as phagocytosis and encapsulation Interestingly, PPO induction varies across different insect developmental stages. For example, in *F. occidentalis*-TSWV interactions, serine proteases are downregulated in pupal stages and PPO was upregulated in the adult stages (Schneweis *et al.* 2017). In contrast, infection of *L. straitellus* with Southern rice black stripe dwarf virus downregulates the PPO in adults (Wang *et al.* 2016). These contradicting results from different systems further support the specificity of virus–arthropod interactions. However, it is noteworthy to mention the spectrum of roles of PPO in fundamental insect physiology, including development, wing formation and egg melanization and hence observed changes might be an indirect effect of virus infection (Lu *et al.* 2014).

7.10. Host and vector manipulation by plant viruses

Parasites have been thought to enhance their own transmission through manipulation of hosts since Hofmann first reported that, instead of hiding in the soil and pupating like non-infected caterpillars, caterpillars infected with baculoviruses climbed to the top of their host tree, died, liquefied and released virions into the environment to infect new hosts (Hofmann 1891). Decades later, following many detailed reports of host manipulation by parasites, a "host manipulation hypothesis" was proposed suggesting that pathogens evolve to alter their hosts in ways that enhance successful transmission (Holmes and Bethel 1972). Since that time, many investigations support the concept that parasites of hosts, from all major taxa known to be parasitized, have evolved to influence their own spread through host manipulation (Poulin 1995; Thomas *et al.* 2005; Lefèvre *et al.* 2009; Hughes 2014; reviewed in van Houte *et al.* 2013; Heil 2016; Mauck *et al.* 2018, 2019; Wilson *et al.* 2019). In the following, we focus on the remarkably adaptive ways in which plant viruses have evolved to manipulate their plant hosts and insect vectors to enhance their own transmission.

7.10.1. *Indirect (plant-mediated) manipulation of insect vectors by plant viruses*

Many pathogens that infect animals, humans or plants rely on arthropod vectors for transmission to new hosts, and these pathogens have evolved remarkably similar strategies to manipulate the efficiency of their spread by these vectors (reviewed in Guiguet *et al.* 2016; Heil 2016). Pathogens may influence vector behavior and population dynamics *indirectly* by manipulating the final host in ways that influence altered attractiveness and/or host suitability to the vector. This strategy is remarkably convergent across animal-, human- and plant-infecting pathogens. For example, infection of animals and humans with the parasite causing malaria, *Plasmodium falciparum*, can enhance their attractiveness to mosquito vectors (Lacroix *et al.* 2005; Cornet *et al.* 2012; De Moraes *et al.* 2014). Similarly, plant virus infection may alter emission of volatile organic compounds (VOCs) from infected hosts, enhancing plant attractiveness to vector insects (Eigenbrode *et al.* 2002; Jiménez-Martínez *et al.* 2004; Mauck *et al.* 2010a; Safari *et al.* 2019). In addition, infection may profoundly affect insect vector populations by improving host quality (nutritionally or by suppressing induced defenses) for vector herbivores (Blua and Perring 1992; Wu *et al.* 2017; Kumar 2019; Li *et al.* 2019; Wu *et al.* 2019). Indeed, there are more than 100 reports detailing instances of plant virus effects on plant hosts that indirectly manipulate vector orientation (host attractiveness), feeding behavior, or population development (reviewed in Mauck *et al.* 2019). Selected examples of plant viruses affecting their plant hosts, in ways that influence insect vector behavior, are shown in Table 7.1.

Mode of transmission	Virus (by family, genus and species)/vector	Virus protein/virion or virus infection	Effect in host plant or insect	Effect of plant alteration on insect vector	Citations
Non-circulative	***Bromoviridae***				
	Cucumovirus: *Cucumber mosaic virus*/*Myzus persicae* (aphid)	2b*	Attenuation of JA signaling due to 2b, JAZ protein interactions that prevent JAZ degradation	Insect vectors are attracted to infected plants	(Wu *et al.* 2017)
	Cucumovirus: *Cucumber mosaic virus*/*Myzus persicae* (aphid)	2b*	Quenches transcriptional responses of plant genes to JA	Increased aphid fecundity and phloem ingestion on infected plants	(Ziebell *et al.* 2011)
	Cucumovirus: *Cucumber mosaic virus*/*Myzus persicae* (aphid)	1a, 2a, 2b*	The trio of proteins interact, ultimately allowing synthesis of the feeding deterrent, 4MI3M	Reduced insect performance, enhanced dispersal from infected plants	(Westwood *et al.* 2013)

Non-circulative (cont.)	***Potyviridae***				
	Potyvirus *Potato virus Y*/ *Myzus persicae* (aphid)	HC-Pro* (in transgenic plants, but not in PVY infected plants)	Inhibition of JA-induced gene expression	Increased aphid fecundity	(Westwood *et al.* 2014)
	Potyvirus *Turnip mosaic virus*/ *Potato virus Y*/ *Myzus persicae* (aphid)	NIA-Pro	Stimulates ethylene production and ET-regulated genes, suppresses callose deposition. When aphids probe, NIA-Pro relocates from the nucleus to the vacuole, and reverts when aphids are removed.	Vector preference for infected plants, greater vector fecundity	(Casteel *et al.* 2014, 2015; Casteel and Falk 2016; Bak *et al.* 2017)
Circurlative, Non-propagative	***Geminiviridae***				
	Begomovirus *Tomato yellow leaf curl China virus*/*Bemisia tabaci* (whitefly) *Begomovirus* *Cabbage leaf curl virus*/ *B. tabaci* (whitefly)	βC1* (satellite DNA) DNA-B, BV1	Satellite or viral proteins, respectively, bind the MYC2 transcription factor resulting in altered leaf development and suppressed activation of terpene synthase and JA-mediated defenses.	Increased attractiveness to vectors, increased survival and fecundity of vector insects	(Yang *et al.* 2008; Zhang *et al.* 2012, Li *et al.* 2014; Liu *et al.* 2017)

Circurlative, Non-propagative (cont.)	***Geminiviridae* cont.**				
	Begomovirus *Tomato yellow leaf curl virus*/*B.tabaci* (whitefly)	C2*	Interacts with ubiquitin to compromise degradation of JAZ1 protein, thus inhibiting JA-mediated defense and expression of MYC2-regulated terpene synthase genes	Increased survival and fecundity of vector insects	(Li *et al.* 2019)
		Virion	Unknown	Shortened development time for males, altered feeding behavior	(Moreno-Delafuente *et al.* 2013; Maluta *et al.* 2014)
	Luteoviridae				
	Luteovirus: *Barley yellow dwarf virus*/*Rhopalosiphum padi* (aphid)**	Virion, *in vitro* feeding of purified virus	Unknown, possibly associated with virus movement across cell types and membranes in the insect	Infective aphids prefer noninfected plants, noninfective aphids prefer virus-infected plants	(Ingwell *et al.* 2012)

Circurlative, Non-propagative (cont.)	*Luteoviridae* cont.				
	Luteovirus: *Barley yellow dwarf virus*/*Rhopalosiphum padi* (aphid)	Virus infection, specific plant components unknown	Changes to headspace volatile organic compounds	Non-viruliferous aphids aggregate on screening above infected leaflets and leaflet models treated with volatiles from infected plants	(Jiménez-Martínez *et al.* 2004)
	Polerovirus: *Potato leaf roll virus*/*Myzus persicae* (aphid)	Virus infection, specific plant components unknown	Changes to headspace volatile organic compounds	Non-viruliferous aphids aggregate on screening above infected leaflets and leaflet models treated with volatiles from infected plants	(Eigenbrode *et al.* 2002)
	Polerovirus: *Potato leaf roll virus*/*Myzus persicae* (aphid)**	Virus infection, specific plant components unknown	Changes to headspace volatile organic compounds	Non-viruliferous aphids settled preferentially on or near infected plants, whereas viruliferous aphids settled on or near non-infected plants.	(Rajabaskar *et al.* 2014)

Circulative, Propagative	***Tospoviridae***				
	Orthotospovirus *Tomato spotted wilt virus/Frankliniella occidentalis* (thrips)**	Virus infection of insect by first instar larval acquisition, rearing on non-infected green bean pods	Unknown mechanism, possibly associated with the outcomes of virus replication in the midgut, tubular salivary glands and associated ducts, foregut and principal salivary gland	Infected males feed more on plants and increase rapid non-ingestion probes, most likely to result in inoculation. Infected females increase predator feeding on mite eggs.	(Stafford *et al.* 2011, Stafford-Banks *et al.* 2014)
	Orthotospovirus *Tomato spotted wilt virus/Frankliniella occidentalis* (thrips)	Virus-infected plants	Unknown mechanism, plants damaged by virus-infected thrips were better hosts than those damaged by virus-free thrips	Improved juvenile survival, faster developmental rate resulting in larger larvae that were invulnerable to predation by mites and *Orius laevigatus*	(Belliure *et al.* 2005; Belliure *et al.* 2008)
	Orthotospovirus *Tomato spotted wilt virus/Frankliniella occidentalis* (thrips)	Virus-infected plants	Unknown mechanism, infected plants were more attractive to the insects and positively affected insect development	Preference for infected plants, faster development (larvae hatched earlier from eggs and pupated more quickly)	(Maris *et al.* 2004)

Circulative, Propagative (cont.)	***Tospoviridae* cont.**				
	Orthotospovirus *Tomato spotted wilt virus*/*Frankliniella occidentalis* (thrips)	NSs*	NSs directly interacts with MYC2 and its two close homologs, MYC3 and MYC4, thus disabling JA-mediated activation of terpene synthase genes	Plant volatile monoterpenes are suppressed and thrips preference and performance are enhanced	(Wu *et al.* 2019)
	Phenuiviridae				
	Tenuivirus *Rice stripe virus*/*Laodelphax striatellus*/planthopper**	Virus maintained continuously in an infected planthopper strain	Virus infection in the insect modifies insect physiology, mechanism thought to include down- and up-regulation of genes in the juvenile hormone and ecdysteroid pathways	Reduced fecundity, possibly offset by shorter duration of nymphal stages linked to down-regulated JHAMT (juvenile hormone pathway) and up-regulated CYP307A1 (ecdysteroid pathway). Greater phloem ingestion and higher body weight in association with increased abundance of yeast-like endosymbionts.	(Wan *et al.* 2015)
	Reoviridae				
	Fijivirus *Southern rice black-streaked dwarf virus*/*Sogatella furcifera*/whitebacked planthopper**	Virus infection of insects by a 2 hour acquisition access by newly hatched nymphs)	Unknown mechanism, possibly associated with the outcomes of virus passage to salivary glands and replication in the insect	More frequent ingestion from the phloem, decreased egg hatchability and fecundity when infected males and females mated	(Xu *et al.* 2014)

Circulative, Propagative (cont.)					
	***Reoviridae* cont.**				
	Fijivirus *Southern rice black-streaked dwarf virus* and *Oryzavirus* *Rice ragged stunt virus*/ *Sogatella furcifera* and *Nilaparvata lugens*/ whitebacked planthopper and brown planthopper, respectively**	Virus infection by rearing insects on virus-infected plants)	Changes to host plant defenses (JA, SA and ET biosynthesis genes): plant defense genes down-regulated (OsAOS1, OsICs, OsACS2), and to host plant volatiles: volatile-biosynthesis genes up-regulated (OsLIS, OsCAS, OsHPL3)	Temporal change in attractiveness to vectors, whereby non-infected insects were more attracted to plants infected with the virus they transmit. Planthoppers infected with each of the viruses preferred non-infected plants.	(Lu *et al.* 2016)
	Phytoreovirus *Rice dwarf virus*/ *Nephotettix cincticeps*/ leafhopper	Virus infection by a 2-day acquisition access period	Unknown mechanism, altered host plant physiology resulting in proposed changes to volatiles	Infected leafhoppers preferred non-infected rice plants, while non-infected leafhoppers preferred infected rice plants	(Wang *et al.* 2018)

Table 7.1. *Selected reports of virus–plant and possible virus–insect interactions resulting in altered vector behavior, development or fecundity in ways that enhance virus spread organized by mode of transmission. Single asterisks indicate viral proteins that are also suppressors of gene silencing in plants. Double asterisks indicate interactions in which the literature suggests direct manipulation of the vector by the virus. JA = Jasmonic Acid Signaling Pathway; SA = Salicylic Acid Signaling Pathway; ET = Ethylene Signaling Pathway*

There are multiple mechanisms by which viruses indirectly benefit insect vectors through manipulation of their plant hosts. The multifunctional capacities of plant viral proteins have been broadly shown (reviewed in Zanardo *et al.* 2019). Among these capacities, plant virus proteins can take advantage of multiple cellular pathways in ways that promote infection in their hosts, while simultaneously modifying expression of symptoms in ways that shift insect vector development and responses (reviewed in Eigenbrode *et al.* 2018; Mauck *et al.* 2018; 2019; Wilson *et al.* 2019). In the following, we discuss how viruses can induce changes to the nutritional quality of hosts, and how they can either enhance (e.g. non-circulative viruses that are non-persistently (NPT) or semipersistently transmitted (SPT) or suppress (e.g. non-circulative or circulative viruses that are non-propagative or propagative in their vector and persistently transmitted (PT)) plant defenses against insect vectors, thus enhancing their own spread.

Many different plant viruses induce changes to the nutritional quality of plant hosts during infection, and plant nutritional status may shift as virus infection progresses. Early infections with non-circulative viruses and sustained infection with circulative viruses result in elevated levels of soluble sugars, starch and some amino acids such that insect vector feeding behavior and fitness are positively influenced and, in the case of aphids, the number of winged individuals increases. In reading the literature, it is important to recognize that there is a temporal pattern to these changes. In some cases, findings by different groups appear to be contradictory; however, the timing of the experiments relative to time of infection, host, virus isolate, insect vector species and population, usually explains these inconsistencies. For example, as infection proceeds, especially with circulative viruses, plants often shift from being an improved host to serving as poor quality hosts, thus favoring vector dispersal.

In contrast, when certain plant hosts are infected with non-circulative viruses, such as cucumber mosaic virus, plants produce volatiles that are highly attractive to insect vectors, but the plants themselves may be deleterious to vector development, encouraging rapid dispersal that favor this mode of transmission. For example, ratios of carbohydrates to amino acids may be reduced compared to non-infected plants, as well as having lower levels of essential amino acids required for insect survival (Mauck *et al.* 2014). In general, many pathways and components of plant–virus–vector interactions can alter plant hosts in ways that influence insect vector behavior and physiology to enhance virus spread. These components are well reviewed and illustrated in Mauck *et al.* 2019, Figure 1), as well as (Bosque-Perez and Eigenbrode 2011; Mauck *et al.* 2018; Wilson *et al.* 2019), and are therefore not further reviewed here. However, among the many possible virus-induced changes to plants, perhaps the most striking and consistent is viral targeting and suppression of highly conserved plant defense pathways, resulting in improved feeding behavior and performance of their insect vectors. Notably, plants have multiple highly conserved host defense pathways that are extremely complex and that interact with one another. To truly understand how viral proteins interfere with plant defenses to enhance their own spread requires knowledge of the pathways and how they interact, including the role of degradation of JAZ proteins and the activities of the many MYC transcriptional factors (reviewed in Broekgaarden *et al.* 2015; Van de Poel *et al.* 2015; Hillmer *et al.* 2017).

Drivers of manipulated plant phenotypes are frequently viral proteins or subviral satellite proteins that also serve as suppressors of gene silencing (Table 7.1). Among these are the 2b protein of cucumber mosaic virus (CMV), βC1 satellite protein of tomato yellow leaf curl China virus (TYLCCV), HC-Pro from potato virus Y, C2 of tomato yellow leaf curl virus, and NSs of tomato spotted wilt virus and tomato zonate spot virus (Table 7.1). In nearly all the reports, these viral proteins interact with plant proteins involved in induced plant defense signaling pathways. The C2 protein of a begomovirus lacking DNA satellites (tomato yellow leaf curl virus) promoted survival and reproduction of whitefly vectors by suppressing ubiquitination of JAZ1, compromising degradation of JAZ1 and thereby preventing downstream jasmonic acid (JA) signaling (Li *et al.* 2019). The role of the transcription factors MYC2 and associated homologs as convergent targets for manipulation by viral proteins, is revealed in a recent publication showing that NSs protein of TSWV and tomato zonate virus interacts directly with MYC2, MYC3 and MYC4, thus disabling JA-mediated activation of terpene synthase genes. In the absence of terpene-based resistance, attraction and performance of thrips vectors were improved on infected plants (Wu *et al.* 2019). Table 7.1 shows multiple examples of viral proteins binding transcription factors and thereby suppressing plant defenses, in particular the jasmonic acid (JA) signaling pathway.

While viral proteins that are also silencing suppressors are commonly found playing these roles, viral proteases (NIa-pro) and viral proteins including replicases, RNA-dependent RNA polymerases and nuclear shuttle proteins are also involved in manipulating plant defense pathways in ways that enhance vector performance or behaviors enhancing transmission, (reviewed in Mauck *et al.* 2019; Wilson *et al.* 2019). Table 7.1 presents many of these interactions in the context of modes of transmission, and more detail about specific viral components and their interactions in defense pathways may be found in reviews by Mauck *et al.* (2019) and Wilson *et al.* (2019).

7.10.2. *Direct manipulation of insect vectors by plant viruses*

Evidence is building for *direct* interactions between parasites and their vectors that result in altered vector behavior or physiology, to enhance biting or probing, movement and/or host choice in ways that enhance pathogen spread. For example, mosquitoes infected with *P. falciparum* may feed longer and bite more people than non-infected mosquitoes (Koella *et al.* 1998; Smallegange *et al.* 2013). Insect vectors infected with circulative, (PT) propagative plant viruses also change feeding behavior in ways that enhance transmission, as well as shifting their feeding preferences. There are a handful of reports about these phenomena that include multiple Hemipteran and Thysanopteran insect vectors and viruses from the families *Tospoviridae* (Stafford *et al.* 2011; Keough *et al.* 2016), *Reoviridae* (Xu *et al.* 2014, Lu *et al.* 2016, Wang *et al.* 2018), and *Phenuiviridae* (Wan *et al.* 2015). For example, western flower thrips males infected with (TSWV) fed longer than their non-infected counterparts, as well as making more probes conducive to virus inoculation (Stafford *et al.* 2011). Interestingly, when insects acquire circulative, non-propagative plant viruses (that do not replicate in their insect vectors), they also shift their behavior in ways that enhance virus transmission. For example, aphids that have acquired various luteovirids (*Barley yellow dwarf virus*, *Potato leaf roll virus*, *Cucurbit aphid-borne yellows virus*) prefer non-infected plants, while aphids that have not yet acquired the viruses prefer infected plants (Ingwell *et al.* 2012; Rajabaskar *et al.* 2014; Carmo-Sousa *et al.* 2016). There are similar reports of changes in vector behavior following acquisition of viruses in the *Gemniviridae* (Liu *et al.* 2013; Moreno-Delafuente *et al.* 2013; Maluta *et al.* 2014; Fereres *et al.* 2016). Interestingly, the finding that viruliferous insects (those that have acquired a plant virus) prefer non-infected plants, while those that have not yet acquired the viruses prefer infected plants is a common theme, suggesting that this phenomenon is a convergent adaptation of plant viruses that rely on insect vectors. The importance of co-evolution between the plant host, virus and vector is frequently underscored by findings that show that vector transmission efficiency, competence of the insect to act as a vector, and mode of transmission all play a role in whether manipulation occurs (Chesnais *et al.* 2017; Mauck *et al.* 2018, 2019).

When taken together, these findings lend support to the "vector manipulation hypothesis" proposing that pathogens directly manipulate their arthropod vectors (Ingwell *et al.* 2012), but more data are needed to establish that criteria for direct manipulation of the insect vectors by the viruses has been met (Blanc and Michalakis 2016; Heil 2016; Mauck *et al.* 2019). Direct vector manipulation requires that alterations to the insect are under genetic control by the pathogen rather than the host (e.g. representing generalized immune responses) (Heil 2016; Mauck *et al.* 2019). A serious challenge to determining this critical aspect is the difficulty in separating direct effects of the viruses from indirect effects of virus infection in plant host sources, even when researchers have attempted this separation by using purified viruses in experiments (Ingwell *et al.* 2012). Multiple studies that show plant host proteins may remain attached to the virion surface during purification, and even play an important role in transmission. For example, a proteomic study of cereal yellow dwarf virus (CYDV) transmission by aphids showed host plant proteins co-purified with the virus, and in the absence of these proteins CYDV was no longer transmitted (Cilia *et al.* 2012). Similar results were shown with cucurbit aphid borne yellows virus in studies with aphid transmission (Bencharki *et al.* 2010). Hence, separating plant host and direct virus effects is a critical challenge not yet met by the scientific community. A direct effect of a plant virus on its vector insect is perhaps best shown with rice stripe virus, a member of the *Phenuiviridae* (Wan *et al.* 2015). This virus replicates in its planthopper vector, *Laodelphax striatellus*, and it also transovarially transmitted, an attribute that can maintain the virus in nature even in the absence of a good plant host. The fecundity of infected insects is reduced by infection; however, this negative effect is apparently offset by shorter duration of nymphal stages linked to down-regulated JHAMT (juvenile hormone pathway) and up-regulated CYPc07A1 (ecdysteroid pathway) in infected adults. Furthermore, infected adults feed more in the phloem and have higher body weight, in association with increased abundance of their yeast-like endosymbiont.

We expect that as the focus on characterizing how viruses directly modify their insect vectors increases, answers to these important, outstanding questions will emerge. For example, there has been much progress in examining how plant viruses affect gene expression in their vector insects. Multiple reports indicate plant viruses in the families *Luteoviridae*, *Gemniviridiae* and *Tospoviridae* can profoundly affect gene expression in their aphid, leafhopper, whitefly and thrips vectors (Brault *et al.* 2010; Zhang *et al.* 2010; Whitfield *et al.* 2011; Badillo-Vargas *et al.* 2012; Kaur *et al.* 2017; Schneweis *et al.* 2017; Geng *et al.* 2018; Hasegawa *et al.* 2018) and, as reviewed earlier in this chapter, can have important impacts on insect innate immunity. While details about how these changes may, or may not, relate to virus manipulation of insect vectors has not been fully revealed; these studies provide the tools and a much needed foundation for carrying this area of study forward. Although most of the investigations describing direct effects of plant viruses on their vectors did not identify functional explanations for the observed behavioral changes, advances in transcriptomics and proteomics will advance this area of study (reviewed in Mauck *et al.* 2019; Wilson *et al* 2019).

7.10.3. *Mode of transmission and virus manipulation of plant hosts leading to enhanced vector transmission*

There is abundant evidence that plant viruses induce changes in their plant hosts which modulate vector behavior and development, ultimately altering virus transmission by the vector (reviewed in Mauck *et al.* 2012; 2018; Eigenbrode *et al.* 2018; Wilson *et al.* 2019). These effects of virus infection on plant host phenotypes show an extraordinary convergence among unrelated viruses whose transmission depends upon very similar vector behaviors, and this convergence centers on mode of transmission rather than virus phylogeny (Mauck *et al.* 2012; Chesnais *et al.* 2017; Mauck *et al.* 2018). Selected reports of virus–plant and possible virus–insect interactions resulting in altered vector behavior, development or fecundity, in ways that enhance virus spread are shown by mode of transmission in Table 7.1, where a double asterisk indicates reports where direct effects are suggested. Infections with insect-borne plant viruses (all modes of transmission) cause many visual and biochemical changes in their hosts that are generally attractive to insect vectors (Blua and Perring 1992; Eigenbrode *et al.* 2002; Mauck *et al.* 2010a; 2010b; Wu *et al.* 2017). In contrast, the mode of transmission is important in determining how virus-mediated changes in host quality alters insect vector responses, such as settling, feeding preferences and population development (Mauck *et al.* 2016; Chesnais *et al.* 2017; Mauck *et al.* 2018).

In the case of non-circulative viruses that rely on aphids remaining on the infected plant for a very short time, virus-mediated visual and volatile cues often deceptively signal a good host (Mauck *et al.* 2010a; 2010b). These authors showed that plants infected by the non-circulative and non-persistent (CMV) produced elevated emissions of a volatile blend otherwise similar to the odor of healthy plants, making them very attractive to aphids even though the infected plants were poor hosts. In this situation, aphid vectors quickly determine lack of host suitability via short sample probes, ideal for NPT virus acquisition, and then quickly leave in search of a more favorable host, carrying the virus along and inoculating it to a new host during sample probing. These relationships are often intricate, for example Blua and Perring (1992) demonstrated that as infection progressed, the NPT *Zucchini yellow mosaic virus* (ZYMV) mediated a change in host suitability for the aphid vector, from improved quality to very poor quality. Hence, during early stages of infection, population growth of aphid vectors and winged aphids increased. Later in the infection, host suitability decreased and large numbers of winged aphids dispersed, likely carrying the virus with them to new hosts. Similarly, while plants infected with the non-circulative SPT *cauliflower mosaic virus* (CaMV) are attractive to aphid vectors, virus infection mediated plant effects that were detrimental to two very efficient aphid vector species, resulting in feeding behavior conducive to virus acquisition and dispersal (Chesnais *et al.* 2017).

In contrast, circulative, PT viruses typically require prolonged feeding for their dissemination and are most often associated with beneficial effects on fitness and behavior of their vector insects (Castle and Berger 1993; Castle *et al.* 1998, Jiménez-Martínez *et al.* 2004; Maris *et al.* 2004; Chesnais *et al.* 2017; Wilson *et al.* 2019). Visual and volatile cues remain important in signaling insect vector attraction to PT viruses (Eigenbrode *et al.* 2002; Jiménez-Martínez *et al.* 2004); and, insect vector settling, feeding, growth rate, longevity and population growth on infected plants are often improved relative to non-infected hosts (Montllor *et al.* 1983; Montllor and Gildow 1986; Colvin *et al.* 2006; Fereres and Moreno 2009; Fereres *et al.* 2016; Eigenbrode *et al.* 2018). Enhanced reproduction and associated increased production of winged aphids on infected plants also promotes a reservoir of insects ready to carry and inoculate viruses to new host plants (Gildow 1980, 1983; Jiu *et al.* 2007; Hodge *et al.* 2010).

These are all examples in which plant viruses alter host plant cues that favor changes to vector feeding behavior with the potential to influence virus transmission, a crucial determinant of virus fitness, supporting the hypothesis that viruses evolve traits that induce or maintain effects on vectors that enhance virus spread (Mauck *et al.* 2018). These effects are ubiquitous, identified for many vector groups, across many plant virus families. The importance of this area of study is reflected in the large number of published reports of plant virus manipulation of host phenotypes that favor vector attraction to infected plants or other life history traits/feeding behavior enhancing virus transmission by vector insects (reviewed in Casteel and Falk 2016; Eigenbrode and Boscue-Perez 2016; Eigenbrode *et al.* 2018; Mauck *et al.* 2018; Mauck *et al.* 2019).

7.11. Summary points

– Plant viruses are transmitted by arthropod vectors by a variety of diverse mechanisms. Transmission is mediated by specific interactions between viral and vector proteins.

– Non-circulative viruses rely on viral capsid proteins and in some cases "helper" proteins encoded by the virus for attachment and retention to the vector stylet or foregut.

– The first non-circulative virus receptor, a cuticle protein, stylin-01 was localized to the acrostyle, a specialized region of the aphid stylet in the common canal.

– Circulative viruses transit through the body of the vector and traverse multiple tissue barriers for successful transmission to occur.

– Aminopeptidase N and ephrin receptors were identified as gut receptors for luteovirid entry into aphid vectors.

– Interestingly, several vector proteins such as cyclophilins and cuticle proteins have been shown to interact with viruses with different transmission strategies.

– Vertically transmitted plant viruses utilize endosymbionts and vitellogenin pathways for transovarial transmission.

– Non-circulative viruses regulate transmission via formation of transmission morphs and can also alter calcium-mediated early plant defense mechanisms.

– Genome resources and "-omics" tools enable higher resolution analysis of plant–virus–vector interactions and functional analysis of vector genes.

– Plant viruses can modulate expression of vector innate immune-associated genes.

– Plant viruses may directly and indirectly manipulate their insect vectors to enhance virus transmission.

– Viral proteins (frequently viral suppressors of gene silencing) manipulate plant phenotypes to increase vector performance on host plants.

7.12. Acknowledgments

We thank our friends and colleagues in the field of plant vector biology for the collegial and helpful environment they foster, which results in the representation of diverse perspectives on these important biological questions.

The authors were funded in part by the following agencies: the French National Institute for Agricultural Research (INRA), the Bill and Melinda Gates Foundation (GCEag, OPP1130147) the French National Research Agency (ANR15-CE20-0011), USDA-NIFA (2016-67013-27492 and 2018-67013-28495) and USDA-FNRI (6034-22000-039-06S).

7.13. References

Ammar, E.-D. and Hogenhout, S.A. (2008). A neurotropic route for Maize mosaic virus (Rhabdoviridae) in its planthopper vector. *Peregrinus maidis. Virus. Res.*, 131(1), 77–85.

Ammar, E.-D., Gargani, D., Lett, J.M., and Peterschmitt, M. (2009). Large accumulations of maize streak virus in the filter chamber and midgut cells of the leafhopper vector. *Cicadulina mbila. Arch. Virol.*, 154(2), 255–262.

Antunes, L.C.M., Han, J., Pan, J., Moreira, C.J.C., Azambuja, P., Borchers, C.H., and Carels, N. (2013). Metabolic signatures of triatomine vectors of *Trypanosoma cruzi* unveiled by metabolomics. *PLOS ONE*, 8(10).

Badillo-Vargas, I.E., Rotenberg, D., Schneweis, D.J., Hiromasa, Y., Tomich, J.M., and Whitfield, A.E. (2012). Proteomic analysis of *Frankliniella occidentalis* and differentially expressed proteins in response to tomato spotted wilt virus infection. *J. Virol.*, 86(16), 8793–8809.

Badillo-Vargas, I.E., Rotenberg, D., Schneweis, B.A., and Whitfield, A.E. (2015). RNA interference tools for the western flower thrips, *Frankliniella occidentalis*. *J. Insect Physiol*, 76, 36–46.

Badillo-Vargas, I.E., Chen, Y., Martin, K.M., Rotenberg, D., and Whitfield, A.E. (2019). Discovery of novel thrips vector proteins that bind to the plant bunyavirus, tomato spotted wilt virus. *J. Virol.*, 93, e00699-19.

Bak, A., Gargani, D., Macia, J.-L., Malouvet, E., Vernerey, M.-S., Blanc, S., and Drucker, M. (2013). Virus factories of cauliflower mosaic virus are virion reservoirs that engage actively in vector transmission. *J. Virol.*, 87(22), 12207–12215.

Beerntsen, B.T., James, A.A., and Christensen, B.M. (2000). Genetics of mosquito vector competence. *Microbiol. Mol. Biol. Rev.*, 64(1), 115–137.

Bencharki, B., Boissinot, S., Revollon, S., Ziegler-Graff, V., Erdinger, M., Wiss, L., Dinant, S., Renard, D., Beuve, M., Lemaitre-Guillier, C., and Brault, V. (2010). Phloem protein partners of Cucurbit aphid borne yellows virus: Possible involvement of phloem proteins in virus transmission by aphids. *Mol. Plant. Microbe. In.,* 23(6), 799–810.

Berthelot, E., Macia, J.-L., Martinière, A., Morisset, A., Gallet, R., Blanc, S., Khelifa, M., and Drucker, M. (2019a). Pharmacological analysis of transmission activation of two aphid-vectored plant viruses, turnip mosaic virus and cauliflower mosaic virus. *Sci. Rep-UK*, 9(1).

Berthelot, E., Ducousso, M., Macia, J.-L., Bogaert, F., Baecker, V., Thébaud, G., Gallet, R., Yvon, M., Blanc, S., Khelifa, M., and Drucker, M. (2019b). Turnip mosaic virus is a second example of a virus using transmission activation for Plantto-plant propagation by aphids. *J. Virol.*, 93(9).

Blanc, S. and Michalakis, Y. (2016). Manipulation of hosts and vectors by plant viruses and impact of the environment. *Curr. Opin. Insect. Sci.*, 16, 36–43.

Blanc, S., Cerutti, M., Usmany, M., Vlak, J.M., and Hull, R. (1993). Biological activity of cauliflower mosaic virus aphid transmission factor expressed in a heterologous system. *Virology*, 192(2), 643–650.

Blanc, S., Dolja, V.V., Garcia-Lampasona, S., Baker, J., Llave, C., Ammar, E.D., and Pirone, T.P. (1998). Mutations in the potyvirus helper component protein: Effects on interactions with virions and aphid stylets. *J. Gen. Virol.*, 79(12), 3119–3122.

Blua, M.J. and Perring, T.M. (1992). Effects of zucchini yellow mosaic virus on colonization and feeding behavior of *Aphis gossypii* (Homoptera: Aphididae) alatae. *Environ. Entomol.*, 21(3), 578–585.

Bosque-Pérez, N.A. and Eigenbrode, S.D. (2011). The influence of virus-induced changes in plants on aphid vectors: Insights from luteovirus pathosystems. *Virus. Res.*, 159(2), 201–205.

Brault, V., Mutterer, J., Scheidecker, D., Simonis, M.T., Herrbach, E., Richards, K., and Ziegler-Graff, V. (2000). Effects of point mutations in the readthrough domain of the beet western yellows virus minor capsid protein on virus accumulation in planta and on transmission by aphids. *J. Virol*, 74(3), 1140–1148.

Brault, V., Bergdoll, M., Mutterer, J., Prasad, V., Pfeffer, S., Erdinger, M., Richards, K.E., and Ziegler-Graff, V. (2003). Effects of point mutations in the major capsid protein of beet western yellows virus on capsid formation, virus accumulation, and aphid transmission. *J. Virol.*, 77(5), 3247–3256.

Brault, V., Périgon, S., Reinbold, C., Erdinger, M., Scheidecker, D., Herrbach, E., Richards, K., and Ziegler-Graff, V. (2005). The polerovirus minor capsid protein determines vector specificity and intestinal tropism in the aphid. *J. Virol.*, 79(15), 9685–9693.

Brault, V., Tanguy, S., Reinbold, C., Le Trionnaire, G., Arneodo, J., Jaubert-Possamai, S., Guernec, G., and Tagu, D. (2010). Transcriptomic analysis of intestinal genes following acquisition of pea enation mosaic virus by the pea aphid *Acyrthosiphon pisum*. *J. Gen. Virol.*, 91(3), 802–808.

Bressan, A. and Watanabe, S. (2011). Immunofluorescence localisation of Banana bunchy top virus (family Nanoviridae) within the aphid vector, *Pentalonia nigronervosa*, suggests a virus tropism distinct from aphid-transmitted luteoviruses. *Virus. Res.*, 155(2), 520–525.

Bricault, C.A. and Perry, K.L. (2013). Alteration of intersubunit acid–base pair interactions at the quasi-threefold axis of symmetry of Cucumber mosaic virus disrupts aphid vector transmission. *Virology*, 440(2), 160–170.

Briddon, R.W., Pinner, M.S., Stanley, J., and Markham, P.G. (1990). Geminivirus coat protein gene replacement alters insect specificity. *Virology*, 177(1), 85–94.

Broekgaarden, C., Caarls, L., Vos, I.A., Pieterse, C.M.J., and Van Wees, S.C.M. (2015). Ethylene: Traffic controller on hormonal crossroads to defense. *Plant Physiol.*, 01020.2015.

Calatayud, P.A., Rahbé, Y., Tjallingii, W.F., Tertuliano, M., and Rü, B. (1994). Electrically recorded feeding behaviour of cassava mealybug on host and non-host plants. *Entomol. Exp. Appl.*, 72(3), 219–232.

Carmo-Sousa, M., Moreno, A., Plaza, M., Garzo, E., and Fereres, A. (2016). Cucurbit aphid-borne yellows virus (CABYV) modifies the alighting, settling and probing behaviour of its vector *Aphis gossypii* favouring its own spread. *Ann. Appl. Biol.*, 169(2), 284–297.

Cassone, Bryan J., Carter, F.M.C., Michel, A.P., Stewart, L.R., and Redinbaugh, M.G. (2014a). Genetic insights into *Graminella nigrifrons* competence for maize fine streak virus infection and transmission. *PLOS ONE*, 9(11), e113529.

Cassone, Bryan J., Wijeratne, S., Michel, A.P., Stewart, L.R., Chen, Y., Yan, P., and Redinbaugh, M.G. (2014b). Virus-independent and common transcriptome responses of leafhopper vectors feeding on maize infected with semi-persistently and persistent propagatively transmitted viruses. *BMC Genomics*, 15(1), 133.

Casteel, C.L. and Falk, B.W. (2016). Plant virus-vector Interactions: More than just for virus transmission. *Curr. Res. Topics Plant Virol.*, 217–240.

Castle, S.J. and Berger, P.H. (1993). Rates of growth and increase of *Myzus persicae* on virus-infected potatoes according to type of virus-vector relationship. *Entomol. Exp. Appl.*, 69(1), 51–60.

Castle, S.J., Mowry, T.M., and Berger, P.H. (1998). Differential settling by *Myzus persicae* (Homoptera: Aphididae) on various virus infected host plants. *Ann. Entomol. Soc. Am.*, 91(5), 661–667.

Cerkvenik, U., Dodou, D., van Leeuwen, J.L., and Gussekloo, S.W.S. (2018). Functional principles of steerable multi-element probes in insects. *Biol. Rev.*, 94(2), 555–574.

Chay, C.A., Gunasinge, U.B., Dinesh-Kumar, S.P., Miller, W.A., and Gray, S.M. (1996). Aphid transmission and systemic plant infection determinants of barley yellow dwarf luteovirus-PAV are contained in the coat protein read through domain and 17-kDa protein, respectively. *Virology*, 219(1), 57–65.

Chen, A.Y.S., Walker, G.P., Carter, D., and Ng, J.C.K. (2011). A virus capsid component mediates virion retention and transmission by its insect vector. *Proc. Natl. Acad. Sci. USA.*, 108(40), 16777–16782.

Chen, Q. Chen, H., Mao, Q., Liu, Q., Shimizu, T., Uehara-Ichiki, T., Wu, Z., Xie, L., Omura, T., and Wei, T. (2012a). Tubular structure induced by a plant virus facilitates viral spread in its vector insect. *PLOS Pathog.*, 8(11), e1003032.

Chen, Y., Cassone, B.J., Bai, X., Redinbaugh, M.G., and Michel, A.P. (2012b). Transcriptome of the plant virus vector *Graminella nigrifrons*, and the molecular interactions of maize fine streak rhabdovirus transmission. *PLOS ONE*, 7, e40613.

Chen, Y., Redinbaugh, M. G., Michel, A. P. (2015). Molecular interactions and immune responses between Maize fine streak virus and the leafhopper vector Graminella nigrifrons through differential expression and RNA interference. *Insect Molecular Biology*, 24(3), 391–401.

Chen, W., Hasegawa, D.K., Kaur, N., Kliot, A., Pinheiro, P.V., Luan, J., Stensmyr, M.C., Zheng, Y., Liu, W., Sun, H., Xu, Y., Luo, Y., Kruse, A., Yang, X., Kontsedalov, S., Lebedev, G., Fisher, T.W., Nelson, D.R., Hunter, W.B., Brown, J.K., Jander, G., Cilia, M., Douglas, A.E., Ghanim, M., Simmons, A.M., Wintermantel, W.M., Ling, K.-S., and Fei, Z. (2016). The draft genome of whitefly *Bemisia tabaci* MEAM1, a global crop pest, provides novel insights into virus transmission, host adaptation, and insecticide resistance. *BMC Biol.*, 14(1), 110.

Chen, Q., Zheng, L., Mao, Q., Liu, J., Wang, H., Jia, D., Chen, H., Wu, W., and Wei, T. (2019a). Fibrillar structures induced by a plant reovirus target mitochondria to activate typical apoptotic response and promote viral infection in insect vectors. *PLOS Pathog.*, 15(1), e1007510.

Chen, Y., Dessau, M., Rotenberg, D., Rasmussen, D.A., and Whitfield, A.E. (2019b). Chapter Two - Entry of bunyaviruses into plants and vectors. In *Adv. Virus. Res.*, Kielian, M., Mettenleiter, T.C., and Roossinck, M.J. (eds). Academic Press (Virus Entry), 65–96.

Chesnais, Q., Couty, A., Uzest, M., Brault, V., and Ameline, A. (2017). Plant infection by two different viruses induce contrasting changes of vectors fitness and behavior. *Insect. Sci.*, 26(1), 86–96.

Cilia, M., Tamborindeguy, C., Fish, T., Howe, K., Thannhauser, T. W. and Gray, S. (2011). Genetics coupled to quantitative intact proteomics links heritable aphid and endosymbiont protein expression to circulative polerovirus transmission. *Journal of Virology*, 85(5), 2148–2166.

Cilia, M., Peter K. A., Bereman, M. S., Howe, K., Fish, T., Smith, D., Gildow, F., MacCoss, M. J., Thannhauer, W. T., Gray, S. M. (2012). Discovery and targeted LC-MS/MS of purified polerovirus reveals differences in the virus-host interactome associated with altered aphid transmission. *PLoS ONE*, 7(10), e48177.

Colvin, J., Omongo, C.A., Govindappa, M.R., Stevenson, P.C., Maruthi, M.N., Gibson, G., Seal, S.E., and Muniyappa, V. (2006). Host–plant viral infection effects on arthropod–vector population growth, development and behaviour: Management and epidemiological implications. *Adv. Virus. Res.*, 419–452.

Cornet, S., Nicot, A., Rivero, A., and Gon, S. (2012). Malaria infection increases bird attractiveness to uninfected mosquitoes. *Ecol. Lett.*, 16(3), 323–329.

Cuesta, R., Yuste-Calvo, C., Gil-Cartón, D., Sánchez, F., Ponz, F., and Valle, M. (2019). Structure of Turnip mosaic virus and its viral-like particles. *Sci. Rep-UK.*, 9(1).

Day, M. and Irzykiewicz, H. (1954). On the mechanism of transmission of non-persistent phytopathogenic viruses by aphids. *Aust. J. Biol. Sci.*, 7(3), 251.

De Moraes, C.M., Stanczyk, N.M., Betz, H.S., Pulido, H., Sim, D.G., Read, A.F., and Mescher, M.C. (2014). Malaria-induced changes in host odors enhance mosquito attraction. *Proc. Natl. Acad. Sci. USA.*, 111(30), 11079–11084.

Deshoux, M., Monsion, B., and Uzest, M. (2018). Insect cuticular proteins and their role in transmission of phytoviruses. *Curr. Opin. Virol*, 33, 137–143.

Deshoux, M., Masson, V., Arafah, K., Voisin, S., Guschinskya, N., van Munster, M., Cayrol, B., Webster, C.G., Rahbé, Y., Blanc, S., Bulet, P., Uzest, M. (2020). Cuticular structure proteomics in the pea aphid Acyrthosiphon pisum reveals new plant virus receptor candidates at the tip of maxillary stylets. *J. Proteome Res.* 19(3):1319-1337.

Dombrovsky, A., Sobolev, I., Chejanovsky, N., and Raccah, B. (2007). Characterization of RR-1 and RR-2 cuticular proteins from *Myzus persicae*. *Comp. Biochem. Phys. B.*, 146(2), 256–264.

Drucker, M. and Then, C. (2015). Transmission activation in non-circulative virus transmission: a general concept? *Curr. Opin. Virol.*, 15, 63–68.

Drucker, M. Froissart, R., Hebrard, E., Uzest, M., Ravallec, M., Esperandieu, P., Mani, J.-C., Pugniere, M., Roquet, F., Fereres, A., and Blanc, S. (2002). Intracellular distribution of viral gene products regulates a complex mechanism of cauliflower mosaic virus acquisition by its aphid vector. *Proc. Natl. Acad. Sci. USA.* 99(4), 2422–2427.

Eigenbrode, S.D. and Bosque-Pérez, N.A. (2016). Chemical ecology of aphid-transmitted plant viruses. *Vector-Mediated Transmission of Plant Pathogens*, 3–19.

Eigenbrode, S.D., Ding, H., Shiel, P., and Berger, P.H. (2002). Volatiles from potato plants infected with potato leafroll virus attract and arrest the virus vector, *Myzus persicae* (Homoptera: Aphididae). *Proc. R. Soc. London.*, 269(1490), 455–460.

Eigenbrode, S.D., Bosque-Pérez, N.A., and Davis, T.S. (2018). Insect-borne plant pathogens and their vectors: Ecology, evolution, and complex interactions. *Annu. Rev. Entomol.*, 63(1), 169–191.

Espinoza, A.M., Medina, V., Hull, R., and Markham, P.G. (1991). Cauliflower mosaic virus gene II product forms distinct inclusion bodies in infected plant cells. *Virology*, 185(1), 337–344.

Fereres, A. and Moreno, A. (2009). Behavioural aspects influencing plant virus transmission by homopteran insects. *Virus. Res.*, 141(2), 158–168.

Fereres, A., Peñaflor, M., Favaro, C., Azevedo, K., L,i, C., Maluta, N., Bento, J., and Lopes, J. (2016). Tomato infection by whitefly-transmitted circulative and non-circulative viruses induce contrasting changes in plant volatiles and vector behaviour. *Viruses*, 8(8), 225.

Fletcher, S.J., Shrestha, A., Peters, J.R., Carroll, B.J., Srinivasan, R., Pappu, H.R., and Mitter, N. (2016). The tomato spotted wilt virus genome is processed differentially in its plant host *Arachis hypogaea* and its thrips vector *Frankliniella fusca*. *Front. Plant. Sci.*, 7, 1349.

Forbes, A.R. (1969). The stylets of the green peach aphid, *Myzus Persicae* (Homoptera: Aphididae). *Can. Entomol.*, 101(1), 31–41.

Franz, A.W.E., van der Wilk, F., Verbeek, M., Dullemans, A.M., van den Heuvel, and J.F.J.M. (1999). Faba bean necrotic yellows virus (genus *Nanovirus*) requires a helper factor for its aphid transmission. *Virology*, 262(1), 210–219.

Gamage, S.M.W., Rotenberg, D., Schneweis, D.J., Tsai, C.-W., and Dietzgen, R.G. (2018). Transcriptome-wide responses of adult melon thrips (*Thrips palmi*) associated with capsicum chlorosis virus infection. *PLOS ONE*, 13(12), e0208538.

Garry, C.E. and Garry, R.F. 2004. Proteomics computational analyses suggest that the carboxyl terminal glycoproteins of Bunyaviruses are class II viral fusion protein (beta-penetrenes). *Theoretical Biology and Medical Modelling*, 1(1), p. 10.

Geng, L., Qian, L.-X., Shao, R.-X., Liu, Y.-Q., Liu, S.-S., and Wang, X.-W. (2018). Transcriptome profiling of whitefly guts in response to Tomato yellow leaf curl virus infection. *Virology. J.*, 15(1).

Gerardo, N.M. Altincicek, B., Anselme, C., Atamian, H., Barribeau, S.M., de Vos, M., Duncan, E.J., Evans, J.D., Gabaldón, T., Ghanim, M., Heddi, A., Kaloshian, I., Latorre, A., Moya, A., Nakabachi, A., Parker, B.J., Pérez-Brocal, V., Pignatelli, M., Rahbé, Y., Ramsey, J.S., Spragg, C.J., Tamames, J., Tamarit, D., Tamborindeguy, C., Vincent-Monegat, C., and Vilcinskas, A. (2010). Immunity and other defenses in pea aphids, *Acyrthosiphon pisum*. *Genome Biol.*, 11(2).

Gildow, F.E. (1980). Increased production of alatae by aphids1 reared on oats infected with barley yellow dwarf virus. *Ann. Entomol. Soc. Am.*, 73(3), 343–347.

Gildow, F.E. (1983). Influence of barley yellow dwarf virus-infected oats and barley on morphology of aphid vectors. *Phytopathology*, 73(8), 1196.

Gildow, F.E. (1985). Transcellular transport of barley yellow dwarf virus into the hemocoel of the aphid vector, *Rhopalosiphum padi*. *Phytopathology (USA)*.

Govier, D.A. and Kassanis, B. (1974). A virus-induced component of plant sap needed when aphids acquire potato virus Y from purified preparations. *Virology*, 61(2), 420–426.

Gray, S.M. and Banerjee, N. (1999). Mechanisms of arthropod transmission of plant and animal viruses. *Microbiol. Mol. Biol. Rev.*, 63(1), 128–148.

Gray, S.M. and Gildow, F.E. (2003). Luteovirus-aphid interactions. *Annu. Rev. Phytopathol.*, 41(1), 539–566.

Grigoras, I., Vetten, H.-J., Commandeur, U., Ziebell, H., Gronenborn, B., and Timchenko, T. (2018). Nanovirus DNA-N encodes a protein mandatory for aphid transmission. *Virology*, 522, 281–291.

Guiguet, A., Dubreuil, G., Harris, M.O., Appel, H.M., Schultz, J.C., Pereira, M.H., and Giron, D. (2016). Shared weapons of blood- and plant-feeding insects: Surprising commonalities for manipulating hosts. *J. Insect. Physiol.*, 84, 4–21.

Guo, B., Lin, J., and Ye, K. (2011). Structure of the autocatalytic cysteine protease domain of potyvirus helper-component proteinase. *J. Biol. Chem.*, 286(24), 21937–21943.

Guschinskaya, N., Ressnikoff, D., Arafah, K., Voisin, S., Bulet, P., Uzest, M., Rahbé, Y. (2020). Insect mouthpart transcriptome unveils extension of cuticular protein repertoire and complex organization. *iScience*, 23(2), 100828.

Habili, N. and Francki, R.I.B. (1974). Comparative studies on tomato aspermy and cucumber mosaic viruses. *Virology*, 61(2), 443–449.

Han, S.H., Kim, J.H., Kim, K., and Lee, S.H. (2019). Selection of lethal genes for ingestion RNA interference against western flower thrips, *Frankliniella occidentalis*, via leaf disc-mediated dsRNA delivery. *Pestic. Biochem. Physiol.*, 161, 47–53.

Harris, K.F. (1977). An ingestion-egestion hypothesis of noncirculative virus transmission. In *Aphids as Virus Vectors*, Harris, K.F. and Maramorosch, K. (eds). Elsevier, 165–220.

Hasegawa, D.K., Chen, W., Zheng, Y., Kaur, N., Wintermantel, W.M., Simmons, A.M., Fei, Z., and Ling, K.-S. (2018). Comparative transcriptome analysis reveals networks of genes activated in the whitefly, *Bemisia tabaci* when fed on tomato plants infected with Tomato yellow leaf curl virus. *Virology*, 513, 52–64.

Hebrard, E., Drucker, M., Leclerc, D., Hohn, T., Uzest, M., Froissart, R., Strub, J-M., Sanglier, S., van Dorsselaer, A., Padilla, A., Labesse, G., and Blanc, S. (2001). Biochemical characterization of the helper component of cauliflower mosaic virus. *J. Virol.*, 75(18), 8538–8546.

Heil, M. (2016). Host manipulation by parasites: Cases, patterns, and remaining doubts. *Front. Ecol. Evol.*, 4.

Hillmer, R.A., Tsuda, K., Rallapalli, G., Asai, S., Truman, W., Papke, M.D., Sakakibara, H., Jones, J.D.G., Myers, C.L., and Katagiri F. (2017). The highly buffered Arabidopsis immune signaling network conceals the functions of its components. *PLOS Genet.*, 13(5), e1006639.

Hodge, S. and Powell, G. (2010). Conditional facilitation of an aphid vector, *Acyrthosiphon pisum*, by the plant pathogen, pea enation mosaic virus. *J. Insect. Sci.*, 10(155), 1–14.

Hofmann, R. (1891). *Insektentötende Pilze mit Besonderer berücksichtigung der "Nonne"*. P. Weber.

Hogenhout, S.A., Ammar, E.-D., Whitfield, A.E., and Redinbaugh, M.G. (2008). Insect vector interactions with persistently transmitted viruses. *Annu. Rev. Phytopathol.*, 46(1), 327–359.

Hoh, F., Uzest, M., Drucker, M., Plisson-Chastang, C., Bron, P., Blanc, S., and Dumas, C. (2010). Structural insights into the molecular mechanisms of cauliflower mosaic virus transmission by its insect vector. *J. Virol*, 84(9), 4706–4713.

Holmes, J.C., Bethel, W.M., Canning, E.U., and Wright, C.A. (1972). Behavioural aspects of parasite transmission. *Zool. J. Linn. Soc*, 51, 123–149.

van Houte, S., Ros, V.I., and van Oers, M.M. (2013). Walking with insects: molecular mechanisms behind parasitic manipulation of host behavior. *Mol. Ecol.*, 22(13), 3458–3475.

Howarth, A.J., Gardner, R.C., Messing, J., and Shepherd, R.J. (1981). Nucleotide sequence of naturally occurring deletion mutants of cauliflower mosaic virus. *Virology*, 112(2), 678–685.

Huet, H., Gal-On, A., Meir, E., Lecoq, H., and Raccah, B. (1994). Mutations in the helper component protease gene of zucchini yellow mosaic virus affect its ability to mediate aphid transmissibility. *J. Gen. Virol.*, 75(6), 1407–1414.

Hughes, D. P. (2014). On the origins of parasite-extended phenotypes. *Integr. Comp. Biol.*, 54(2), 210–217.

Huo, Y., Liu, W., Zhang, F., Chen, X., Li, L., Liu, Q., Zhou, Y., Wei, T., Fang, R., and Wang, X. (2014). Transovarial transmission of a plant virus is mediated by vitellogenin of its insect vector. *PLOS Pathog.*, 10(3), e1003949.

Imler, J.-L. and Zheng, L. (2004). Biology of Toll receptors: Lessons from insects and mammals. *J. Leukoc. Biol.*, 75(1), 18–26.

Ingwell, L.L., Eigenbrode, S.D., and Bosque-Pérez, N.A. (2012). Plant viruses alter insect behavior to enhance their spread. *Sci. Rep-UK.*, 2(1).

Jaouannet, M., Rodriguez, P.A., Thorpe, P., Lenoir, C.J.G., MacLeod, R., Escudero-Martinez, C., and Bos, J.I.B. (2014). Plant immunity in plant-aphid interactions. *Front. Plant. Sci.*, 5.

Jia, D., Mao, Q., Chen, H., Wang, A., Liu, Y., Wang, H., Xie, L., and Wei, T. (2014). Virus-induced tubule: A vehicle for rapid spread of virions through basal lamina from midgut epithelium in the insect vector. *J. Virol.,* 88(18), 10488–10500.

Jia, D., Mao, Q., Chen, Y., Liu, Y., Chen, Q., Wu, W., Zhang, X., Chen, H., Li, Y., and Wei, T. (2017). Insect symbiotic bacteria harbour viral pathogens for transovarial transmission, *Nat. Microbial.*, 2(5), 17025.

Jia, D., Chen, Q., Mao, Q., Zhang, X., Wu, W., Chen, H., Yu, X., Wang, Z., and Wei, T. (2018). Vector mediated transmission of persistently transmitted plant viruses. *Curr. Opin. Virol.*, 28, 127–132.

Jiménez-Martínez, E.S., Bosque-Pérez, N.A., Berger, P.H., Zemetra, R.S., Ding, H., and Eigenbrode, S.D. (2004). Volatile cues influence the response of *Rhopalosiphum padi* (Homoptera: Aphididae) to barley yellow dwarf virus–infected transgenic and untransformed wheat. *Environ. Entomol.*, 33(5), 1207–1216.

Jiu, M., Zhou, X.-P., Tong, L., Xu, J., Yang, X., Wan, F.-H., and Liu, S.-S. (2007). Vector-virus mutualism accelerates population increase of an invasive whitefly. *PLOS ONE*, 2(1), e182.

Johnson, D.D. and Walker, G.P. (1999). Intracellular punctures by the adult whitefly *Bemisia argentifolii* on DC and AC electronic feeding monitors. *Entomol. Exp. Appl.*, 92(3), 257–270.

Junglen, S. (2016). Evolutionary origin of pathogenic arthropod-borne viruses – A case study in the family Bunyaviridae. *Curr. Opin. Insect. Sci.*, 16, 81–86.

Kanakala, S. and Ghanim, M. (2016). Implication of the whitefly *Bemisia tabaci* cyclophilin B protein in the transmission of Tomato yellow leaf curl virus. *Front. Plant. Sci.*, 7, 1702.

Kanakala, S., Kontsedalov, S., Lebedev, G., and Ghanim, M. (2019). Plant-Mediated Silencing of the Whitefly *Bemisia tabaci* Cyclophilin B and Heat Shock Protein 70 impairs insect development and virus transmission. *Front. Physiol.*, 10, 557.

Kaper, J.M. and Geelen, J.L.M.C. (1971). Studies on the stabilizing forces of simple RNA viruses. *J. Mol. Biol.*, 56(2), 277–294.

Kaplan, I.B., Lee, L., Ripoll, D.R., Palukaitis, P., Gildow, F., and Gray, S.M. (2007). Point mutations in the potato leafroll virus major capsid protein alter virion stability and aphid transmission. *J. Gen. Virol.*, 88(Pt 6), 1821–1830.

Kassanis, B. and Govier, D.A. (1971). The role of the helper virus in aphid transmission of potato aucuba mosaic virus and potato virus C. *J. Gen. Virol.*, 13(2), 221–228.

Katayama, S., Wei, T., Omura, T., Takagi, J., and Iwasaki, K. (2007). Three-dimensional architecture of virus-packed tubule. *J. Electron. Microsc., (Tokyo)*, 56(3), 77–81.

Kaur, N., Chen, W., Zheng, Y., Hasegawa, D.K., Ling, K.-S., Fei, Z., and Wintermantel, W.M. (2017). Transcriptome analysis of the whitefly, Bemisia tabaci MEAM1 during feeding on tomato infected with the crinivirus, Tomato chlorosis virus, identifies a temporal shift in gene expression and differential regulation of novel orphan genes. *BMC Genomics*, 18(1).

Kennedy, J.S., Day, M.F., and Eastop, V.F. (1962). *A Conspectus of Aphids as Vectors of Plant Viruses. Commonwealth Institute of Entomology*. The Eastern Press Ltd.

Keough, S., Han, J., Shuman, T., Wise, K., and Nachappa, P. (2016). Effects of soybean vein necrosis virus on life history and host preference of its vector, *Neohydatothrips variabilis*, and evaluation of vector status of *Frankliniella tritici* and *Frankliniella fusca*. *J. Econ. Entomol.*, 109(5), 1979–1987.

Kežar, A. Kavčič, L., Polák, M., Nováček, J., Gutiérrez-Aguirre, I., Žnidarič, M.T., Coll, A., Stare, K., Gruden, K., Ravnikar, M., and Pahovnik, D. (2019). Structural basis for the multitasking nature of the potato virus Y coat protein. *Sci. Adv.*, 5(7), eaaw3808.

Killiny, N., Harper, S.J., Alfaress, S., El Mohtar, C., and Dawson, W.O. (2016). Minor coat and heat shock proteins are involved in the binding of citrus tristeza virus to the foregut of its aphid vector, *Toxoptera citricida*. *App. Environ. Microbiol.*, 82(21), 6294–6302.

Killiny, N., Hijaz, F., El-Shesheny, I., Alfaress, S., Jones, S.E., and Rogers, M.E. (2017). Metabolomic analyses of the haemolymph of the Asian citrus psyllid *Diaphorina citri*, the vector of huanglongbing. *Physiol. Entomol.*, 42(2), 134–145.

Kingsolver, M.B., Huang, Z., and Hardy, R.W. (2013). Insect antiviral innate immunity: Pathways, effectors, and connections. *J. Mol. Biol.*, 425(24), 4921–4936.

Kistler, K.E., Vosshall, L.B., and Matthews, B.J. (2015). Genome engineering with CRISPR-Cas9 in the mosquito *Aedes aegypti*. *Cell. Rep.*, 11(1), 51–60.

Koella, J.C., Sørensen, F.L., and Anderson, R.A. (1998). The malaria parasite, *Plasmodium falciparum*, increases the frequency of multiple feeding of its mosquito vector, *Anopheles gambiae*. *P. R. Soc. Lon. B.*, 265(1398), 763–768.

Kumar, R.V. (2019). Plant antiviral immunity against geminiviruses and viral counter-defense for survival. *Front. Microbiol.*, 10.

Lacroix, R., Mukabana, W.R., Gouagna, L.C., and Koella, J.C. (2005). Malaria infection increases attractiveness of humans to mosquitoes. *PLOS Biol.*, 3(9), e298.

Lan, H., Wang, H., Chen, Q., Chen, H., Jia, D., Mao, Q., and Wei, T. (2016). Small interfering RNA pathway modulates persistent infection of a plant virus in its insect vector. *Sci. Rep-UK.*, 6(1).

Le Trionnaire, G., Tanguy, S., Hudaverdian, S., Gleonnec, F., Richard, G., Cayrol, B., Monsion, B., Pichon, E., Deshoux, M., Webster, C., and Uzest, M. (2019). An integrated protocol for targeted mutagenesis with CRISPR-Cas9 system in the pea aphid. *Insect Biochem. Mol. Biol.*, 110, 34–44.

Leclerc, D., Burri, L., Kajava, A.V., Mougeot, J.-L., Hess, D., Lustig, A., Kleemann, G., and Hohn, T. (1998). The open reading frame III product of cauliflower mosaic virus forms a tetramer through a N-terminal coiled-coil. *J. Biol. Chem.*, 273(44), 29015–29021.

Lee, J.H., Choi, J.Y., Tao, X.Y., Kim, J.S., Kim, W., and Je, Y.H. (2013). Transcriptome analysis of the small brown planthopper, *Laodelphax striatellus* carrying rice stripe virus. *Plant. Pathol. J.*, 29(3), 330–337.

Lefèvre, T., Lebarbenchon, C., Gauthier-Clerc, M., Missé, D., Poulin, R., and Thomas, F. (2009). The ecological significance of manipulative parasites. *Trends. Ecol. Evol.,* 24(1), 41–48.

Leh, V. (1999). Aphid transmission of cauliflower mosaic virus requires the viral PIII protein. *EMBO J.*, 18(24), 7077–7085.

Leh, V., Jacquot, E., Geldreich, A., Haas, M., Blanc, S., Keller, M., and Yot, P. (2001). Interaction between the open reading frame III product and the coat protein is required for transmission of cauliflower mosaic virus by aphids. *J. Virol.*, 75(1), 100–106.

Li, J., Andika, I.B., Shen, J., Lv, Y., Ji, Y., Sun, L., and Chen, J. (2013). Characterization of rice black-streaked dwarf virus- and rice stripe virus-derived siRNAs in singly and doubly infected insect vector *Laodelphax striatellus. PLOS ONE*, 8(6).

Li, J., Liang, X., Wang, X., Shi, Y., Gu, Q., Kuo, Y.-W., Falk, B.W., and Yan, F. (2016). Direct evidence for the semipersistent transmission of Cucurbit chlorotic yellows virus by a whitefly vector. *Sci. Rep-UK.*, 6(1).

Li, P., Liu, C., Deng, W.-H., Yao, D.-M., Pan, L.-L., Li, Y.-Q., Liu, Y.-Q., Liang, Y., Zhou, X-P., and Wang, X.-W. (2019). Plant begomoviruses subvert ubiquitination to suppress plant defenses against insect vectors. *PLOS. Pathog.*, 15(2), e1007607.

Liang, Y. and Gao, X.-W. (2017). The cuticle protein gene MPCP4 of *Myzus persicae* (Homoptera: Aphididae) plays a critical role in cucumber mosaic virus acquisition. *J Econo. Entomol.*, 110(3), 848–853.

Linz, L.B., Liu, S., Chougule, N.P., and Bonning, B.C. (2015). *In vitro* evidence supports membrane alanyl aminopeptidase N as a receptor for a plant virus in the pea aphid vector. *J. Virol.*, 89(22), 11203–11212.

Liu, S., He, X., Park, G., Josefsson, C., and Perry, K.L. (2002). A conserved capsid protein surface domain of cucumber mosaic virus is essential for efficient aphid vector transmission. *J. Virol.*, 76(19), 9756–9762.

Liu, B., Preisser, E.L., Chu, D., Pan, H., Xie, W., Wang, S., Wu, Q., Zhou, X., Zhang, Y. (2013). Multiple forms of vector manipulation by a plant-infecting virus: *Bemisia tabaci* and tomato yellow leaf curl virus. *J. Virol.*, 87(9), 4929–4937.

Liu, B., Qin, F., Liu, W., and Wang, X. (2016). Differential proteomics profiling of the ova between healthy and rice stripe virus-infected female insects of *Laodelphax striatellus*. *Sci. Rep-UK.*, 6(1).

Liu, W., Gray, S., Huo, Y., Li, L., Wei, T., and Wang, X. (2015). Proteomic analysis of interaction between a plant virus and its vector insect reveals new functions of hemipteran cuticular protein. *Mol Cell Proteomics*, 14(8), 2229–2242.

Lu, A., Zhang, Q., Zhang, J., Yang, B., Wu, K., Xie, W., Luan, Y.-X., and Ling, E. (2014). Insect prophenoloxidase: the view beyond immunity. *Front. Physiol.*, 5.

Lu G., Li S., Zhou, C., Qian, X., Xiang, Q., Yang, T., Wu, J., Zhou, X., Zhou, Y., Ding, XS., and Tao, X. (2019). Tenuivirus utilizes its glycoprotein as a helper component to overcome insect midgut barriers for its circulative and propagative transmission. *PLOS Pathog.*, 15(3), e1007655.

Luan, J.-B., Li, J.-M., Varela, N., Wang, Y.-L., Li, F.-F., Bao, Y.-Y., Zhang, C.-X., Liu, S.-S., and Wang, X.-W. (2011). Global analysis of the transcriptional response of whitefly to Tomato yellow leaf curl China virus reveals the relationship of coevolved adaptations. *J. Virol.*, 85(7), 3330–3340.

Lung, M.C.Y. and Pirone, T.P. (1973). Studies on the reason for differential transmissibility of cauliflower mosaic virus isolates by aphids. *Phytopathology*, 63(9), 4.

Lung, M.C.Y. and Pirone, T.P. (1974). Acquisition factor required for aphid transmission of purified cauliflower mosaic virus. *Virology*, 60(1), 260–264.

Maluta, N.K.P., Garzo, E., Moreno, A., Lopes, J.R.S., and Fereres, A. (2014). Tomato yellow leaf curl virus benefits population growth of the Q biotype of *Bemisia tabaci* (Gennadius) (Hemiptera: Aleyrodidae). *Neotrop. Entomol.*, 43(4), 385–392.

Mao, Q Wu, W., Liao, Z., Li, J., Jia, D., Zhang, X., Chen, Q., Chen, H., Wei, J., and Wei, T. (2019). Viral pathogens hitchhike with insect sperm for paternal transmission. *Nat. Commun*, 10(1), 1–10.

Maris, P.C., Joosten, N.N., Goldbach, R.W., and Peters, D. (2004). Tomato spotted wilt virus infection improves host suitability for its vector *Frankliniella occidentalis*. *Phytopathology*, 94(7), 706–711.

Markham, P.G., Pinner, M.S., Raccah, B., and Hull, R. (1987). The acquisition of a caulimovirus by different aphid species: comparison with a potyvirus. *Ann. Appl. Biol.*, 111(3), 571–587.

Martin, J.A. and Wang, Z. (2011). Next-generation transcriptome assembly. *Nat Rev. Genet.*, 12(10), 671–682.

Martin, B., Fereres, A., Tjallingii, W.F., and Collar, J.L. (1997). Intracellular ingestion and salivation by aphids may cause the acquisition and inoculation of non-persistently transmitted plant viruses. *J. Gen. Virol.*, 78(10), 2701–2705.

Martin, K.M., Barandoc-Alviar, K., Schneweis, D.J., Stewart, C.L., Rotenberg, D., and Whitfield, A.E. (2017). Transcriptomic response of the insect vector, *Peregrinus maidis*, to maize mosaic rhabdovirus and identification of conserved responses to propagative viruses in hopper vectors. *Virology*, 509, 71–81.

Martinière, A. Bak, A., Macia, J.L., Lautredou, N., Gargani, D., Doumayrou, J., Garzo, E., Moreno, A., Fereres, A., Blanc, S., and Drucker, M. (2013). A virus responds instantly to the presence of the vector on the host and forms transmission morphs. *Elife*, 2, e00183.

Mathers, T.C., Chen, Y., Kaithakottil, G., Legeai, F., Mugford, S.T., Baa-Puyoulet, P., Bretaudeau, A., Clavijo, B., Colella, S., Collin, O., and Dalmay, T. (2017). Rapid transcriptional plasticity of duplicated gene clusters enables a clonally reproducing aphid to colonise diverse plant species. *Genome Biol.*, 18(1).

Matthews, B.J. Dudchenko, O., Kingan, S.B., Koren, S., Antoshechkin, I., Crawford, J.E., Glassford, W.J., Herre, M., Redmond, S.N., Rose, N.H., and Weedall, G.D., *et al.* (2018). Improved reference genome of *Aedes aegypti* informs arbovirus vector control. *Nature*, 563(7732), 501–507.

Mauck, K.E., De Moraes, C.M., and Mescher, M.C. (2010a). Deceptive chemical signals induced by a plant virus attract insect vectors to inferior hosts. *Proc. Natl. Acad. Sci. USA.*, 107(8), 3600–3605

Mauck, K.E., De Moraes, C.M., and Mescher, M.C. (2010b). Effects of Cucumber mosaic virusinfection on vector and non-vector herbivores of squash. *Commun. Integr. Biol.*, 3(6), 579–582.

Mauck, K., Bosque-Pérez, N.A., Eigenbrode, S.D., De Moraes, C.M., and Mescher, M.C. (2012). Transmission mechanisms shape pathogen effects on host-vector interactions: Evidence from plant viruses. *Funct. Ecol.*, 26(5), 1162–1175.

Mauck, K.E., De Moraes, C.M., and Mescher, M.C. (2014). Biochemical and physiological mechanisms underlying effects of Cucumber mosaic virus on host-plant traits that mediate transmission by aphid vectors. *Plant Cell. Environ.*, 37(6), 1427–1439.

Mauck, K.E., Chesnais, Q., and Shapiro, L.R. (2018). Evolutionary determinants of host and vector manipulation by plant viruses. *Environ. Virol. Virus. Ecol.*, 189–250.

Mauck, K.E., Kenney, J., and Chesnais, Q. (2019). Progress and challenges in identifying molecular mechanisms underlying host and vector manipulation by plant viruses. *Curr. Opin. Insect. Sci.*, 33, 7–18.

McLean, D.L. and Kinsey, M.G. (1968). Probing behavior of the pea aphid, *Acyrthosiphon pisum*. II. Comparisons of salivation and ingestion in host and non-host plant leaves. *Ann. Entomol. Soc. Am.*, 61(3), 730–739.

Medina, V., Pinner, M.S., Bedford, I.D., Achon, M.A., Gemeno, C., and Markham, P.G. (2006). Immunolocalization of Tomato yellow leaf curl Sardinia virus in natural host plants and its vector *Bemisia tabaci*. *J. Plant. Pathol.*, 299–308.

Megahed, E.-S. and Pirone, T.P. (1966). *Comparative Tramsmission Of Cucumber Mosaic Virus Acquired By Aphids From Plants Or Through A Membrane*. Amer Phytopathological Soc.

Miyazaki, N., Higashiura, A., Higashiura, T., Akita, F., Hibino, H., Omura, T., Nakagawa, A., and Iwasaki, K. (2015). Electron microscopic imaging revealed the flexible filamentous structure of the cell attachment protein P2 of Rice dwarf virus located around the icosahedral 5-fold axes. *J. Biochem.*, 159(2), 181–190.

Montero-Astúa, M., Rotenberg, D., Leach-Kieffaber, A., Schneweis, B.A., Park, S., Park, J.K., German, T.L., and Whitfield, A.E. (2014). Disruption of vector transmission by a plant-expressed viral glycoprotein. *Mol. Plant. Microbe. Interact.*, 27(3), 296–304.

Montero-Astúa, M., Stafford-Banks, C.A., Badillo-Vargas, I.E., Rotenberg, D., Ullman, D.E., and Whitfield, A.E. (2016a). Tospovirus-thrips biology. In *Vector-Mediated Transmission of Plant Pathogens*, Mauricio Montero-Astúa *et al.* (eds). The American Phytopathological Society (General Plant Pathology), 289–308.

Montero-Astúa, M., Ullman, D.E., and Whitfield, A.E. (2016b). Salivary gland morphology, tissue tropism and the progression of tospovirus infection in *Frankliniella occidentalis*. *Virology*, 493, 39–51.

Montllor, C.B. and Gildow, F.E. (1986). Feeding responses of two grain aphids to barley yellow dwarf virus-infected oats. *Entomol. Exp. Appl*, 42(1), 63–69.

Montllor, C.B., Campbell, B.C., and Mittler, T.E. (1983). Natural and induced differences in probing behavior of two biotypes of the greenbug, schizaphis graminum, in relation to resistance in sorghum. *Entomol. Exp. Appl.*, 34(1), 99–106.

Moreno, A., Hebrard, E., Uzest, M., Blanc, S., and Fereres, A. (2005). A single amino acid position in the helper component of cauliflower mosaic virus can change the spectrum of transmitting vector species. *J. Virol.*, 79(21), 13587–13593.

Moreno-Delafuente, A., Garzo, E., Moreno, A., and Fereres, A. (2013). A plant virus manipulates the behavior of its whitefly vector to enhance its transmission efficiency and spread. *PLOS ONE*, 8(4), e61543.

Mulot, M., Monsion, B., Boissinot, S., Rastegar, M., Meyer, S., Bochet, N., and Brault, V. (2018). Transmission of turnip yellows virus by *Myzus persicae* is reduced by feeding aphids on double-stranded RNA targeting the ephrin receptor protein. *Front. Microbiol.*, 9.

Nakagawa, A. Miyazaki, N., Taka, J., Naitow, H., Ogawa, A., Fujimoto, Z., Mizuno, H., Higashi, T., Watanabe, Y., Omura, T., Cheng, R.H., and and Tsukihara, T. (2003). The atomic structure of rice dwarf virus reveals the self-assembly mechanism of component proteins. *Structure*, 11(10), 1227–1238.

Ng, J.C.K. (2004). Quantitative parameters determining whitefly (*Bemisia tabaci*) transmission of Lettuce infectious yellows virus and an engineered defective RNA. *J. Gen. Virol.*, 85(9), 2697–2707.

Ng, J.C.K. and Perry, K.L. (1999). Stability of the aphid transmission phenotype in cucumber mosaic virus. *Plant Pathol.*, 48(3), 388–394.

Ng, J.C.K., Liu, S., and Perry, K.L. (2000). Cucumber mosaic virus mutants with altered physical properties and defective in aphid vector transmission. *Virology*, 276(2), 395–403.

O'Neill, J.R. (2019). An overview of mass spectrometry-based methods for functional proteomics. *I Functional Proteomics: Methods and Protocols*, Wang, X. and Kuruc, M. (eds). Springer, 179–196.

Peng, Y.H., Wang, Y., Gal-On, A., Huet, H., Kadoury, D., and Raccah, B. (1998). Mutations in the HC-Pro gene of zucchini yellow mosaic potyvirus: Effects on aphid transmission and binding to purified virions. *J. Gen. Virology.*, 79(4), 897–904.

Perry, K.L. (2001). Cucumoviruses. *Virus-Insect-Plant Interactions*, 167–180.

Perry, K.L., Zhang, L., Shintaku, M.H., and Palukaitis, P. (1994). Mapping determinants in cucumber mosaic virus for transmission by *Aphis gossypii*. *Virology*, 205(2), 591–595.

Perry, K.L., Zhang, L., and Palukaitis, P. (1998). Amino acid changes in the coat protein of cucumber mosaic virus differentially affect transmission by the aphids *Myzus persicae* and *Aphis gossypii*. *Virology*, 242(1), 204–210.

Peter, K.A., Liang, D., Palukaitis, P., and Gray, S.M. (2008). Small deletions in the potato leafroll virus readthrough protein affect particle morphology, aphid transmission, virus movement and accumulation. *J. Gen. Virol.*, 89(8), 2037–2045.

Pirone, T.P. and Blanc, S. (1996). Helper-dependent vector transmission of plant viruses. *Ann. Rev. Phytopathol.*, 34(1), 227–247.

Pirone, T.P. and Megahed, E.-S. (1966). Aphid transmissibility of some purified viruses and viral RNA's. *Virology*, 30(4), 631–637.

Plisson, C., Uzest, M., Drucker, M., Froissart, R., Dumas, C., Conway, J., Thomas, D., Blanc, S., and Bron, P. (2005). Structure of the mature p3-virus particle complex of cauliflower mosaic virus revealed by cryo-electron microscopy. *J. Mol. Biol.*, 346(1), 267–277.

Pollard, D.G. (1969). Directional control of the stylets in phytophagous Hemiptera, in *Proc. R. Soc. London. A.*, 173–185.

Pollard, D.G. (1973). Plant penetration by feeding aphids (Hemiptera, Aphidoidea): A review. *Bulletin of Entomological Research*, 62(4), 631–714.

Poulin, R. (1995). 'Adaptive' changes in the behaviour of parasitized animals: A critical review. *Int. J. Parasitol.*, 25(12), 1371–1383.

Powell, G. (2005). Intracellular salivation is the aphid activity associated with inoculation of non-persistently transmitted viruses. *Journal of General Virology*, 86(2), 469–472.

Powell, G., Maniar, S.P., Pickett, J.A., and Hardie, J. (1999). Aphid responses to non-host epicuticular lipids. *Proceedings of the 10th International Symposium on Insect-Plant Relationships*, 115–123.

Qin, F., Liu, W., Wu, N., Zhang, L., Zhang, Z., Zhou, X., and Wang, X. (2018). Invasion of midgut epithelial cells by a persistently transmitted virus is mediated by sugar transporter 6 in its insect vector. *PLOS Pathog.* 14(7).

Rajabaskar, D., Bosque-Pérez, N.A., and Eigenbrode, S.D. (2014). Preference by a virus vector for infected plants is reversed after virus acquisition. *Virus. Res.*, 186, 32–37.

Reinbold, C., Herrbach, E., and Brault, V. (2003). Posterior midgut and hindgut are both sites of acquisition of Cucurbit aphid-borne yellows virus in *Myzus persicae* and *Aphis gossypii*. *J. Gen. Virol.*, 84(12), 3473–3484.

Rotenberg, D., Krishna Kumar, N.K., Ullman, D.E., Montero-Astúa, M., Willis, D.K., German, T.L., and Whitfield, A.E. (2009). Variation in tomato spotted wilt virus titer in *Frankliniella occidentalis* and its association with frequency of transmission. *Phytopathology*, 99(4), 404–410.

Rotenberg, D., Baumann, A. A., Ben-Mahmoud, S., Christiaens, O., Dermauw, W., Ioannidis, P., Jacobs, C. G. C., Jentzsch, I. M. V., Oliver, J. E., Poelchau, M. F., Rajarapu, S. P., Schneweis, D. J., Snoeck, S., Taning, C. N. T., Wei, D., Windana-Gamage, S. M. K., Hughes, D. S. T., Murali, S. C., Bailey, S., Bejerman, N. E., Holmes, C. J., Jennings, E. C., Rosendale, A. J., Rosselot, A., Hervery, K., Schneweis, B., Cheng, S., Childers, C., Simão, F. A., Dietzgen, R. G., Chao, H., Dinh, H., Doddapaneni, H., Dugan, S., Han, Y., Lee, S. L., Muzny, D. M., Qu, J., Worley, K. C., Benoit, J. B., Friedrich, M., Jones, J. W., Panfilio, K. A., Park, Y., Robertson, H. M., Smagghe, G., Ullman, D. E., Van der Zee, M., Van Leeuwen, T., Veenstra, J. A., Waterhouse, R. M., Weirauch, M. T., Werren, J. H., Whitfield, A. E., Zdobnov, E. M., Gibbs, R. A., Richards, S. (2020). Genome-enabled insights into the biology of thrips as crop pests. *bioRxiv*, 2020.02.12.941716.

Ruiz-Ferrer, V., Boskovic, J., Alfonso, C., Rivas, G., Llorca, O., Lopez-Abella, D., and Lopez-Moya, J.J. (2005). Structural analysis of tobacco etch potyvirus HC-Pro oligomers involved in aphid transmission. *J. Virol.*, 79(6), 3758–3765.

Safari, M., Ferrari, M.J., and Roossinck, M.J. (2019). Manipulation of aphid behavior by a persistent plant virus. *J. Virol.*, 93(9).

Schneweis, D.J., Whitfield, A.E., and Rotenberg, D. (2017). Thrips developmental stage-specific transcriptome response to tomato spotted wilt virus during the virus infection cycle in *Frankliniella occidentalis*, the primary vector. *Virology*, 500, 226–237.

Scholthof, K.B.G., Adkins, S., Czosnek, H., Palukaitis, P., Jacquot, E., Hohn, T., Hohn, B., Saunders, K., Candresse, T., Ahlquist, P., Hemenway, C., and Foster, G.D. (2011). Top 10 plant viruses in molecular plant pathology. *Mol. Plant. Pathol.*, 12(9), 938–954.

Sharakhov, I.V. and Sharakhova, M.V. (2015). Heterochromatin, histone modifications, and nuclear architecture in disease vectors. *Curr. Opin. Insect Sci.*, 10, 110–117.

Shrestha, A., Champagne, D.E., Culbreath, A.K., Rotenberg, D., Whitfield, A.E., and Srinivasan, R. (2017). Transcriptome changes associated with Tomato spotted wilt virus infection in various life stages of its thrips vector, *Frankliniella fusca* (Hinds). *J. Gen. Virol.*, 98(8), 2156–2170.

Smallegange, R.C., van Gemert, G.-J., van de Vegte-Bolmer, M., Gezan, S., Takken, W., Sauerwein, R.W., and Logan, J.G. (2013). Malaria infected mosquitoes express enhanced attraction to human odor. *PLOS ONE*, 8(5), e63602.

Smith, T.J., Chase, E., Schmidt, T., and Perry, K.L. (2000). The structure of cucumber mosaic virus and comparison to cowpea chlorotic mottle virus. *J. Virol.*, 74(16), 7578–7586.

Stafford, C.A., Walker, G.P., and Ullman, D.E. (2011). Infection with a plant virus modifies vector feeding behavior. *Proc. Natl. Acad. Sci. USA*. 108(23), 9350–9355.

Stafford, C. A., Walker, G. P. and Ullman, D. E. (2012). Hitching a ride: Vector feeding and virus transmission. *Communicative & Integrative Biology*, 5(1), 43–49.

Stafford-Banks, C.A., Rotenberg, D., Johnson, B.R., Whitfield, A.E., and Ullman, D.E. (2014). Analysis of the salivary gland transcriptome of *Frankliniella occidentalis*. *PLOS ONE*, 9(4), e94447.

Stavolone, L., Villani, M.E., Leclerc, D., and Hohn, T. (2005). A coiled-coil interaction mediates cauliflower mosaic virus cell-to-cell movement. *Proc. Natl. Acad. Sci. USA*. 102(17), 6219–6224.

Stewart, L.R., Medina, V., Tian, T., Turina, M., Falk, B.W., and Ng, J.C.K. (2010). A mutation in the lettuce infectious yellows virus minor coat protein disrupts whitefly transmission but not in planta systemic movement. *J. Virol.*, 84(23), 12165–12173.

Sylvester, E.S. (1956). Beet yellows virus transmission by the green peach aphid. *J. Econo. Entomol.*, 49(6), 789–800.

Tamborindeguy, C., Bereman, M.S., DeBlasio, S., Igwe, D., Smith, D.M., White, F., MacCoss, M.J., Gray, S.M., and Cilia, M. (2013). Genomic and proteomic analysis of *Schizaphis graminum* reveals cyclophilin proteins are involved in the transmission of cereal yellow dwarf virus. *PLOS ONE*, 8(8), e71620.

Thomas, F., Adamo, S., and Moore, J. (2005). Parasitic manipulation: Where are we and where should we go? *Behav. Process.*, 68(3), 185–199.

Tian, T., Crawford, B., Falk, B.W., Yeh, H.H., and Rubio, L. (1999). Lettuce infectious yellows virus: *In vitro* acquisition analysis using partially purified virions and the whitefly *Bemisia tabaci*. *J. Gen. Virol.*, 80(5), 1111–1117.

Ullman, D.E., Westcot, D.M., Hunter, W.B., and Mau, R.F.L. (1989). Internal anatomy and morphology of *Frankliniella occidentalis* (Pergande) (Thysanoptera: Thripidae) with special reference to interactions between thrips and tomato spotted wilt virus. *Int. J. Insect. Morphol. Embryol.*, 18(5), 289–310.

Uzest, M., Gargani, D., Drucker, M., Hebrard, E., Garzo, E., Candresse, T., Fereres, A., and Blanc, S. (2007). A protein key to plant virus transmission at the tip of the insect vector stylet. *Proc. Natl. Acad. Sci. USA.*, 104(46), 17959–17964.

Uzest, M., Gargani, D., Dombrovsky, A., Cazevieille, C., Cot, D., and Blanc, S. (2010). The 'acrostyle': A newly described anatomical structure in aphid stylets. *Arthropod. Struct. Dev.*, 39(4), 221–229.

Valli, A.A., Gallo, A., Rodamilans, B., López-Moya, J.J., and García, J.A. (2017). The HCPro from thePotyviridaefamily: an enviable multitasking Helper Component that every virus would like to have. *Mol. Plant. Path.*, 19(3), 744–763.

Wan, G., Jiang, S., Wang, W., Li, G., Tao, X., Pan, W., Sword, G.A., and Chen, F. (2015). Rice stripe virus counters reduced fecundity in its insect vector by modifying insect physiology, primary endosymbionts and feeding behavior. *Sci. Rep-UK.*, 5(1).

Wang, H., Wu, K., Liu, Y., Wu, Y., and Wang, X. (2015). Integrative proteomics to understand the transmission mechanism of Barley yellow dwarf virus-GPV by its insect vector *Rhopalosiphum padi, Sci. Rep-UK.*, 5, 10971.

Wang, L., Tang, N., Gao, X., Guo, D., Chang, Z., Fu, Y., Akinyemi, I.A., and Wu, Q. (2016). Understanding the immune system architecture and transcriptome responses to southern rice black-streaked dwarf virus in *Sogatella furcifera. Sci. Rep-UK.*, 6(1).

Wang, Q., Li, J., Dang, C., Chang, X., Fang, Q., Stanley, D., and Ye, G. (2018). Rice dwarf virus infection alters green rice leafhopper host preference and feeding behavior. *PLOS ONE*, 13(9), e0203364.

Wang, Y., Mao, Q., Liu, W., Mar, T., Wei, T., Liu, Y., and Wang, X. (2014). Localization and distribution of wheat dwarf virus in its vector leafhopper, *Psammotettix alienus. Phytopathol.*, 104(8), 897–904.

Watanabe, S. and Bressan, A. (2013). Tropism, compartmentalization and retention of banana bunchy top virus (Nanoviridae) in the aphid vector *Pentalonia nigronervosa. J. Gen. Virol.*, 94(1), 209–219.

Watson, M.A., Roberts, F.M., and Russell, E.J. (1939). A comparative study of the transmission of Hyoscyamus virus 3, potato virus Y and cucumber virus 1 by the vectors *Myzus persicae* (Sulz), *M. circumflexus* (Buckton), and *Macrosiphum gei* (Koch). *Proc. R. Soc. London B.*, 127(849), 543–576.

Webster, C.G., Thillier, M., Pirolles, E., Cayrol, B., Blanc, S., and Uzest, M. (2017). Proteomic composition of the acrostyle: Novel approaches to identify cuticular proteins involved in virus-insect interactions. *Insect. Sci.*, 24(6), 990–1002.

Webster, C.G., Pichon, E., Van Munster, M., Monsion, B., Deshoux, M., Gargani, D., Calevro, F., Jiménez, J., Moreno, A., Krenz, B., Thompson, J.R., Perry, K.L., Fereres, A., Blanc, S., and Uzest, M. (2018). Identification of plant virus receptor candidates in the stylets of their aphid vectors. *J. Virol.*, 92(14).

Wei, T., Kikuchi, A., Moriyasu, Y., Suzuki, N., Shimizu, T., Hagiwara, K., Chen, H., Takahashi, M., Ichiki-Uehara, T., and Omura, T. (2006). The spread of Rice dwarf virus among cells of its insect vector exploits virus-induced tubular structures. *J. Virol.*, 80(17), 8593–8602.

Wei, T., Chen, H., Ichiki-Uehara, T., Hibino, H., and Omura, T. (2007). Entry of Rice dwarf virus into cultured cells of its insect vector involves clathrin-mediated endocytosis. *J. Virol.*, 81(14), 7811–7815.

Wei, T., Shimizu, T., and Omura, T. (2008). Endomembranes and myosin mediate assembly into tubules of Pns10 of Rice dwarf virus and intercellular spreading of the virus in cultured insect vector cells. *Virology*, 372(2), 349–356.

Wenger, J.A., Cassone, B.J., Legeai, F., Johnston, J.S., Bansal, R., Yates, A.D., Coates, B.S., Pavinato, V.A.C., and Michel, A. (2017). Whole genome sequence of the soybean aphid, *Aphis glycines. Insect. Biochem. Mol Biol.*, 102917.

Whitfield, A.E., Ullman, D.E., and German, T.L. (2004). Expression and characterization of a soluble form of tomato spotted wilt virus glycoprotein GN. *J. Virol.*, 78(23), 13197–13206.

Whitfield, A.E., Ullman, D.E., and German, T.L. (2005). Tomato spotted wilt virus glycoprotein GC is cleaved at acidic pH. *Virus. Res.*, 110(1), 183–186.

Whitfield, A.E., Kumar, N.K.K., Rotenberg, D., Ullman, D.E., Wyman, E.A., Zietlow, C., Willis, D.K., and German, T.L. (2008). A soluble form of the tomato spotted wilt virus (TSWV) glycoprotein GN (GN-S) inhibits transmission of TSWV by *Frankliniella occidentalis. Phytopathology*, 98(1), 45–50.

Whitfield, A.E., Rotenberg, D., Aritua, V., and Hogenhout, S.A. (2011). Analysis of expressed sequence tags from Maize mosaic rhabdovirus-infected gut tissues of *Peregrinus maidis* reveals the presence of key components of insect innate immunity. *Insect Mol. Biol.*, 20(2), 225–242.

Whitfield, A.E., Falk, B.W., and Rotenberg, D. (2015). Insect vector-mediated transmission of plant viruses. *Virology*, 479, 278–289.

Whitfield, A.E., Huot, O.B., Martin, K.M., Kondo, H., and Dietzgen, R.G. (2018). Plant rhabdoviruses-their origins and vector interactions, *Curr. Opin. Virol.*, 33, 198–207.

Willis, J.H. (2010). Structural cuticular proteins from arthropods: Annotation, nomenclature, and sequence characteristics in the genomics era. *Insect. Biochem. Mol. Biol*, 40(3), 189–204.

Wilson, J. R., DeBlasio, S. L., Alexander, M. M., and Heck, M. (2019). Looking through the lens of 'Omics Technologies: Insights into the transmission of insect vector-borne plant viruses. *Curr. Issues. Mol. Biol.*, 34, 113–144.

Wu, Q., Luo, Y., Lu, R., Lau, N., Lai, E.C., Li, W.-X., and Ding, S.-W. (2010). Virus discovery by deep sequencing and assembly of virus-derived small silencing RNAs. *Proc. Natl. Acad. Sci. USA.*, 107(4), 1606–1611.

Wu, D., Qi, T., Li, W.X., Tian, H., Gao, H., Wang, J., Ge, J., Yao, R., Ren, C., Wang, X.B., Liu, Y., Kang, L., Ding, S, -W., and Xie, D. (2017). Viral effector protein manipulates host hormone signaling to attract insect vectors. *Cell Res.*, 27(3), 402–415.

Wu, W., Huang, L., Mao, Q., Wei, J., Li, J., Zhao, Y., Zhang, Q., Jia, D., and Wei, T. (2019a). Interaction of viral pathogen with porin channels on the outer membrane of insect bacterial symbionts mediates their joint transovarial transmission. *Philos. T. R. Soc. B.*, 374(1767), 20180320.

Wu, X., Xu, S., Zhao, P., Zhang, X., Yao, X., Sun, Y., Fang, R., and Ye, J. (2019b). The Orthotospovirus nonstructural protein NSs suppresses plant MYC-regulated jasmonate signaling leading to enhanced vector attraction and performance. *PLOS Pathog.*, 15(6), e1007897.

Xue, J. Zhou, X., Zhang, C.X., Yu, L.L., Fan, H.W., Wang, Z., Xu, H.J., Xi, Y., Zhu, Z.R., Zhou, W.W., Pan, P.L., Li. B.L., Colbourne, J.K., Noda, H., Suetsugu, Y., Kobayashi, T., Zheng, Y., Liu, S., Zhang, R., Liu, Y., Luo, Y.D., Fang, D.M., Chen, Y., Zhan, D.L., Lv, X.D., Cai, Y., Wang, Z.B., Huang, H.J., Cheng, R.L., Zhang, X.C., Lou, Y.H., Yu, B., Zhuo, J.C., Ye, Y.X., Zhang, W.Q., Shen, Z.C., Yang, H.M., Wang, J., Wang, J., Bao, Y.Y., and Cheng, J.-A. (2014). Genomes of the rice pest brown planthopper and its endosymbionts reveal complex complementary contributions for host adaptation. *Genome Biol.*, 15(12), 521.

Xue, W.-H., Xu, N., Yuan, X.-B., Chen, H.-H., Zhang, J.-L., Fu, S.-J., Zhang, C.-X., and Xu, H.-J. (2018). CRISPR/Cas9-mediated knockout of two eye pigmentation genes in the brown planthopper, *Nilaparvata lugens* (Hemiptera: Delphacidae). *Insect. Biochem. Mol. Biol.*, 93, 19–26.

Yao, J., Rotenberg, D., Afsharifar, A., Barandoc-Alviar, K., and Whitfield, A.E. (2013). Development of RNAi Methods for *Peregrinus maidis*, the Corn Planthopper. *PLOS ONE*, 8(8), e70243.

Zamora, M., Méndez-López, E., Agirrezabala, X., Cuesta, R., Lavín, J.L., Sánchez-Pina, M.A., Aranda, M.A., and Valle, M. (2017). Potyvirus virion structure shows conserved protein fold and RNA binding site in ssRNA viruses. *Science Adv.*, 3(9), eaao2182.

Zanardo, L.G., de Souza, G.B., and Alves, M.S. (2019). Transcriptomics of plant–virus interactions: a review. *Theo. Exp. Plant. Phys.*, 31(1), 103–125.

Zhang, F., Guo, H., Zheng, H., Zhou, T., Zhou, Y., Wang, S., Fang, R., Qian, W., and Chen, X. (2010). Massively parallel pyrosequencing-based transcriptome analyses of small brown planthopper (*Laodelphax striatellus*), a vector insect transmitting rice stripe virus (RSV). *BMC Genomics*, 11(1).

Zhang, T., Feng, W., Ye, J., Li, Z., and Zhou, G. (2018). Metabolomic changes in *Sogatella furcifera* under Southern rice black-streaked dwarf virus infection and temperature stress. *Viruses*, 10(7), 344.

Zhao, J., Chi, Y., Zhang, X.-J., Lei, T., Wang, X.-W., and Liu, S.-S. (2019). Comparative proteomic analysis provides new insight into differential transmission of two begomoviruses by a whitefly. *Virology. J.*, 16(1).

Zhao, W., Lu, L., Yang, P., Cui, N., Kang, L., and Cui, F. (2016). Organ-specific transcriptome response of the small brown planthopper toward rice stripe virus. *Insect. Biochem. Mol. Biol.*, 70, 60–72.

8

Evolution and Origin of Human Viruses

Rachele CAGLIANI, Alessandra MOZZI,
Chiara PONTREMOLI, Manuela SIRONI
IRCCS Eugenio Medea, Bosisio Parini, Italy

8.1. Introduction

Like all living organisms, humans are constantly exposed to a huge diversity of viruses. Over 260 known viral species can infect humans and most of them also infect other mammals (Olival *et al.* 2017). Some of these diseases have been known for centuries, whereas others have emerged (or have been recognized) only recently. Human viruses are genetically highly diverse, and new genotypes, strains and species can evolve rapidly (over time periods of years or decades). The rate of viral evolution is determined by viral intrinsic features, but is also influenced by host factors (e.g. genetics, antiviral drug use), by human population interactions among themselves and by anthropic changes to the natural environment.

Most likely, a number of viruses remain to be identified and to be associated with human diseases. Indeed, several human conditions have an unknown etiology, and for some of these, a viral cause is suspected (Mokili *et al.* 2012). In recent years, the advent of metagenomic approaches has opened new avenues of research and has already resulted in the identification of unknown disease-associated human viruses (e.g. redondoviruses (Abbas *et al.* 2019)) or in the characterization of the genetic

---

For a color version of all figures in this book, see www.iste.co.uk/saleh/virology.zip.

Virology,
coordinated by María-Carla SALEH and Félix AUGUSTO REY. © ISTE Ltd 2021

diversity of known viral species (Chiu and Miller 2019). Indeed, metagenomics has revolutionized the field of virology to such a level that the International Committee on Taxonomy of Viruses (ICTV) has made the decision to classify new viral species (or higher taxa) solely on the basis of metagenomic data, provided that appropriate checks on data integrity have been performed (Simmonds *et al.* 2017). Importantly, metagenomic approaches and, more generally, next-generation sequencing technologies have allowed the characterization of the viral diversity hosted by many other organisms, both eukaryotic and prokaryotic. In turn, this has allowed researchers to trace the evolutionary history of human viruses back in time and to define the key events that led to the origins of some of the most dangerous human pathogens.

8.2. Origin and ancient evolutionary history of human viruses

8.2.1. *Origin and ancient evolutionary history of human-infecting RNA viruses*

RNA viruses comprise a major part of the eukaryotic virome, whereas few prokaryotic RNA viruses are known. In contrast, the prokaryotic virome is largely dominated by dsDNA viruses (Dolja and Koonin 2018). The reasons for this difference are unknown, but it is thought that the expansion of the RNA virome in eukaryotes was facilitated by the emergence of a compartmentalized cytoplasm with internal membranes that provided a suitable environment for RNA replication. Conversely, the nuclear envelope may represent a barrier for the access of DNA viruses to the host replication and transcription machineries (Dolja and Koonin 2018).

Recently, the ICTV ratified the establishment of the realm *Riboviria* to include viruses with single-stranded positive- or negative-sense genomic RNA and double-stranded genomic RNA that use cognate RNA-directed RNA polymerases (RdRps) for replication (Walker *et al.* 2019). The realm thus comprises genetically diverse viruses that infect prokaryotic and eukaryotic hosts, and belong to three classes of Baltimore classification (III–V)[1]. The creation of this realm, which represents the highest taxonomic rank permitted by the ICTV code, was mainly motivated by the observation that all these viruses are likely to form a monophyletic lineage based on the RdRp palm domain phylogeny (Figure 8.1).

1. https://talk.ictvonline.org/taxonomy/.

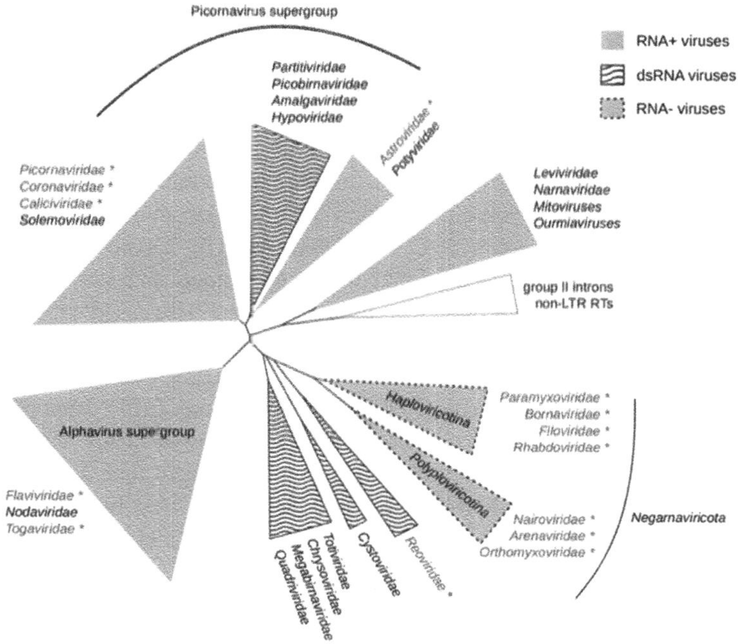

Figure 8.1. Phylogenetic tree determined using RdRp palm domains. *The tree was obtained from previous data (Wolf* et al. *2018). Representative viral families are indicated. Families that include important human pathogens are shown in red with an asterisk. dsRNA, double-stranded RNA viruses; RNA- and RNA+, single-stranded negative-sense and positive-sense RNA viruses, respectively*

Indeed, phylogenetic analysis indicated that double-stranded RNA viruses evolved, on at least two occasions, from positive-sense RNA viruses, whereas negative-sense RNA viruses evolved from a distinct group of dsRNA viruses (Wolf *et al.* 2018, 2019; Kuhn *et al.* 2019) (Figure 8.1). Although not universally accepted (Holmes and Duchene 2019), this scenario is in line with the observation that positive-sense RNA viruses infect both eukaryotic and prokaryotic hosts, whereas all known negative-sense RNA viruses infect eukaryotes only, which is consistent with their relatively recent emergence (Wolf *et al.* 2018). RNA virus RdRps also show homology to the reverse transcriptases (RTs) of group II introns and other non-LTR (Long Terminal Repeat) retrotransposons. Group II introns are common in prokaryotes and they are thought to be the ancestors of the RTs of all other retrotransposons and retroviruses/pararetroviruses (recently classified as the order *Ortervirales*) (Koonin and Dolja 2014; Krupovic *et al.* 2018) (see below).

The gene encoding the RdRp is the only universal feature among RNA viruses. It was thus suggested that the last common ancestor of eukaryotic RNA viruses encoded only the RdRp and, possibly, a single jelly-roll (SJR) capsid protein, which is the most common type of capsid protein among positive-sense RNA viruses (Wolf *et al.* 2018). Extensive structural comparisons also indicated that the ancestral SJR viral protein was probably acquired from a cellular protein. In fact, cellular SJR proteins are ubiquitous and some of them are able to form multi-subunit virus-like particles (Krupovic and Koonin 2017). Because the SJR capsid is not universal among RNA viruses, the hypothesis that the last common ancestor of eukaryotic RNA viruses encoded such a protein also implies that the SJR-based capsid was lost multiple times and replaced with other capsid folds. In fact, structural analyses indicated that, most likely, several other capsid proteins emerged through the acquisition of cellular or viral genes (Krupovic and Koonin 2017). Importantly, the gain of one or more helicases is thought to have allowed RNA virus genome expansion, as the majority of viruses with genomes longer than 6 kb encode a helicase protein (Wolf *et al.* 2018). In general, gene gain must have been a major evolutionary force during the diversification of RNA viruses and it is thought to have involved module exchange among diverse viruses and acquisition of cellular genes (Koonin *et al.* 2015b; Shi *et al.* 2016; Wolf *et al.* 2018) (see also below). As a result, extant viral genomes encode diverse proteins, some of them specific for distinct viral families or genera. For instance, several viruses in the picornavirus-like superfamily encode a chymotrypsin-like protease (3CPro) and a superfamily 3 helicase (S3H) which are thought to have been independently acquired from a cellular gene and a plasmid homolog, respectively (Koonin *et al.* 2008; Kazlauskas *et al.* 2019).

Recent meta-transcriptomic approaches indicated that a large diversity of RNA viruses is present in invertebrates and cold-blooded vertebrates (Shi *et al.* 2016; 2018). Analyses of these viral genomes showed that RNA viruses evolved by frequent recombination events, lateral gene transfers (among viruses or between a virus and its hosts), gene gains and losses, as well as genome rearrangements. Overall, these analyses also indicated that most viral families that infect mammals have counterparts in aquatic vertebrates and in invertebrates, clearly implying that they are evolutionarily very ancient (Shi *et al.* 2016; 2018). In particular, the analysis of RNA viruses in fish, amphibians and reptiles revealed that the macroevolution of most viral families was driven by virus–host co-divergence, although host switching events were also common. Thus, the origin of most human-infecting RNA viruses can be traced back to aquatic metazoans and paralleled the emergence of terrestrial animals (Shi *et al.* 2018).

Extant human-infecting RNA viruses are highly diverse in terms of genome size and organization, as well as of infection cycle and transmission dynamics. Their pathogenetic potential is also highly variable ranging from dangerous human pathogens with high fatality rates, such as Ebola virus (EBOV) (family *Filoviridae*) and rabies virus (family *Rhabdoviridae*), to viruses that usually cause no or little symptoms in our species (e.g. several members of the *Togaviridae* family). All RNA viruses, however, share a common feature: an error-prone RdRp, which determines high rates of sequence change and fast adaptation to new hosts/environments (Holmes 2003) (see below). Such high error rates are also thought to limit genome size in RNA viruses, because, as genome size increases, the number of deleterious mutations also increases to a point that affects fitness (Holmes 2003). Thus, human RNA virus genomes have an average length of 9–10 kb, with only a few families deviating from this size (Belshaw *et al.* 2007; Simon-Loriere *et al.* 2013). The most notable exception is represented by the *Coronaviridae* family, whose members represent the largest vertebrate-infecting RNA viruses identified to date (Forni *et al.* 2017). The family includes recently emerged human pathogens such as severe acute respiratory syndrome coronavirus (SARS-CoV) and Middle East respiratory syndrome coronavirus (MERS-CoV), both with genomes longer than 29 kb. All coronaviruses have non-segmented positive-sense RNA genomes with similar organization and encode about 20 proteins common to all family members, plus a variable number of accessory proteins (Forni *et al.* 2017). Increased replication fidelity most likely contributed to genome expansion in coronaviruses. In fact, a major step that allowed genome expansion was the acquisition of a set of RNA-processing enzymes (e.g. an RNA 3'-to-5' exoribonuclease and an endoribonuclease) that improved the low fidelity of RNA replication (Forni *et al.* 2017). However, features distinct from replication fidelity, such as a peculiar genome organization and a processive replication complex, may have also contributed to genome expansion in this family (Forni *et al.* 2017). Whatever the underlying mechanisms, genome expansion allowed coronaviruses to acquire and maintain genes encoding diverse accessory proteins, which in turn contribute to virus adaptation to specific hosts, as well as to the suppression of immune responses (Forni *et al.* 2017). In fact, coronaviruses have a high propensity for host shifts: both MERS-CoV and SARS-CoV originated in bats and were transmitted to humans via intermediate hosts (camels and small carnivores, respectively).

Members of the *Reoviridae* family also have relatively large genome (around 23 kb). However, their genome organization is totally different from that of coronaviruses, as the reovirus dsRNA genome is segmented. The *Reoviridae* family includes several clinically important human pathogens, the best studied of which are rotaviruses. Rotaviruses cause life-threatening gastroenteritis in infants and, before the introduction of vaccination campaigns, these viruses were responsible for substantial morbidity and mortality (Tate *et al.* 2012). The rotavirus A genome consists of 11 segments. Genome segmentation is common to other important human viruses, including members of the *Orthomyxoviridae* (e.g. influenza virus), *Arenaviridae* (e.g. Lassa virus), *Phenuiviridae* (e.g. Rift Valley fever virus) and *Nairoviridae* (e.g. Crimean-Congo hemorrhagic fever virus) families (McDonald *et al.* 2016). The origin of genome segmentation is still a matter of debate, and different theories have been proposed to explain how this feature evolved (McDonald *et al.* 2016). One hypothesis is that segmented viruses evolved as a result of polyploidy or diploidy followed by diversification of the additional genomes. Another possibility is that segmentation resulted from the accidental packaging of RNA genomes from different viruses. Of course, these theories are not mutually exclusive as there is no indication that segmentation arose only once during RNA virus evolution. Conversely, recent studies have indicated that, over long evolutionary timescales, the evolution of segmented genomes, the loss of segmentation or a change in the number of segments has occurred (Shi *et al.* 2016; 2018). Most mammalian segmented viruses have the capacity to exchange genome segments during co-infection. This process, referred to as reassortment, clearly provides an additional strategy for genetic diversification (McDonald *et al.* 2016). Reassortment is thought to be advantageous in several respects, for instance by allowing acquisition of new antigenic combinations (see below). Several data, however, also indicate that reassortment can decrease viral fitness by disrupting interactions among co-evolving proteins encoded by different segments (McDonald *et al.* 2016). Interestingly, reassortment has been exploited as a strategy to produce live-attenuated vaccines against influenza A viruses and rotavirus A (McDonald *et al.* 2016).

Viruses belonging to a few other viral families have longer than average genome sizes; these include EBOV and the Marburg virus (family *Filoviridae*), both with genomes around 19 kb. Infection with both viruses is thought to be zoonotic and to be transmitted to humans by a reservoir host, most likely bats (Malvy *et al.* 2019). Until recently, filoviruses had only been described in mammals. However, the above-mentioned meta-transcriptomic approaches revealed that filoviruses with very similar genome organization as the mammal-infecting viruses exist in fish (Shi *et al.* 2018). This finding suggests that similar unbiased analyses in other animal hosts may further expand knowledge on the diversity of this viral family, which represents an epitome of the potentially devastating and largely unpredictable consequences of virus spillovers to our species.

It should be noted, however, that there is no general correlation between genome complexity or length and pathogenetic potential or overall disease burden in human populations. Human viruses with small and simple genomes can cause (or have caused in the human population throughout history) a huge burden in terms of morbidity and mortality. Examples include influenza A virus (family *Orthomyxoviridae*, segmented negative-sense linear RNA genome of ~13 kb) and hepatitis C virus (family *Flaviviridae*, non-segmented positive-sense linear RNA genome of ~10 kb) (see below), as well as a number of arthropod-borne viruses. Another interesting example is the measles virus, a member of the *Paramyxoviridae* family. The virus has a non-segmented negative-sense RNA linear genome of about 16 kb. The measles virus is considered the most contagious virus that infects humans (Nambulli *et al.* 2016) and, although infection is often self-resolving, complications are relatively common, with fatality rates of 1–2 out of 1000 infected people[2]. Infection with measles generates lifelong immunity, meaning that, for the virus to persist, it must frequently encounter new susceptible individuals. This clearly depends on the population reaching a certain threshold density or interconnection with other communities. Thus, data on human population history, which was initially characterized by small and close communities, suggest that the virus could not be maintained in our species until a few thousand years ago, and that, necessarily, it was introduced from another species (Keeling and Grenfell 1997; Nambulli *et al.* 2016). To date, the closest relative of measles virus, rinderpest virus, was identified in ruminants (Nambulli *et al.* 2016), suggesting that the close contact between humans and domestic animals originated the human virus in the 11th–12th Centuries (Furuse *et al.* 2010). These observations clearly imply that the evolutionary history of human viruses is tightly linked to the evolution of our species, not only in terms of genetic make-up, but also in a wider ecological and cultural perspective (see also below).

8.2.2. *Origin and ancient evolutionary history of human-infecting reverse-transcribing viruses*

Reverse-transcribing viruses, which synthesize a copy of genomic DNA from an RNA template, include important human pathogens, such as human immunodeficiency virus (HIV, order *Ortervirales*, family *Retroviridae*) and hepatitis B virus (HBV, family *Hepadnaviridae*) (Krupovic *et al.* 2018). These viruses, as well as a plethora of diverse retroelements identified in prokaryotes and eukaryotes, share a single common feature: the presence of an RT gene (Koonin and Dolja 2014). Retroelements are poorly represented in prokaryotes, whereas they are extremely common in eukaryotes, with some eukaryotic genomes being replete with different retroelement copies. Evolutionary analyses have indicated that all viral RTs are likely to have a common root, and it was proposed that group II introns represent the ancestors of all eukaryotic retroelements (Koonin and Dolja 2014). As in the case of RNA viruses, gene gains and losses are thought to have promoted the subsequent diversification of eukaryotic retroelements. Such early events also led

2. https://www.cdc.gov/vaccines/pubs/pinkbook/meas.html#measles.

different retroelements to adopt diverse genome organizations and infection cycles. Thus, retroviruses, as well as other viruses that have never been detected in mammals (families *Belpaoviridae*, *Metaviridae*, *Pseudoviridae*), have single-stranded RNA genomes with LTRs and commonly integrate into the host genomes as part of their replication cycles (Baltimore class VI). Conversely, members of the *Hepadnaviridae* and *Caulimoviridae* (plant-infecting viruses) families encapsidate a circular double-stranded DNA (dsDNA) genome (with no LTR) and do not actively integrate into host chromosomes (Krupovic *et al.* 2018). However, the RT of hepadnaviruses is the most divergent among those of the other viral families and these viruses possess an unrelated capsid protein. For these reasons, hepadnaviruses are not classified in the same order (*Ortervirales*) as the other reverse-transcribing eukaryotic viruses (Krupovic *et al.* 2018).

Retroviruses are further divided into two recognized subfamilies (*Orthoretrovirinae* and *Spumaretrovirinae*). Spumaviruses (also known as foamy viruses) are considered the most primitive retroviral form (Hayward *et al.* 2015). Retroviruses have been detected in all major groups of vertebrates but, at present, not in invertebrates (Hayward *et al.* 2015; Xu *et al.* 2018), suggesting that these viruses originated and diversified with their vertebrate hosts. Indeed, evidence for this hypothesis was recently obtained: using endogenous viruses integrated in the genomes of amphibians and fish, it was estimated that spumaretroviruses originated in the ocean at least 450 million years ago (Aiewsakun and Katzourakis 2017). Together with evidence supporting the co-speciation history of vertebrates and foamy viruses (Xu *et al.* 2018), these estimates indicate that spumaretroviruses (and thus all retroviruses) emerged with their jawed vertebrate hosts and underwent parallel water-to-land transition (Aiewsakun and Katzourakis 2017). In line with the long-standing association with their hosts, foamy viruses infect a wide variety of mammals including cats, cows, horses, bats and non-human primates (NHPs), but they are usually non-pathogenic. The zoonotic transmission of foamy viruses to humans has been documented in several circumstances and it is usually asymptomatic (Pinto-Santini *et al.* 2017).

Conversely, the *Orthoretrovirinae* subfamily includes two prominent human pathogens, namely HIV (human immunodeficiency virus, genus *Lentivirus*) and human T-cell leukemia virus (HTLV, genus *Deltaretrovirus*), both of which have an Old-World NHP origin. HTLV was discovered in 1980, before HIV, and its geographic distribution is characterized by clusters of high endemicity that include the Southwestern part of Japan, sub-Saharan Africa and South America (Gessain and Cassar 2012). To date, four types of HTLV (HTLV-1 to -4) have been described in humans and for three of them their simian counterparts (STLVs) have been identified (Peeters *et al.* 2014). Among non-primate mammals, only cattle are known to be infected by a virus related to HTLV (bovine leukemia virus) (Polat *et al.* 2017). HTLVs and STLVs are formally classified in three species referred to as *primate T-lymphotropic virus (PTLV)-1* to *-3*. For HTLV-4, a simian counterpart has yet to be identified, but the virus was detected in African bushmeat hunters, hinting at a wildlife (primate) origin (Wolfe *et al.* 2005b). Molecular dating estimated that HTLV-4 originated in the same period as our species (Switzer *et al.* 2009), suggesting that early humans may have acquired the virus from an unknown primate reservoir. STLV-1 and STLV-3 infect a wide variety of primate species, indicating that they have low species-specificity and can easily be transmitted among primates living in the same geographic area (Peeters *et al.* 2014). In fact, human- and NHP-derived sequences of PTLV-1 and PTLV-3 are interspersed or closely related in phylogenetic trees, suggesting that multiple zoonotic events account for the genetic diversity of HTLV-1 and -3 in humans (Peeters *et al.* 2014). The origin of HTLV-2 is less clear, but the virus is thought to have emerged in Central Africa (Vandamme *et al.* 1998). Thus, the origin of HTLV as a human pathogen narrates a tale of frequent cross-species transmission from primates. However, the evolutionary events that led to the origin of these viruses in primates (and cattle) are presently unknown. Recently, the first endogenous deltaretrovirus was discovered in the genomes of minopterid bats, indicating that these viruses already existed in a time interval between 20 and 45 million years ago (Farkasova *et al.* 2017).

In analogy to deltaretroviruses, lentiviruses are rarely found as endogenous elements in mammalian genomes. Recently, though, the identification of endogenous lentiviruses in the genomes of lagomorphs (Katzourakis *et al.* 2007), lemurs (Gifford *et al.* 2008), ferrets (Cui and Holmes 2012b) and colugos (Han and Worobey 2015) allowed reconstruction of the ancient history of these viruses and indicated that they originated at least 14 million years ago (Han and Worobey 2015). In particular, phylogenetic analysis of the colugo lentivirus showed that it is a sister group of all known lentiviruses, either endogenous or actively replicating, and that it possesses primitive genomic features of lentiviruses such as the lack of *vif* (*viral infectivity factor*) and *rev* (*viral protein R*) genes, which were probably acquired at a later stage of lentiviral evolution (Han and Worobey 2015).

In addition to humans, extant actively replicating lentiviruses infect a relatively limited range of mammals, including NHP (simian immunodeficiency virus, SIV), cats (feline immunodeficiency virus), horses (equine infectious anemia virus) and ruminants (bovine immunodeficiency virus, caprine arthritis encephalitis virus). Phylogenetic analyses suggest that most lentiviruses have co-evolved for a long time with their mammalian hosts (Foley *et al.* 2016). However, cross-species transmission events have also occurred and these originated HIV-1 and HIV-2, as detailed below.

Finally, as anticipated above, hepadnaviruses, a group of reverse-transcribing dsDNA viruses to which HBV belongs, retain the reverse transcription stage in the reproduction cycle, without integration in the host genome (Nassal 2008). These viruses also encode a capsid protein unrelated to that of reverse-transcribing viruses in the order *Ortervirales* (Krupovic *et al.* 2018).

Hepadnaviruses were detected in several mammals (NHPs, rodents and bats), birds, fish and reptiles (Rasche *et al.* 2016; de Carvalho Dominguez Souza *et al.* 2018). Recently, Lauber *et al.* discovered a family of fish viruses (nackednaviruses) which lack the envelope protein gene, but share genomic features with hepadnaviruses (Lauber *et al.* 2017). This finding, together with the discovery of endogenous hepadnavirus elements integrated in the genome of birds and reptiles (Cui and Holmes 2012a; Liu *et al.* 2012; Gilbert *et al.* 2014; Suh *et al.* 2014), indicated that the *Hepadnaviridae* family originated some hundred million years ago. The identification of nackednaviruses led Lauber and coworkers to suggest that hepadnaviruses evolved from non-enveloped ancestors and that acquisition of the envelope protein occurred *de novo*, allowing a major transition in viral lifestyle. The same authors also estimated that the genotypes of human HBV (including isolates from NHPs) emerged during the last 30 million years (Lauber *et al.* 2017). The more recent history of HBV in human populations seems to be highly complex. HBV is one of the few human viruses that could be sequenced from archaeological human samples. These include a 16th Century Asian mummy (Kahila Bar-Gal *et al.* 2012), European Neolithic remains (Krause-Kyora *et al.* 2018) and Eurasian samples of different ages (800–4500 years old) (Muhlemann *et al.* 2018). Analysis of these ancient viral genomes indicated that some HBV lineages of the past went extinct and that the geographic distribution of ancient samples does not match the modern genotype distribution (Muhlemann *et al.* 2018). Thus, a high level of uncertainty concerning the ultimate origin of HBV still exists. Moreover, these observations imply that, in the absence of archeological samples or of genome-integrated fossil elements, analysis of extant viral lineages can provide only partial information on the evolutionary events that shaped viral diversity and distribution.

8.2.3. *Origin and ancient evolutionary history of human-infecting DNA viruses*

Similarly to RNA viruses, DNA viruses display an extraordinary diversity of genome architectures and sizes, replication mechanisms, virion structures and host tropisms. Their genomes may consist of double-stranded DNA (dsDNA) or single-stranded DNA (ssDNA).

Most single-stranded DNA viruses are small in size (few kilobases) and their genomes typically encode less than 10 genes; conversely, double-stranded DNA viruses exhibit a wide range of genome sizes (~5 to ~2.5 × 10^3 Kb) and a highly variable gene content (from 5 up to ~2600 genes) (Koonin *et al.* 2015a). For example, "giant viruses" (dsDNA viruses recently discovered in eukaryotes) display physical dimensions and genome sizes exceeding those of many bacteria and archaea, as well as of some parasitic unicellular eukaryotes (Raoult *et al.* 2004; Philippe *et al.* 2013).

The reconstruction of the evolutionary history of DNA viruses represents a much debated topic. Different studies have pointed out how the evolution of viruses in eukaryotes was driven by a complex network of gene exchanges between different selfish elements (e.g. bacteriophages, transposable elements, linear DNA plasmids), as well as by a parallel acquisition of host genes (Koonin *et al.* 2015b). According to the latest reports, eukaryotic ssDNA viruses evolved via a fusion of genes from prokaryotic rolling circle-replicating plasmids and positive-sense RNA viruses (Kazlauskas *et al.* 2019). Concerning dsDNA viruses, the majority originated from a specific group of bacteriophages, called *Tectiviridae*. However, some dsDNA virus families originated in independent instances: herpesviruses originated from a distinct bacteriophage root, whereas polyoma/papillomaviruses (PVs) are assumed to derive from ssDNA viruses and ultimately from plasmids (Koonin *et al.* 2015b) (Figure 8.2).

Extant DNA viruses include many important mammalian and human pathogens, such as herpesviruses, poxviruses, PVs and polyomaviruses (PyVs). Herpesviruses are a large group of dsDNA viruses (order *Herpesvirales*) that infect many vertebrate host species (Davison *et al.* 2009). They are extremely widespread among humans, encompassing herpes simplex virus-1 and -2 (HSV-1 and HSV-2, major causes of orolabial and genital herpes, respectively), varicella zoster virus (VZV, the cause of chickenpox and shingles), Epstein–Barr virus (EBV, implicated in several lymphoproliferative diseases, including mononucleosis and cancers) and human cytomegalovirus (HCMV, the most common infectious cause of birth defects). Herpesviruses are highly common and causative of severe diseases also in other mammals (e.g. elephantid herpesvirus 1, alcelaphine herpesvirus 1, ovine herpesvirus 2) (Sood *et al.* 2013; Long *et al.* 2016), and can be strongly detrimental in the context of animal farming, as for bovine herpesvirus type 1 (BoHV-1), causative of infectious bovine rhinotracheitis (IBR), a highly contagious disease which affects the respiratory tract and the reproductive system of calves (Muylkens *et al.* 2007).

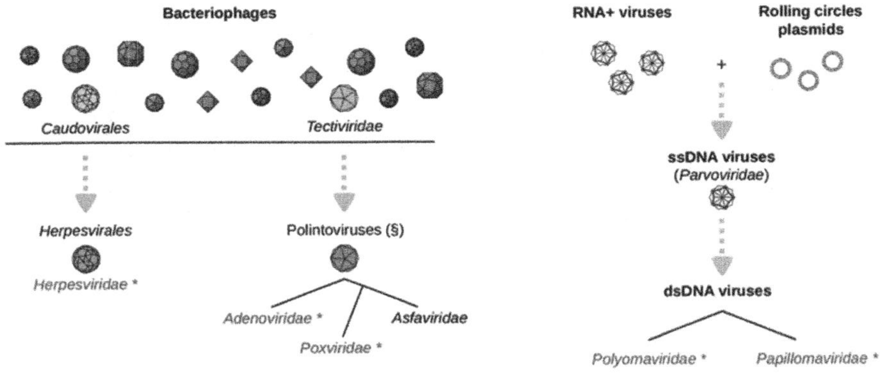

Figure 8.2. Origin of mammalian-infecting dsDNA viruses. *Families that include important human pathogens are shown in red with an asterisk. (§) Currently unclassified by the ICTV because the existence of these virions is yet to be validated*

Unlike the majority of eukaryotic dsDNA viruses, herpesviruses show a distinct morphogenetic connection to bacteriophages belonging to the order *Caudovirales* (Figure 8.2). Indeed, in both orders, the homologous major capsid proteins, which are the key component of the viral capsid, share the same protein fold, known as the Hong Kong 97 (HK97)-fold, so called after the X-ray resolution of the first coat protein structure of this group of bacteriophages (Conway *et al.* 1995). Moreover, members of the two orders show similarities in capsid assembly and maturation mechanisms, as well as in several other proteins (Baker *et al.* 2005; Selvarajan Sigamani *et al.* 2013; Rixon and Schmid 2014). Because the host range of herpesviruses is limited to animals, their bacteriophage ancestor is assumed to derive from a bacterial symbiont of early animals (Koonin *et al.* 2015b).

All herpesviruses that infect humans belong to the *Herpesviridae* family, which is divided into three subfamilies (*Alpha-*, *Beta-* and *Gammaherpesvirinae*). Ample evidence suggests that herpesviruses have co-diverged with their hosts for millions of years. Indeed, in each of the three herpesvirus subfamilies, many features of branching and approximate relative scale in the virus tree show congruence with features in the phylogeny of corresponding host organisms (McGeoch and Cook 1994; McGeoch *et al.* 1995, 2000). Due to the long-time co-divergence with their host, herpesviruses are generally species-specific (Davison *et al.* 2009). For example, in the *Gammaherpesvirinae* subfamily, the origin of three genera is well defined: percaviruses are considered to have originated in perissodactyls (mainly equids) and carnivores (felids and mustelids), macaviruses in artiodactyls, and lymphocryptoviruses in primates (McGeoch *et al.* 2005; Ehlers *et al.* 2008).

A completely different evolutionary scenario was proposed for poxviruses (family *Poxviridae*). In analogy to several other dsDNA viruses infecting eukaryotes, poxviruses were suggested to derive from Polintons, large DNA transposons that are common in eukaryotic genomes (Krupovic and Koonin 2015). In turn, Polintons possibly evolved from tectiviruses (which infect bacteria) and gave rise to several groups of large and giant eukaryotic dsDNA viruses, which were proposed to be included in a new order (*Megavirales*) (Colson *et al.* 2013) (Figure 8.2).

The *Poxviridae* family includes several species that infect humans and their domestic animals, such as variola virus (VARV), the causative agent of smallpox. Other mammal-infecting poxviruses are also pathogenic to humans, although with mild effects; these include vaccinia virus (VACV), cowpox virus (CPXV) and monkeypox virus (MPXV) (Fenner 2000; Moss 2001; Shchelkunov *et al.* 2006). Phylogenetic and evolutionary studies of the genetic structures of poxviruses suggested that VARV is a relatively young virus that emerged approximately 3000–4000 years ago in East Africa, in accordance with archeological, epidemiological and historical records regarding the appearance of smallpox as a human disease (Babkin and Babkina 2012, 2015; Firth *et al.* 2010; Hughes *et al.* 2010).

Phylogenetic analyses also supported the view that transfers from one host species to another have been a recurrent feature of the evolution of poxviruses, rejecting the idea of a long-standing co-evolution with their hosts (Hughes *et al.* 2010). CPXV possesses the broadest host range and all genes characteristic of orthopoxviruses. It is thus thought that the common progenitor of orthopoxviruses was CPXV-like and possibly able to infect a wide range of hosts. Presumably, the CPXV-like ancestor spread mainly in rodent populations, which are currently the natural reservoirs of the majority of orthopoxviruses (Essbauer *et al.* 2010), before coming into contact with camels, which were introduced in Africa 3500–4500 years ago as a new species (Uerpmann and Uerpmann 2002; Babkin and Babkina 2015). During a relatively short time span, further divergence due to subsequent adaptation of the ancestral virus to new hosts (i.e. camel, naked sole gerbil and human) led to the emergence of three new highly specialized viruses: the camelpox virus (CMLV), the taterapox virus (TATV) and VARV, respectively (Babkin and Babkina 2015).

PVs and PyVs constitute two closely related families of dsDNA viruses with small, circular genomes. These viruses present an origin completely distinct from that of all larger dsDNA viruses, as they seem to be derivatives of ssDNA viruses, which in turn originated by the recombination of bacterial rolling circle–replicating plasmids and positive-sense ssRNA viruses (Krupovic 2013) (Figure 8.2). Indeed, ssDNA viruses present the same Rep-dependent replication mechanism typical of bacterial plasmids, as well as a single jelly-roll capsid protein found in ssRNA viruses (Koonin *et al.* 2015b; Kazlauskas *et al.* 2019). Both PVs and PyVs show striking homologies in the beta-barrel jelly-roll capsids and in some replicative proteins typical of these ssDNA viruses (i.e. circoviruses, nanoviruses, geminiviruses and parvoviruses) (Koonin *et al.* 2015b). In particular, very recent analyses on the SF3 helicase domains confirmed parvoviruses as the likely viral ancestors, although the detailed evolutionary events have not been elucidated yet (Kazlauskas *et al.* 2019).

PVs infect stratified squamous epithelia in a wide range of vertebrates, spanning all amniotes (Rector and Van Ranst 2013). Members of this family are causative agents of malignant tumors, such as cervical cancer, while others are associated with benign proliferative lesions (Doorbar *et al.* 2015). Conversely, despite their name, only a few PyVs are causative of cancer. Human PyVs can induce several diseases in immunocompromised patients, whereas they seem to be quite harmless in healthy individuals; less is known about the pathogenicity of mammalian-infecting PyVs.

For both PVs and PyVs, an ancient association with their hosts was assumed, suggesting co-divergence as the major driver in their evolutionary history (Geoghegan *et al.* 2017; Moens *et al.* 2017). However, with the expansion of viral sequences detected in an ever increasing number of animal species, some deviations from a simplistic model of co-divergence between virus and host have emerged. The discovery of highly divergent PyV sequences associated with fish and arthropods indicated that additional evolutionary processes occurred at different timescales, including host switching, intra-host viral speciation, incomplete lineage sorting and recombination (Sharp and Simmonds 2011; Buck *et al.* 2016). Indeed, although PyVs have been gradually co-evolving with their animal hosts for at least half a billion years, phylogenetic analyses performed on individual gene sequences suggest ancient recombination events involving distantly related ancestral PyV lineages that contribute to shape the genetic diversity observed in modern avian and mammalian PyVs (Buck *et al.* 2016).

Some exception to the strict co-divergent model was also reported for PVs. Different lines of evidence support the view that, before extensive virus–host co-speciation, an ancestral generalist PV went through a "niche sorting" process, giving rise to some niche-specialized viruses (Bravo and Alonso 2007; Van Doorslaer 2013). The same process of specific host niche adaptation was confirmed to be the first stage of evolution of primate PVs. After this sorting, primate PVs co-evolved with their host for at least 40 million years (Chen *et al.* 2018). Interestingly, recent evidence also suggested that some PV lineages were transmitted from Neanderthals to modern non-African humans through sexual interaction (Pimenoff *et al.* 2017; Chen *et al.* 2018).

Other DNA viruses that infect humans belong to the *Adenoviridae* (dsDNA) family, as well as to the *Anelloviridae* and *Parvoviridae* (both ssDNA) families. Adenoviruses, which are thought to share their ancient evolutionary history with poxviruses (Figure 8.2) have linear genomes of 30–35 kb in length. These viruses have been detected in numerous and diverse vertebrate species, but never in invertebrates (Harrach *et al.* 2019). Human-infecting adenoviruses (HAdVs) belong to different species, all of them classified in the Mastadenovirus genus[3]. HAdVs are increasingly recognized as causative of clinical illnesses, including pneumonia, pharyngitis, bronchiolitis, meningitis, hemorrhagic cystitis, hepatitis and conjunctivitis. The general view is that adenoviruses coevolved with their vertebrate hosts, although host-switch events most likely occurred several times (Harrach *et al.* 2019). Recent evidence also indicated that the zoonotic transmission of adenoviruses may not be rare, especially between humans and NHPs (Borkenhagen *et al.* 2019).

Adenoviruses can also function as helper viruses for some members of the *Parvoviridae* family, which includes both autonomous and non-autonomous viruses. Parvoviruses have small (~4 to 6 kb) linear genomes, but a large host spectrum. Indeed, these viruses infect both vertebrates and invertebrates, suggesting that they originated before the radiation of the major animal phyla (Koonin *et al.* 2015b). In humans, several autonomous parvoviruses cause disease (e.g. B19 parvovirus, which is responsible for fifth disease, transient aplastic crisis, and other conditions), whereas adeno-associated parvoviruses are generally non pathogenic (Söderlund-Venermo 2019). For both parvoviruses and anelloviruses, the latter comprising several species of mammalian viruses with small genomes, the advent of next-generation sequencing and metagenomic approaches has unveiled a large viral genetic diversity in humans and other animals (Söderlund-Venermo 2019; Focosi *et al.* 2016). Thus, the relevance of these and other ssDNA viruses (e.g. redondoviruses (Abbas *et al.* 2019)) for human health remains to be comprehensively evaluated.

8.3. Sources of viral genetic diversity

The huge biological diversity that characterizes viruses suggests the presence of different strategies to generate genetic change. Mutation, recombination, duplication, gene gain and gene loss account for virus genetic diversity, providing the raw materials for evolutionary change.

3. https://talk.ictvonline.org/taxonomy/.

Mutations are the ultimate source of genetic variation and mainly result from errors during replication. Most mutations are "silent" and do not result in any obvious changes to the fitness of progeny virions, but others confer advantages that increase the fitness of the viruses in the environment (e.g. by facilitating the evasion of within-host immune responses or herd immunity in the host population, or by promoting host switches).

Mutation rates (the rates at which mutations are generated per genome per round of replication) can vary across microorganisms, spanning up to several orders of magnitude. In particular, viruses are the biological systems with the widest variation in mutation rates, ranging between 10^{-4} (RNA viruses) and 10^{-8} (dsDNA viruses) substitutions per nucleotide per cell infection (s/n/c) (Sanjuan and Domingo-Calap 2016). Viral mutation rates are strongly correlated with the nucleic acid backbone: in general, RNA viruses have a higher mutation rate than retroviruses, which in turn change faster than DNA viruses. These differences in the mutation rate are in part attributable to the fidelity of the viral polymerase, as well as to the ability of a virus to correct errors through proofreading and post-replicative repairs (Duffy *et al.* 2008). Indeed, the higher per-site mutation rates of RNA viruses can be accounted for by RdRps that, lacking proofreading activity, are unable to correct mistakes during replication. On the other hand, lower DNA virus mutation rates (ranging between 10^{-8} and 10^{-6} s/n/c) are due to DNA polymerase proofreading activity and to the presence of error-correcting domains.

Single nucleotide mutations in a protein-coding gene may produce either a synonymous codon mutation (silent mutation) or an amino acid change, which could alter the functionality of the protein (non-synonymous codon mutation). Usually, non-synonymous mutations are deleterious. Nevertheless, in dynamic environments, they may provide an adaptive potential. In this scenario, viruses evolve by maintaining a balance between the establishment of the highest mutation rate that allows the best adaptation in specific environmental conditions and an excessive accumulation of harmful mutations that would lead to the loss of viability (Holmes 2003).

As anticipated above, mutation rates are also thought to constrain viral genome size: in order to reduce the probability of generating lethal mutations, viruses with higher mutation rates tend to have reduced genome lengths. Indeed, RNA viruses are characterized by smaller genomes than DNA viruses. However, exceptions have been reported, including coronaviruses (see above). In this perspective, single-stranded DNA viruses may also represent an exception, as mutation rates for some of them have been reported to be similar to those of ssRNA viruses (Sanjuan and Domingo-Calap 2016). Recently, Domingo-Calap and colleagues (Domingo-Calap *et al.* 2018) showed that BK PyV, a small double-stranded DNA virus, evolves as fast as single-stranded ones and RNA viruses. Moreover, they demonstrated a significant accumulation of coding substitutions within predicted HLA (Human leukocyte antigen)-C-bound viral peptides, suggesting viral evolution to evade HLA-C-mediated immune responses.

Recombination is another mechanism with great impact on viral evolution. In fact, even though mutation provides the first source of genetic diversity, the ability to combine different mutations has the potential to generate high viral diversity (Arenas *et al.* 2018). Recombination is widespread in viruses and occurs when at least two viral genomes co-infect the same host cell and exchange genetic material. Several factors affect recombination rates; these include circulation of different strains in the same geographical area, frequency of coinfection, genetic homology and viral load (Perez-Losada *et al.* 2015). Two types of recombination are recognizable (Austermann-Busch and Becher 2012; Scheel *et al.* 2013): homologous (involving the same sites in both parental strands) and illegitimate recombination (involving different sites). In all cases, these processes do not create new mutation, but introduce new combinations of existing ones, whose fate will ultimately be determined by natural selection and genetic drift. On one hand, recombination acts by preventing the accumulation of harmful mutations; on the other hand, it facilitates the introduction of genetic variants with different biological features (Arenas *et al.* 2018). This process is common in both DNA and RNA viruses and was associated with several evolutionary aspects, such as emergence of new viral species, generation of diverse viral genotypes, evasion of host immunity, vaccine escape and resistance to antivirals. Recombination events were also shown to contribute to the expansion of host viral ranges, alteration of transmission vector specificities, increases in virulence and pathogenesis, and modification of tissue tropism (Chopra *et al.* 2003; Woodford and Ellington 2007; Martin *et al.* 2011; Simon-Loriere and Holmes 2011; Arenas *et al.* 2018).

In many viruses, such as HIV-1, HSV-1 and HBV, recombination is an important element to generate genetic variability with possible clinical implications (Bowden *et al.* 2004; Araujo 2015; Arenas *et al.* 2016). In HBV, for example, recombination explains much of the great diversity among isolates. Indeed, recombination events played a major role in the emergence and establishment of novel genotypes with an enhanced virulence and could represent a critical mechanism by which HBV evades detection by the immune system and develops resistance to antiviral therapies (Simmonds and Midgley 2005; Araujo 2015; Castelhano *et al.* 2017). Moreover, evidence of recombination between human genotypes and NHP-derived HBV isolates has been reported, suggesting an involvement of this process also in cross-species transmission of some HBV variants (Vartanian *et al.* 2002; Magiorkinis *et al.* 2005; Simmonds and Midgley 2005).

As anticipated above, in RNA viruses with segmented genomes, a different mechanism (known as shuffling or reassortment) occurs to generate new combinations of genetic variants. Reassortment is the process by which viruses swap gene segments and it occurs when two distinct strains co-infect one cell. In analogy to recombination, reassortment through gene swapping has the potential to generate significant viral diversity, making possible rapid evolution under selective pressure (e.g. adaptation to new host environments, evasion of host immune responses, acquisition of resistance to antiviral drugs). However, the recombination process can occur anywhere in the viral genome (also in the middle of a gene) and it has the potential to also generate non-functional chimeric fusion protein. Conversely, in reassortment, entire genes or sets of genes are exchanged by swapping of segments, thus maintaining coding potential and protein integrity. Nevertheless, new gene/gene set combinations are not necessarily viable or fit. Reassortment is common to different virus families (i.e. *Cystoviridae*, *Orthomyxoviridae* and *Reoviridae*) (McDonald *et al.* 2016). Many segmented viruses, such as hantaviruses and mammarenaviruses, show a relatively low level of reassortment, while others display high rates; among the latter, influenza A virus and rotavirus A are the most studied and characterized (McDonald *et al.* 2016; Simmonds and Midgley 2005). In both these viruses, gene reassortment (hemagglutinin (HA) and neuraminidase (NA) for influenza A viruses, VP4 and VP7 for rotavirus A), in combination with a gradual accumulation of nucleotide mutations and amino acid substitutions, results in the emergence of new antigenic variants (antigenic shift), which in turn are associated with the evasion of host immunity and sometimes to the occurrence of epidemics (see also below) (McDonald *et al.* 2016; Petrova and Russell 2018).

Viral genomes are also dynamic in terms of gene content and, consequently, genome size. This genomic variability is associated with the ability of viruses to lose (gene loss) or acquire (gene gain) genes. In the latter case, three possible mechanisms are involved: (1) gene duplication (within genome) followed by extensive sequence divergence; (2) recombination between genes, which produces novel protein products; and (3) horizontal gene transfer (HGT). Again, these events are important for adapting to the host and for subverting the host antiviral response.

HGT provides gene flow between evolutionarily distant lineages. Several cases of host-to-virus gene transfer are known, and in a number of these, the transferred gene is involved in immune responses (e.g. genes encoding cytokines, chemokines and interferon regulatory factors). The immune-related factor might be used by the virus to trick the immune response of the host to gain immune evasion or the acquired gene could contribute to adaptation to a specific host. Examples of cellular genes acquired by the host have been described in detail in several RNA viruses such as coronaviruses, picornaviruses, potyviruses and nidoviruses (Forni *et al.* 2017; Wolf *et al.* 2018). In large eukaryotic double-stranded DNA viruses, high proportions of cellular genes result from host-to-virus HGT (Holzerlandt *et al.* 2002; Hughes and Friedman 2003; Filee and Chandler 2010). For example, it was estimated that at least 10% of giant virus genes and from 9% to 30% of herpesvirus genes likely originated from eukaryotic or prokaryotic genomes (Holzerlandt *et al.* 2002; Hughes and Friedman 2005; Moreira and Brochier-Armanet 2008).

The cellular genes that can be identified in a large proportion of viral genomes most likely result from relatively ancient HGTs, while genes that appear in a subset of viruses probably are the result of relatively recent gene transfer, suggesting that this process has had and continues to play an important role in defining gene composition in viral genomes.

8.4. Viral evolution and host range

The number of host species that a virus can infect constitutes its host range. The amplitude of the host range is determined by several factors. Among human viruses, there are very contrasting examples of host specificity. Several viruses possess such a narrow host range that they are adapted to infect only one species (e.g. VARV, HCV, HSV-1 and HSV-2, HCMV). On the contrary, there are viruses, such as rabies virus or hepatitis E virus, that infect a very wide host spectrum, even spanning species from different mammalian orders (Figure 8.3). In general, the level of host specialization depends both on the ability of the virus to adapt to new host species and on the opportunity of transmission. A wide host range is likely the result of different host shift events in the evolutionary history of the virus, whereas single-host specialization usually reflects long-standing host–virus interactions that often result in a balanced relationship between viral infectivity and host fitness (Woolhouse *et al.* 2001).

The process of host shifting is commonly driven by genetic changes (e.g. nucleotide substitutions, genetic reassortments) that allow a virus to overcome the barriers imposed by the new host. Indeed, appropriate cell-type recognition, cell entry/fusion, trafficking within the cell, genome replication, gene expression, viral maturation and evasion of the host immune system are all fundamental steps that are required for a virus to establish an efficient infection (Parrish *et al.* 2008). As RNA viruses tend to have higher mutation rates than DNA viruses, they have a higher propensity for host shifting. Indeed, most RNA viruses infecting humans are (or were) acquired from an animal source (i.e. they are zoonotic in origin) (see below) (Olival *et al.* 2017).

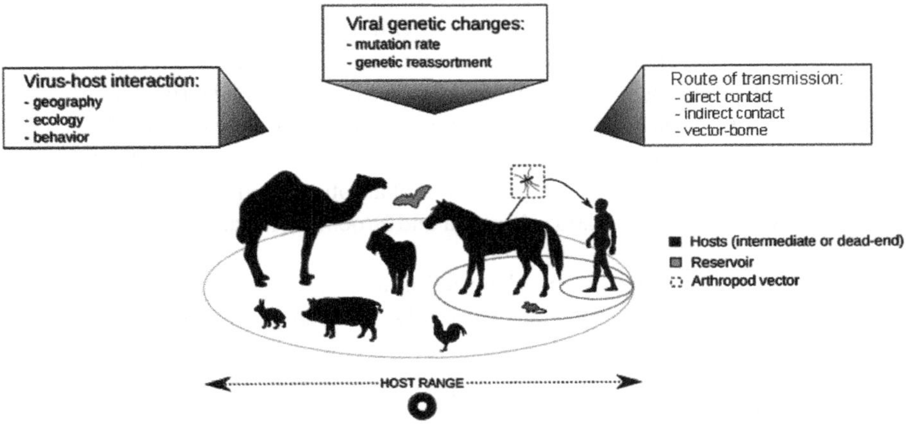

Figure 8.3. *Factors contributing to the amplitude of the viral host range*

Host shifts can sometimes occur with limited sequence change. An exemplar case is that of canine parvovirus type 2 (CPV-2, family *Parvoviridae*), which emerged in the 1970s. It has been widely assumed that CPV-2 originated from feline panleukopenia virus (FPV) due to two mutations in the capsid gene that allowed the recognition of canine transferrin receptor (TfR) (Chang *et al.* 1992). However, a phylogenetic analysis on a broader sampling of parvoviruses hosted by domestic and wild carnivore species challenged this view, suggesting that CPV and FPV independently evolved from a common ancestor (Allison *et al.* 2013). Nevertheless, in both lineages, few changes in the TfR binding domain define viral species-specificity, irrespective of the genetic background. Thus, these changes are adaptive and play a key role in defining the host range (Allison *et al.* 2014). Intriguingly, analysis of the evolutionary history of TfR revealed that the ancestors of dogs were likely infected by a parvovirus millions of years ago until they evolved resistance (Kaelber *et al.* 2012). This clearly implies that viruses, once eliminated from a host, can evolve and regain the infectivity for that species, and as well can shape the genetic diversity of host species. Another striking example in which limited viral changes underlie host shifts is represented by Venezuelan equine encephalitis virus (VEEV, family *Togaviridae*), which causes a severe febrile illness that may progress to fatal encephalitis in equines. The virus can also be transmitted to humans (as a vector-borne disease) with similar clinical manifestation. The last major outbreaks have taken place in Central and South America in the early 1990s, with over 100,000 cases reported in humans (Weaver *et al.* 1996). The VEEV epidemic strain (serotype IC) emerged from an equine-avirulent strain circulating among rodents, after a single mutation in the E2 envelope glycoprotein, which allowed the virus to replicate efficiently in equine species which, in turn, serve as an amplifier host for human infections (Anishchenko *et al.* 2006).

The maintenance of a high viral performance in the original host and the balance between virulence and transmission potential are important trade-offs during host shifts. Specifically, the maintenance of a wider host range depends on viral pathogenicity and epidemiology. High levels of virulence can be maladaptive because the window of time for viral transmission is reduced. Epitomal in this respect is the case of myxoma virus (family *Poxviridae*). In South American *Sylvilagus* rabbits, this virus causes benign localized cutaneous fibroma. However, when (in the 1950s) myxoma virus jumped into European rabbits (*Oryctolagus cuniculus*) as a new naive host species, it caused a lethal disseminated disease known as myxomatosis. The virulence was so high that it reached a mortality rate of 99.8% in Australia, leading the rabbit population close to extinction. As a consequence, attenuated strains with higher transmissibility out-competed the most virulent strain (Kerr 2012).

Clearly, a necessary condition to host shifts is the opportunity of contact between different host species, which depends on geographical, ecological and behavioral determinants. Over millennia, human activities have led to significant environmental changes which have disrupted the dynamic balance between viruses and host species, leading to the emergence of new infectious diseases (Woolhouse *et al.* 2005). Modifications in host distribution and host ecologies were primarily driven by urbanization and industrialization. These processes affected the distribution, the status and genetics of many host species through domestication, wildlife trade, habitat destruction, human population expansion and migrations. All these processes contributed to exponentially increased contact rates between host species, thus fostering the opportunity of viral exchange between multiple host species, including humans (Woolhouse *et al.* 2005; Kock 2015).

A major event during human evolutionary history was the transition from the nomadic lifestyle typical of early hunter-gatherer populations to sedentary farming communities, also committed to animal husbandry. This allowed human population expansion and increased the chances of transmission of several viral and non-viral pathogens. Human settlements with an ever increasing population density and close proximity to livestock holdings intensified human-to-human and animal-to-human contacts, fostering viral transmission among different hosts. As an example, the evolution of different hepatitis E virus genotypes was shown to occur at the same time as the initial domestication of some animal hosts (pigs, camels and rabbits) (Forni *et al.* 2018a).

Viruses can be transmitted between hosts through different transmission routes: direct contact (i.e. physical contact or *in utero*), indirect contact (e.g. food contamination, environmental contamination, farming practice, contacts with reservoirs/intermediate hosts) and by vectors (e.g. via biting arthropods, including mechanical transmission) (Figure 8.3). The route of transmission strongly influences the range of species that can be reached by the virus. In the viral host spectrum, some species can act as dead-end hosts with limited epidemiological impact, some can be intermediate hosts that can drive species jumps and some can act as reservoirs in which the virus persists. Among mammalian reservoirs, bats are recognized to be the natural host of many common human and animal viruses such as mumps and rabies viruses. They have also been identified as the natural reservoir of several viruses causing severe emerging diseases in the past decades. In addition to EBOV, SARS-CoV and MERS-CoV, we mention Nipah virus (NiV) and Hendra virus (HeV) (Han *et al.* 2015).

With about 1240 recognized species, bats represent the second largest and most diversified order of mammals. This order (Chiroptera) originated about 52.5 million years ago (Teeling *et al.* 2005). Interestingly, a similarly ancient origin was also traced back for some zoonotic viruses hosted by bat species (e.g. lyssaviruses (Badrane and Tordo 2001)) suggesting a long history of co-speciation. The ability of bats to fly and to cover long distances during seasonal migration, as well as the longevity of several species and their social behavior, can strongly favor viral dispersal (Han *et al.* 2015). To date, little is known about the immunological mechanisms adopted by bats to control viral infection without evident pathological effects. A recent comparative genome analysis revealed that flight-induced adaptations led to changes in genes shared by DNA damage pathways and by the innate immune system; such changes may have inadvertently affected bat immune function (Zhang *et al.* 2013). A preliminary report on adaptive immunity in fruit-eating bats (only two samples) revealed that the distribution of B-, T- and NK-cells in the blood, bone marrow and spleen is different from that of other mammalian species (Periasamy *et al.* 2019). However, this issue needs further investigation.

The drivers of recurrent spillovers in humans of bat-borne infectious diseases can be identified in the increased interactions between humans and bats due to the obligate sharing of habitats and food resources caused by human population growth and environmental exploitation (Han *et al.* 2015). The routes of viral transmission from bats to humans can be multiple. The transmission by intermediate host species, which are in close contact with humans and can amplify and transmit the virus, is certainly a possibility. These species can, for example, make contact with bat-borne viruses by the consumption of fruits discarded by bats and contaminated by their saliva and/or urine. Intermediate hosts were reported for SARS-CoV (small carnivores), MERS-CoV (dromedary camels), and in some localized outbreaks of NiV (pigs) and EBOV (great apes) (Chua *et al.* 2000; Leroy *et al.* 2005; Wang and Eaton 2007). However, transmission by direct contact or by aerosol transmission in caves, where bats roost, can also represent a mode of infection (Timen and Koopmans, 2008; Leroy *et al.* 2009; Kuzmin *et al.* 2010). A case of direct transmission by a bat's scratch has been reported for Duvenhage virus (DUVV, family *Rhabdoviridae*) in an elderly man from South Africa (Paweska *et al.* 2006). Moreover, the handling and the consumption of bat as bush-meat, a common practice in some regions of Africa, can result in transmission, as reported for the 2007 Ebola outbreak in the Democratic Republic of Congo (Leroy *et al.* 2009).

Like bats, rodents (order Rodentia) are also important reservoirs of a number of human zoonotic viruses that significantly impact public health. Interestingly, rodents share many features with bats that are supposed to affect the reservoir potential: an ancient evolutionary origin, a highly diverse number of species, peridomestic habitats and low metabolite rate during torpor (Luis *et al.* 2013). The transmission to humans prevalently occurs via inhalation of aerosolized rodent urine, saliva and feces (Kruger *et al.* 2011). Among viruses that have rodent reservoirs, we mention hantaviruses (family *Hantaviridae*), which are causative of hantavirus pulmonary syndrome (HCPS) and hemorrhagic fever with renal syndrome (HFRS) (Kruger *et al.* 2011), and arenaviruses (family *Arenaviridae*), which are the etiological agents of Lassa fever and other hemorrhagic fevers transmitted in Latin America (Charrel and de Lamballerie 2010).

For viruses that use vectors as a transmission route, the vector itself may determine whether there are opportunities for inter-species transmission, imposing further constraints on the host range. Nevertheless, in recent times, a rapid progression of vector-borne virus outbreaks (e.g. West Nile virus (WNV), Dengue virus (DENV), Chikungunya virus (CHIKV), yellow fever virus (YFV) and Zika virus (ZIKV)) was recorded worldwide (Ruckert and Ebel 2018). Vectors are typically arthropod species that can transmit infectious diseases between humans or from animals to humans. Many of these vectors are bloodsucking insects; they contribute to viral dispersal by ingesting the virus during a blood meal from an infected host, and by subsequently injecting it into new hosts during the following blood meals. Mosquitoes and ticks species are the most important vectors of arthropod-borne human viruses (Kock 2015). Thus, vector-borne viruses have to optimize their fitness in at least three players: the reservoir, the recipient host and the vector. There are many barriers that the virus, ingested during feeding, must cross within the arthropod vector in order to be transmitted; in competent vectors, the virus undergoes replication, usually in gut tissues, disseminates to secondary sites of replication, including the salivary glands, and is ultimately released into salivary secretions, so that it can be inoculated into the host during subsequent blood meals (Ruckert and Ebel 2018). Each of these barriers represents a bottleneck that shapes the genetic diversity of the virus, with possible effects on replication rate, host tropism and/or disease severity.

The power of a vector to drive viral spread is also influenced by the feeding behavior and the host preference. Many insects are generalist feeders, but some specialize on a single host species. For example, mosquitoes of the *Aedes aegypti* species feed almost exclusively on human hosts (Scott and Takken 2012) and are extremely efficient vectors for viruses that replicate to high titers in humans, such as Dengue (DENV, family *Flaviviridae*), Chikungunya (CHIKV, family *Togaviridae*) and ZIKA (ZIKV, family *Flaviviridae*) viruses. In contrast, several mosquito species within the genus *Culex* tend to prefer birds, making them efficient vectors for West Nile virus (WNV, family *Flaviviridae*) or other zoonotic vector-borne viruses (Farajollahi *et al.* 2011).

The dynamics of human-infecting vector-borne virus emergence are influenced by several factors, including environmental and demographic changes caused by human activities and viral adaptation to new vector species that colonize periurban and urban settings. For example, vector switching from sylvatic mosquito species (i.e. *Ae. africanus*, *Ae. furcifer*), circulating between non-human primates, to peridomestic vectors such as *Ae. aegypti* and *Ae. albopictus*, which are typical of urban areas of the tropics and the subtropics, provided access to humans as new naive host species for ZIKV and CHIKV (Eastwood *et al.* 2017; Mayer *et al.* 2017). The subsequent exchange of both viruses and vectors across continents due to travels and trades has then resulted in large outbreaks, as recently reported for ZIKV (Kindhauser *et al.* 2016; Mayer *et al.* 2017). This virus was first isolated in 1947 from a rhesus macaque monkey in Uganda (Dick *et al.* 1952). In 1954, the first human infection in a 10 year old Nigerian female was reported (Macnamara 1954). Since the 1970s, after a vector switch from sylvatic to domestic mosquito species, ZIKV rapidly expanded its geographic distribution worldwide, leading to several outbreaks in human populations. From Africa, ZIKV was introduced to Yap Island from Southeast Asia in 2007, spread eastward to French Polynesia and other Pacific Islands in 2013–2014 and, in a single year, it reached Latin America and disseminated further to North America (Kindhauser *et al.* 2016). Recent insight into ZIKV epidemiology has also indicated that person-to-person transmission can occur, both vertically, from mother to fetus, and horizontally through sexual transmission (Kindhauser *et al.* 2016).

In conclusion, the host range is the result of an intricate network of factors that contribute to maintain a delicate balance among the virus, its hosts and/or vectors, and the environment. The disruption and/or the unbalance of these relationships can result in the rapid emergence of severe outbreaks, as observed in the past decades.

8.5. Recent evolution of human RNA viruses – selected examples

RNA viruses are among the most prominent emerging pathogens (Ludwig *et al.* 2003) and they account for about 64% of human disease viruses (Woolhouse and Adair 2013). As anticipated above, most emerging human diseases caused by RNA viruses originated in animals (J Woolhouse *et al.* 2013; Woolhouse and Adair 2013; Zeller and Andersen 2018). Moreover, several animal RNA viruses cause non-specific febrile illness in infected people, even though humans are not the primary hosts. Transmission of these infections has been reported on (1) direct contact of human objects with infected animals (e.g. foot-and-mouth disease, which infects cloven-hoofed animals); (2) handling of infected organisms in the laboratory or farm (e.g. bluetongue virus and Newcastle disease virus, which affect ruminants and birds, respectively); (3) transmission by vectors (e.g. African horse sickness virus, louping ill virus); and (4) ingestion of contaminated food or water (caliciviruses such as swine vesicular exanthema virus, feline calicivirus, rabbit hemorrhagic disease virus).

It is commonly thought that many animal viruses, at some point, will spill over into humans, but then fail to establish onward transmission in our species. This process has been termed "viral chatter", whereby pathogens are repeatedly spread from animals to humans, often without further transmission (Wolfe *et al.* 2005a). For example, sequence data of HIV-1 and HIV-2 suggest that several prior transmission events into human populations occurred over the last century before these viruses emerged globally (see below) (Hahn *et al.* 2000). Other pathogens, such as avian influenza virus and Hendra virus, have also led to several small epidemics with little or no evidence of human-to-human transmission (Wolfe *et al.* 2005a)). Viral chatter is likely to be of great importance in the emergence of viruses, as it is believed to increase viral diversity and, as a consequence, the probability that a virus evolves the ability to infect and spread in human populations (Wolfe *et al.* 2005a; Morse *et al.* 2012; Kreuder Johnson *et al.* 2015).

Other animal viruses have great epidemic and even pandemic potential in humans. The best known example is most likely accounted for by influenza viruses (family *Orthomyxoviridae*), which are among the most successful and unpredictable human pathogens (Vijaykrishna *et al.* 2010; Steel and Lowen 2014; Pinsent *et al.* 2016; Rajao *et al.* 2019). According to the World Health Organization estimates, influenza viruses annually cause 1 billion cases of flu, 3–5 million cases of severe illness, and 300,000–500,000 deaths worldwide. Influenza viruses are classified into four genera in the *Orthomyxoviridae* family. Influenza A virus (genus *Alphainfluenzavirus*) and influenza B virus (genus *Betainfluenzavirus*) are responsible for classical flu symptoms and cause seasonal epidemics. Influenza C virus (genus *Gammainfluenzavirus*) is of little clinical relevance (generally asymptomatic), whereas the ability of influenza D virus (genus *Deltainfluenzavirus*) to infect humans is still unclear (Ghebrehewet *et al.* 2016). Influenza viruses have a single-stranded RNA genome with eight (A and B) or seven (C and D) segments, each carrying one or a few genes. When two influenza viruses of the same type infect the same cell at the same time, they can mix their segments. In particular, pigs are well-known "mixing vessels" for influenza A virus. Pig cells can be recognized and infected by both human and bird influenza viruses (as well as pig viruses) (Ma *et al.* 2008). Mixed infections in pigs can thus result in the release of viruses that contain a mixture of genetic material from the human and bird virus. This kind of swap is common for influenza viruses in nature. For example, the H1N1 influenza strain ("swine flu") that caused a pandemic in 2009 was generated by multiple reassortment events of viral lineages that had circulated in swine for years and that were originally acquired from humans and birds (Smith *et al.* 2009; Vijaykrishna *et al.* 2010).

Routine flu vaccination is the most important intervention for preventing illness and severe complications. Antiviral drugs are also available and play an important role for selected patients who have not been immunized or who do not develop protective immunity from the vaccine. Most drugs target NA, a virion surface protein that, along with HA, allows virus entry in the host cell. As surface antigens, HA and NA evolve rapidly (Chen and Holmes 2010; Bhatt *et al.* 2011). This tendency of all influenza viruses to vary and acquire changes in surface proteins allows them to bypass the barrier imposed by immunization and is responsible for the rise and spread of drug-resistant strains. Genetic changes can take place according to two distinct mechanisms:

– *Antigenic drift:* this is a gradual modification of the sequence of amino acids that make up proteins capable of stimulating an immune response. This phenomenon affects both A and B viruses (but in A viruses, it is more pronounced and frequent) and is responsible for seasonal epidemics. In fact, the new variants become sufficiently unrecognizable to antibodies to make a large number of individuals susceptible to infection (Shil *et al.* 2011; Webster and Govorkova 2014; Sandbulte *et al.* 2015).

– *Antigenic shift*: this is a phenomenon restricted to influenza A viruses and consists of the appearance of a viral strain with surface proteins (HA and/or NA) to which most of the population has no immunity. Antigenic shifts have the potential to cause pandemics; they are due to reassortments between different viruses or to the direct transmission of non-human viruses to humans (the most recent example is the one that occurred in Hong Kong in 1997) (Stephenson and Democratis 2006; Shao *et al.* 2017; Allen and Ross 2018). So, the source of the new subtypes is always animal viruses.

Zoonotic diseases clearly differ in their transmission potential. After the transmission event, some viruses are capable of sustained human-to-human transmission, possibly fueling epidemics or large outbreaks. A clear example is influenza A. Other viruses have limited ability to transmit themselves via human-to-human contact, and the frequency with which human cases occur is basically due to the number and frequency of spillover events. Two viral hemorrhagic fevers, caused by EBOV (family *Filoviridae*) and by Lassa virus (family *Arenaviridae*), which partially share their geographic ranges in Sub-Saharan Africa and display similar disease presentation, exemplify this concept, as detailed below.

As briefly mentioned above, EBOV is a zoonotic virus with high pathogenetic potential. Overall, case fatality rates for EBOV reach 80%[4]. EBOV is classified by the ICTV as *Zaire ebolavirus*, in the *Ebolavirus* genus, which includes four other species: *Bundibugyo ebolavirus* (Bundibugyo virus), *Reston ebolavirus* (Reston virus), *Sudan ebolavirus* (Sudan virus) and *Taï Forest ebolavirus* (Taï Forest virus). Among ebolaviruses, only Bundibugyo, Sudan and EBOVs have been associated with disease outbreaks in humans. EBOV disease in humans was first described in Zaire (now the Democratic Republic of Congo) in 1976. For the next 30 years, only rare recognized ebolavirus outbreaks among humans have occurred, and all of these took place in West-Central Africa. Outbreaks have been shown to coincide with significant mortality among the wildlife in surrounding areas (Feldmann and Geisbert 2011; Pigott *et al.* 2014; Holmes *et al.* 2016), and it is thought that people first acquire infection through the handling of infected animal carcasses (gorillas, chimpanzees, duikers, bats) (Alexander *et al.* 2015; Osterholm *et al.* 2015). On one hand, survey data indicate that the detrimental effects of ebolavirus infection on wildlife populations, mainly chimpanzees and gorillas, over the last two decades have been much more dramatic and widespread than the frequency of human outbreaks would suggest (Walsh *et al.* 2003; Bermejo *et al.* 2006; Caillaud *et al.* 2006; Lahm *et al.* 2007). On the other hand, the rarity of human outbreaks emphasizes the very low probability of transmission to humans.

4. https://www.cdc.gov/vhf/ebola/history/distribution-map.html.

Between 2013 and 2016, the largest ever-recorded EBOV outbreak occurred in West Africa: more than 28,000 cases and 11,000 deaths were registered[5]. The West African outbreak was caused by an EBOV variant referred to as "Makona", after the Makona River, close to the border between Liberia, Guinea and Sierra Leone, where the epidemic originated (Kuhn *et al.* 2014). This EBOV epidemic occurred in a time when next-generation sequencing technologies were developed enough to allow large-scale real-time EBOV genome sequencing. Thus, over 1,500 full-length EBOV Makona sequences were obtained, corresponding to approximately 5% of recorded cases (Holmes *et al.* 2016; Malvy *et al.* 2019). Analysis of these sequences allowed detailed reconstruction of EBOV epidemiology and molecular evolution. It was thus established that the epidemic was caused by a single spillover infection from an unknown animal source, which probably occurred in a small village in Guinea (Baize *et al.* 2014; Gire *et al.* 2014). Subsequent human-to-human transmission spread EBOV Makona to larger villages and towns, eventually crossing the borders with Liberia, Sierra Leone and Nigeria. Due to intercontinental travel, secondary infections were also reported in Spain and in the USA (Malvy *et al.* 2019). Thus, human-mediated transmission had a central role in fueling the epidemic. Despite the unusual size of the outbreak, no evidence suggests that the mode of EBOV Makona transmission (contact with infected body fluids) differs from that of other EBOV strains (Holmes *et al.* 2016; Malvy *et al.* 2019). Conversely, the role of EBOV Makona genetic variants on viral transmissibility is more controversial (Holmes *et al.* 2016; Malvy *et al.* 2019). In particular, the emergence of an alanine to valine exchange in the viral glycoprotein at position 82 (Ala82Val) was proposed to represent an adaptation to our species (or to primates in general) and to enhance EBOV Makona entry in human cells. However, the role of this variant (and others that characterize the Makona strains) has not been definitively established (Holmes *et al.* 2016; Malvy *et al.* 2019). It is worth noting that, while we are writing these lines, another EBOV outbreak is claiming lives in the Democratic Republic of the Congo. The outbreak was declared on August 1, 2018 and, to date, the number of cases has grown to 3,000, with more than 2,000 deaths (as of September 4, 2019)[6]. The responsible viral strain (provisionally referred to as "Ituri") does not cluster phylogenetically with EBOV Makona and represents a distinct introduction into the human population (McMullan *et al.* 2019). EBOV Ituri does not carry the Ala82Val substitution and its epidemiology is still under investigation.

The occurrence of two large outbreaks in a time span of few years might be indicative of dangerous changes in the ecology of ebolaviruses in their wild

5. https://www.cdc.gov/vhf/ebola/history/distribution-map.html.

6. https://www.who.int/emergencies/diseases/ebola/drc-2019.

reservoirs. However, this tendency seems to be shared with another zoonotic virus, Lassa virus (LASV, family *Arenaviridae*), which is mainly transmitted in West Africa. In 2018, Nigeria signaled the largest ever number of Lassa fever cases, with a 10-fold increase compared to the number of infections reported in previous years. Lassa fever, an acute hemorrhagic illness first described in 1969 in the town of Lassa (Nigeria), develops in up to 20% of individuals infected with LASV[7]. The main viral reservoir is the Natal multimammate mouse (*Mastomys natalensis*), although other rodents may also host LASV (Pontremoli *et al.* 2019). In analogy to EBOV, the epidemiology of LASV in West Africa was investigated using next-generation sequencing approaches. Several efforts across different countries and timeframes pointed to the same conclusion – i.e. that most Lassa fever cases are directly acquired from rodents and human-to-human transmission, although documented, plays a very limited role in infection spread (Andersen *et al.* 2015; Siddle *et al.* 2018; Kafetzopoulou *et al.* 2019). This pattern was also found to apply to the 2018 outbreak, and genome sequencing revealed that the increase in human cases is not due to the emergence of a specific viral strain with higher transmissibility (Siddle *et al.* 2018; Kafetzopoulou *et al.* 2019). Although reassuring, these observations imply that either the surge in Lassa fever cases is at least partially due to improved awareness and diagnostic capabilities or to changes in the size and features of the rodent reservoir population.

As a consequence of epidemiological and ecological factors, the burden imposed by EBOV and LASV has, up to now, remained mainly confined to West and Central Africa. Very different is the story of HIV-1 and HIV-2, which also originated from wildlife in Africa, but eventually spread worldwide. Despite long-term and frequent human exposure to SIV (simian immunodeficiency virus)-infected monkeys in Africa, only a few cross-species transmission events are known and only four have resulted in significant human-to-human transmission that gave rise to the different HIV-1 groups that infect our species today. Phylogenetic analyses have revealed that the closest relatives of HIV-1 are SIVs infecting wild-living chimpanzees (*Pan troglodytes troglodytes*) and gorillas (*Gorilla gorilla gorilla*) in West Central Africa. In particular, chimpanzees acquired two forms of SIV from African monkeys (SIVrcm from red-capped mangabey and a clade of SIVs infecting several *Cercopithecus* species, including greater spot-nosed, mustached and mona monkeys) that recombined to produce a virus with a unique genome structure (SIVcpz) (Sharp and Hahn 2010). SIVgor in gorillas and at least two of the four lineages of HIV-1 in humans, the pandemic HIV-1 strain (group M or Main) and a rare strain found only in a few Cameroonian people (group N), arose later by independent cross-species transmission from chimpanzees (Figure 8.4).

---

7. http://www.who.int/mediacentre/factsheets/fs179/en/.

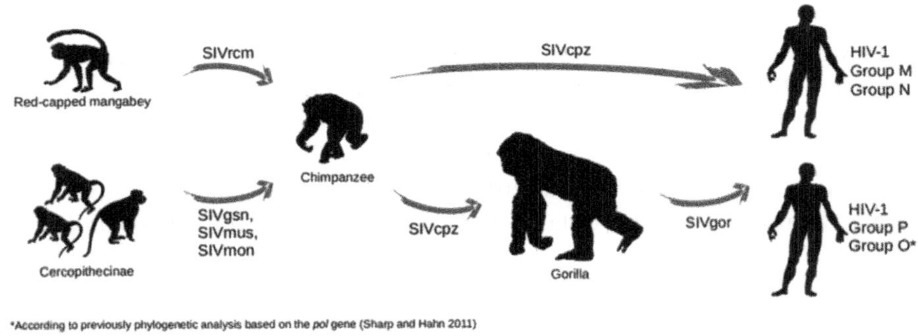

Figure 8.4. *Zoonotic origin of HIV-1. Known cross-species transmissions that gave rise to HIV-1 lineages*

With respect to the other two human lineages, a very rare HIV-1 strain, group P, is clearly derived from SIVgor strains from Cameroon. In contrast, the primate ancestor of HIV-1 group O, a strain infecting 100,000 people mostly from Cameroon as well as from neighboring countries, is unknown, as no ape viruses that are particularly closely related to this group have been identified yet. According to phylogenetic analysis on the *pol* gene sequence, it seems, however, more likely that HIV-1 group O has a gorilla origin (Sharp and Hahn 2011) (Figure 8.4). Exactly when the zoonoses occurred is not known, but using HIV-1 sequences preserved in human biological samples along with estimates of viral mutation rates, researchers calculated that the jump from chimpanzees to humans probably happened during the late 19th or early 20th Century, a time of rapid urbanization and colonization in equatorial Africa (Korber *et al.* 2000; Salemi *et al.* 2001; Lemey *et al.* 2004). Subsequent molecular clock analyses estimated the date of the most recent common ancestor of HIV group M to be approximately 1908 (1884–1924) and approximately 1920 (1890–1940) for the origin of the HIV group O. Group N seems to result from a more recent cross-species transmission, with the most recent common ancestor estimated to be around 1963 (1948–1977) (Wertheim and Worobey 2009). Finally, for HIV group P, which was only discovered in 2009, estimates on dates are uncertain (Plantier *et al.* 2009; Sauter *et al.* 2011; Vallari *et al.* 2011).

As for HIV-2, its origin is also zoonotic and the virus derived from SIVsmm (from sooty mangabey) (Hirsch *et al.* 1989; Gao *et al.* 1992;Lemey *et al.* 2003; Peeters *et al.* 2014). Since its first isolation, at least nine distinct lineages of HIV-2 have been identified, each of which appears to represent an independent host shift. Of these, only groups A and B, whose most recent common ancestor was estimated to date around 1945 (Lemey *et al.* 2003), are pandemic. Humans are probably dead-end hosts for the other HIV-2 groups (C to I) (Santiago *et al.* 2005).

Whereas the evolutionary origins of HIV-1 and HIV-2 are relatively clear and established, the events that generated another extremely common and deadly human pathogen, hepatitis C virus (HCV, family *Flaviviridae*), are still unknown. The host range of HCV is restricted to humans, and the virus is genetically heterogeneous: eight major genotypes (HCV-1 to -8) have been described and these are further divided into at least 90 subtypes[8]. Several of these genotypes and subtypes were identified and classified only recently (Borgia *et al.* 2018; Hedskog *et al.* 2019), suggesting that a considerable proportion of HCV diversity remains undescribed. Due to historical reasons and to the transmission pattern, a small number of "epidemic" HCV subtypes (1a, 1b, 3a and 2a) account for most infections worldwide (Messina *et al.* 2015; Simmonds 2013). Indeed, analysis of HCV diversity indicated that the HCV epidemic started recently, in the 1930s–1940s, as a consequence of practices that determined parenteral or percutaneous exposure (e.g. blood transfusion, vaccination campaigns, intravenous drug injection) (Pybus *et al.* 2001; Simmonds 2013; Jackowiak *et al.* 2014). However, the distribution of the non-epidemic genotypes and subtypes is geographically structured in Sub-Saharan Africa and Asia.

Hepaciviruses related to HCV have been described in several mammals (Burbelo *et al.* 2012; Drexler *et al.* 2013; Baechlein *et al.* 2015; Corman *et al.* 2015; El-Attar *et al.* 2015; Kapoor *et al.* 2011, 2013; Quan *et al.* 2013; Scheel *et al.* 2015; Walter *et al.* 2016). Although the largest diversity of hepacivirus species is hosted by rodents and bats, the closest known relative of HCV was identified in horses (equine hepacivirus, EHV). Recent analyses estimated that extant HCV genotypes originated around 3,000 years ago, whereas the origin of EHV seems to be more recent (Forni *et al.* 2018b). Thus, current evidence suggests that the cross-species transmission of EHV to humans is unlikely to have originated HCV. Even the opposite situation seems to be contradicted by phylogenetic analyses (Pybus and Theze 2016). Thus, an alternative possibility is that EHV and HCV were independently transmitted to their present-day hosts by a third reservoir, with subsequent divergence in human and horse populations

8.6. Conclusion

A wealth of knowledge has accumulated on the evolutionary history of human viruses. Researchers have uncovered the origin of several pathogens that have

8. https://talk.ictvonline.org/ictv_wikis/flaviviridae/w/sg_flavi/56/hcv-classification.

represented or still represent major causes of disability, morbidity and mortality. The timing and circumstances that caused some of the largest and most deadly pandemics of the 21st Century, namely those caused by HIV-1/HIV-2 and by HCV, have been reconstructed by the analysis of viral sequences and by the application of molecular dating tools. Field surveys have uncovered the diversity of viruses hosted by wild and domestic animals that may serve as reservoirs of human pathogens and, for the first time, viral disease outbreaks have been studied in real time by the sequencing of viral samples. The hope is that, in the near future, these advances will translate into direct interventions to improve surveillance programs, develop outbreak prediction/containment measures and assist the selection of future vaccines and drugs.

8.7. References

Abbas, A.A., Taylor, L.J., Dothard, M.I., Leiby, J.S., Fitzgerald, A.S., Khatib, L.A., Collman, R.G., and Bushman, F.D. (2019). Redondoviridae, a family of small, circular DNA viruses of the human oro-respiratory tract associated with periodontitis and critical illness. *Cell Host Microbe*, 25, 719–729.e4.

Aiewsakun, P. and Katzourakis, A. (2017). Marine origin of retroviruses in the early Palaeozoic Era. *Nat Commun*, 8, 13954.

Alexander, K.A., Sanderson, C.E., Marathe, M., Lewis, B.L., Rivers, C.M., Shaman, J., Drake, J.M., Lofgren, E., Dato, V.M., Eisenberg, M.C., and Eubank, S. (2015). What factors might have led to the emergence of Ebola in West Africa? *PLoS Negl Trop Dis*, 9, e0003652.

Allen, J.D. and Ross, T.M. (2018). H3N2 influenza viruses in humans: Viral mechanisms, evolution, and evaluation. *Hum Vaccin Immunother*, 14, 1840–1847.

Allison, A.B., Kohler, D.J., Fox, K.A., Brown, J.D., Gerhold, R.W., Shearn-Bochsler, V.I., Dubovi, E.J., Parrish, C.R., and Holmes, E.C. (2013). Frequent cross-species transmission of parvoviruses among diverse carnivore hosts. *J Virol*, 87, 2342–2347.

Allison, A.B., Kohler, D.J., Ortega, A., Hoover, E.A., Grove, D.M., Holmes, E.C., and Parrish, C.R. (2014). Host-specific parvovirus evolution in nature is recapitulated by *in vitro* adaptation to different carnivore species. *PLoS Pathog*, 10, e1004475.

Andersen, K.G., Shapiro, B.J., Matranga, C.B., Sealfon, R., Lin, A.E., Moses, L.M., Folarin, O.A., Goba, A., Odia, I., Ehiane, P.E., Momoh, M., England, E.M., Winnicki, S., Branco, L.M., Gire, S.K., Phelan, E., Tariyal, R., Tewhey, R., Omoniwa, O., Fullah, M., Fonnie, R., Fonnie, M., Kanneh, L., Jalloh, S., Gbakie, M., Saffa, S., Karbo, K., Gladden, A.D., Qu, J., Stremlau, M., Nekoui, M., Finucane, H.K., Tabrizi, S., Vitti, J.J., Birren, B., Fitzgerald, M., McCowan, C., Ireland, A., Berlin, A.M., Bochicchio, J., Tazon-Vega, B., Lennon, N.J., Ryan, E.M., Bjornson, Z., Milner, D.A. Jr, Lukens, A.K., Broodie, N., Rowland, M., Heinrich, M., Akdag, M., Schieffelin, J.S., Levy, D., Akpan, H., Bausch, D.G., Rubins, K., McCormick, J.B., Lander, E.S., Gunther, S., Hensley, L., Okogbenin, S., Viral Hemorrhagic Fever Consortium, Schaffner, S.F., Okokhere, P.O., Khan, S.H., Grant, D.S., Akpede, G.O., Asogun, D.A., Gnirke, A., Levin, J.Z., Happi, C.T., Garry, R.F., and Sabeti, P.C. (2015). Clinical sequencing uncovers origins and evolution of lassa virus. *Cell*, 162, 738–750.

Anishchenko, M., Bowen, R.A., Paessler, S., Austgen, L., Greene, I.P., and Weaver, S.C. (2006). Venezuelan encephalitis emergence mediated by a phylogenetically predicted viral mutation. *Proc Natl Acad Sci USA*, 103, 4994–4999.

Araujo, N.M. (2015). Hepatitis B virus intergenotypic recombinants worldwide: An overview. *Infect Genet Evol*, 36, 500–510.

Arenas, M., Lorenzo-Redondo, R., and Lopez-Galindez, C. (2016). Influence of mutation and recombination on HIV-1 *in vitro* fitness recovery. *Mol Phylogenet Evol*, 94, 264–270.

Arenas, M., Araujo, N.M., Branco, C., Castelhano, N., Castro-Nallar, E., and Perez-Losada, M. (2018). Mutation and recombination in pathogen evolution: Relevance, methods and controversies. *Infect Genet Evol*, 63, 295–306.

Austermann-Busch, S. and Becher, P. (2012). RNA structural elements determine frequency and sites of nonhomologous recombination in an animal plus-strand RNA virus. *J Virol*, 86, 7393–7402.

Babkin, I.V. and Babkina, I.N. (2012). A retrospective study of the orthopoxvirus molecular evolution. *Infect Genet Evol*, 12, 1597–1604.

Babkin, I.V. and Babkina, I.N. (2015). The origin of the variola virus. *Viruses*, 7, 1100–1112.

Badrane, H. and Tordo, N. (2001). Host switching in Lyssavirus history from the Chiroptera to the Carnivora orders. *J Virol*, 75, 8096–8104.

Baechlein, C., Fischer, N., Grundhoff, A., Alawi, M., Indenbirken, D., Postel, A., Baron, A.L., Offinger, J., Becker, K., Beineke, A., Rehage, J., and Becher, P. (2015). Identification of a novel hepacivirus in domestic cattle from Germany. *J Virol*, 89, 7007–7015.

Baize, S., Pannetier, D., Oestereich, L., Rieger, T., Koivogui, L., Magassouba, N., Soropogui, B., Sow, M.S., Keita, S., De Clerck, H., Tiffany, A., Dominguez, G., Loua, M., Traore, A., Kolie, M., Malano, E.R., Heleze, E., Bocquin, A., Mely, S., Raoul, H., Caro, V., Cadar, D., Gabriel, M., Pahlmann, M., Tappe, D., Schmidt-Chanasit, J., Impouma, B., Diallo, A.K., Formenty, P., Van Herp, M., and Gunther, S. (2014). Emergence of Zaire Ebola virus disease in Guinea. *N Engl J Med*, 371, 1418–1425.

Baker, M.L., Jiang, W., Rixon, F.J., and Chiu, W. (2005). Common ancestry of herpesviruses and tailed DNA bacteriophages. *J Virol*, 79, 14967–14970.

Belshaw, R., Pybus, O.G., and Rambaut, A. (2007). The evolution of genome compression and genomic novelty in RNA viruses. *Genome Res*, 17, 1496–1504.

Bermejo, M., Rodriguez-Teijeiro, J.D., Illera, G., Barroso, A., Vila, C., and Walsh, P.D. (2006). Ebola outbreak killed 5000 gorillas. *Science*, 314, 1564.

Bhatt, S., Holmes, E.C., and Pybus, O.G. (2011). The genomic rate of molecular adaptation of the human influenza A virus. *Mol Biol Evol*, 28, 2443–2451.

Borgia, S.M., Hedskog, C., Parhy, B., Hyland, R.H., Stamm, L.M., Brainard, D.M., Subramanian, M.G., McHutchison, J.G., Mo, H., Svarovskaia, E., and Shafran, S.D. (2018). Identification of a novel Hepatitis C virus genotype from Punjab, India: Expanding classification of Hepatitis C virus into 8 genotypes. *J Infect Dis*, 218, 1722–1729.

Borkenhagen, L.K., Fieldhouse, J.K., Seto, D., and Gray, G.C. (2019). Are adenoviruses zoonotic? A systematic review of the evidence. *Emerg Microbes Infect*, 8, 1679–1687.

Bowden, R., Sakaoka, H., Donnelly, P., and Ward, R. (2004). High recombination rate in herpes simplex virus type 1 natural populations suggests significant co-infection. *Infect Genet Evol*, 4, 115–123.

Bravo, I.G. and Alonso, A. (2007). Phylogeny and evolution of papillomaviruses based on the E1 and E2 proteins. *Virus Genes*, 34, 249–262.

Buck, C.B., Van Doorslaer, K., Peretti, A., Geoghegan, E.M., Tisza, M.J., An, P., Katz, J.P., Pipas, J.M., McBride, A.A., Camus, A.C., McDermott, A.J., Dill, J.A., Delwart, E., Ng, T.F., Farkas, K., Austin, C., Kraberger, S., Davison, W., Pastrana, D.V., and Varsani, A. (2016). The ancient evolutionary history of polyomaviruses. *PLoS Pathog*, 12, e1005574.

Burbelo, P.D., Dubovi, E.J., Simmonds, P., Medina, J.L., Henriquez, J.A., Mishra, N., Wagner, J., Tokarz, R., Cullen, J.M., Iadarola, M.J., Rice, C.M., Lipkin, W.I., and Kapoor, A. (2012). Serology-enabled discovery of genetically diverse hepaciviruses in a new host. *J Virol*, 86, 6171–6178.

Caillaud, D., Levrero, F., Cristescu, R., Gatti, S., Dewas, M., Douadi, M., Gautier-Hion, A., Raymond, M., and Menard, N. (2006). Gorilla susceptibility to Ebola virus: The cost of sociality. *Curr Biol*, 16, R489–91.

de Carvalho Dominguez Souza, B.F., Konig, A., Rasche, A., de Oliveira Carneiro, I., Stephan, N., Corman, V.M., Roppert, P.L., Goldmann, N., Kepper, R., Muller, S.F., Volker, C., de Souza, A.J.S., Gomes-Gouvea, M.S., Moreira-Soto, A., Stocker, A., Nassal, M., Franke, C.R., Rebello Pinho, J.R., Soares, M.D.C.P., Geyer, J., Lemey, P., Drosten, C., Netto, E.M., Glebe, D., and Drexler, J.F. (2018). A novel Hepatitis B virus species discovered in capuchin monkeys sheds new light on the evolution of primate hepadnaviruses. *J Hepatol*, 68, 1114–1122.

Castelhano, N., Araujo, N.M., and Arenas, M. (2017). Heterogeneous recombination among Hepatitis B virus genotypes. *Infect Genet Evol*, 54, 486–490.

Chang, S.F., Sgro, J.Y., and Parrish, C.R. (1992). Multiple amino acids in the capsid structure of canine parvovirus coordinately determine the canine host range and specific antigenic and hemagglutination properties. *J Virol*, 66, 6858–6867.

Charrel, R.N., and de Lamballerie, X. (2010). Zoonotic aspects of arenavirus infections. *Vet Microbiol*, 140, 213–220.

Chen, R. and Holmes, E.C. (2010). Hitchhiking and the population genetic structure of avian influenza virus. *J Mol Evol*, 70, 98–105.

Chen, Z., DeSalle, R., Schiffman, M., Herrero, R., Wood, C.E., Ruiz, J.C., Clifford, G.M., Chan, P.K.S., and Burk, R.D. (2018). Niche adaptation and viral transmission of human papillomaviruses from archaic hominins to modern humans. *PLoS Pathog*, 14, e1007352.

Chiu, C.Y. and Miller, S.A. (2019). Clinical metagenomics. *Nat Rev Genet*, 20, 341–355.

Chopra, I., O'Neill, A.J., and Miller, K. (2003). The role of mutators in the emergence of antibiotic-resistant bacteria. *Drug Resist Updat*, 6, 137–145.

Chua, K.B., Bellini, W.J., Rota, P.A., Harcourt, B.H., Tamin, A., Lam, S.K., Ksiazek, T.G., Rollin, P.E., Zaki, S.R., Shieh, W., Goldsmith, C.S., Gubler, D.J., Roehrig, J.T., Eaton, B., Gould, A.R., Olson, J., Field, H., Daniels, P., Ling, A.E., Peters, C.J., Anderson, L.J., and Mahy, B.W. (2000). Nipah virus: A recently emergent deadly paramyxovirus. *Science*, 288, 1432–1435.

Colson, P., De Lamballerie, X., Yutin, N., Asgari, S., Bigot, Y., Bideshi, D.K., Cheng, X.W., Federici, B.A., Van Etten, J.L., Koonin, E.V., La Scola, B., and Raoult, D. (2013). Megavirales, a proposed new order for eukaryotic nucleocytoplasmic large DNA viruses. *Arch Virol*, 158, 2517–2521.

Conway, J.F., Duda, R.L., Cheng, N., Hendrix, R.W., and Steven, A.C. (1995). Proteolytic and conformational control of virus capsid maturation: The bacteriophage HK97 system. *J Mol Biol*, 253, 86–99.

Corman, V.M., Grundhoff, A., Baechlein, C., Fischer, N., Gmyl, A., Wollny, R., Dei, D., Ritz, D., Binger, T., Adankwah, E., Marfo, K.S., Annison, L., Annan, A., Adu-Sarkodie, Y., Oppong, S., Becher, P., Drosten, C., and Drexler, J.F. (2015). Highly divergent hepaciviruses from African cattle. *J Virol*, 89, 5876–5882.

Cui, J. and Holmes, E.C. (2012a). Endogenous hepadnaviruses in the genome of the budgerigar (*Melopsittacus undulatus*) and the evolution of avian hepadnaviruses. *J Virol*, 86, 7688–7691.

Cui, J. and Holmes, E.C. (2012b). Endogenous lentiviruses in the ferret genome. *J Virol*, 86, 3383–3385.

Davison, A.J., Eberle, R., Ehlers, B., Hayward, G.S., McGeoch, D.J., Minson, A.C., Pellett, P.E., Roizman, B., Studdert, M.J., and Thiry, E. (2009). The order Herpesvirales. *Arch Virol*, 154, 171–177.

Dick, G.W., Kitchen, S.F., and Haddow, A.J. (1952). Zika virus. I. Isolations and serological specificity. *Trans R Soc Trop Med Hyg*, 46, 509–520.

Dolja, V.V. and Koonin, E.V. (2018). Metagenomics reshapes the concepts of RNA virus evolution by revealing extensive horizontal virus transfer. *Virus Res*, 244, 36–52.

Domingo-Calap, P., Schubert, B., Joly, M., Solis, M., Untrau, M., Carapito, R., Georgel, P., Caillard, S., Fafi-Kremer, S., Paul, N., Kohlbacher, O., Gonzalez-Candelas, F., and Bahram, S. (2018). An unusually high substitution rate in transplant-associated BK polyomavirus *in vivo* is further concentrated in HLA-C-bound viral peptides. *PLoS Pathog*, 14, e1007368.

Doorbar, J., Egawa, N., Griffin, H., Kranjec, C., and Murakami, I. (2015). Human papillomavirus molecular biology and disease association. *Rev Med Virol*, 25(1), 2–23.

Drexler, J.F., Corman, V.M., Muller, M.A., Lukashev, A.N., Gmyl, A., Coutard, B., Adam, A., Ritz, D., Leijten, L.M., van Riel, D., Kallies, R., Klose, S.M., Gloza-Rausch, F., Binger, T., Annan, A., Adu-Sarkodie, Y., Oppong, S., Bourgarel, M., Rupp, D., Hoffmann, B., Schlegel, M., Kummerer, B.M., Kruger, D.H., Schmidt-Chanasit, J., Setien, A.A., Cottontail, V.M., Hemachudha, T., Wacharapluesadee, S., Osterrieder, K., Bartenschlager, R., Matthee, S., Beer, M., Kuiken, T., Reusken, C., Leroy, E.M., Ulrich, R.G., and Drosten, C. (2013). Evidence for novel hepaciviruses in rodents. *PLoS Pathog*, 9, e1003438.

Duffy, S., Shackelton, L.A., and Holmes, E.C. (2008). Rates of evolutionary change in viruses: Patterns and determinants. *Nat Rev Genet*, 9, 267–276.

Eastwood, G., Sang, R.C., Guerbois, M., Taracha, E.L.N., and Weaver, S.C. (2017). Enzootic circulation of Chikungunya virus in East Africa: Serological evidence in non-human Kenyan primates. *Am J Trop Med Hyg*, 97, 1399–1404.

Ehlers, B., Dural, G., Yasmum, N., Lembo, T., de Thoisy, B., Ryser-Degiorgis, M.P., Ulrich, R.G., and McGeoch, D.J. (2008). Novel mammalian herpesviruses and lineages within the Gammaherpesvirinae: Cospeciation and interspecies transfer. *J Virol*, 82, 3509–3516.

El-Attar, L.M., Mitchell, J.A., Brooks Brownlie, H., Priestnall, S.L., and Brownlie, J. (2015). Detection of non-primate hepaciviruses in UK dogs. *Virology*, 484, 93–102.

Essbauer, S., Pfeffer, M., and Meyer, H. (2010). Zoonotic poxviruses. *Vet Microbiol*, 140, 229–236.

Farajollahi, A., Fonseca, D.M., Kramer, L.D., and Marm Kilpatrick, A. (2011). "Bird biting" mosquitoes and human disease: A review of the role of Culex pipiens complex mosquitoes in epidemiology. *Infect Genet Evol*, 11, 1577–1585.

Farkasova, H., Hron, T., Paces, J., Hulva, P., Benda, P., Gifford, R.J., and Elleder, D. (2017). Discovery of an endogenous Deltaretrovirus in the genome of long-fingered bats (Chiroptera: Miniopteridae). *Proc Natl Acad Sci USA*, 114, 3145–3150.

Feldmann, H. and Geisbert, T.W. (2011). Ebola haemorrhagic fever. *Lancet*, 377, 849–862.

Fenner, F. (2000). Adventures with poxviruses of vertebrates. *FEMS Microbiol Rev*, 24, 123–133.

Filee, J. and Chandler, M. (2010). Gene exchange and the origin of giant viruses. *Intervirology*, 53, 354–361

Firth, C., Kitchen, A., Shapiro, B., Suchard, M.A., Holmes, E.C., and Rambaut, A. (2010). Using time-structured data to estimate evolutionary rates of double-stranded DNA viruses. *Mol Biol Evol*, 27, 2038–2051.

Focosi, D., Antonelli, G., Pistello, M., and Maggi, F. (2016). Torquetenovirus: The human virome from bench to bedside. *Clin Microbiol Infect*, 22, 589–593.

Foley, B.T., Leitner, T., Paraskevis, D., and Peeters, M. (2016). Primate immunodeficiency virus classification and nomenclature: Review. *Infect Genet Evol*, 46, 150–158.

Forni, D., Cagliani, R., Clerici, M., and Sironi, M. (2017). Molecular evolution of human Coronavirus genomes. *Trends Microbiol*, 25, 35–48.

Forni, D., Cagliani, R., Clerici, M., and Sironi, M. (2018a). Origin and dispersal of Hepatitis E virus. *Emerg Microbes Infect*, 7, 11-017-0009-6.

Forni, D., Cagliani, R., Pontremoli, C., Pozzoli, U., Vertemara, J., De Gioia, L., Clerici, M., and Sironi, M. (2018b). Evolutionary analysis provides insight into the origin and adaptation of HCV. *Front Microbiol*, 9, 854.

Furuse, Y., Suzuki, A., and Oshitani, H. (2010). Origin of measles virus: Divergence from rinderpest virus between the 11th and 12th Centuries. *Virol J*, 7, 52.

Gao, F., Yue, L., White, A.T., Pappas, P.G., Barchue, J., Hanson, A.P., Greene, B.M., Sharp, P.M., Shaw, G.M., and Hahn, B.H. (1992). Human infection by genetically diverse SIVSM-related HIV-2 in West Africa. *Nature*, 358, 495–499.

Geoghegan, J.L., Duchene, S., and Holmes, E.C. (2017). Comparative analysis estimates the relative frequencies of co-divergence and cross-species transmission within viral families. *PLoS Pathog*, 13, e1006215.

Gessain, A. and Cassar, O. (2012). Epidemiological aspects and world distribution of HTLV-1 infection. *Front Microbiol*, 3, 388.

Ghebrehewet, S., MacPherson, P., and Ho, A. (2016). Influenza. *BMJ*, 355, i6258.

Gifford, R.J., Katzourakis, A., Tristem, M., Pybus, O.G., Winters, M., and Shafer, R.W. (2008). A transitional endogenous lentivirus from the genome of a basal primate and implications for lentivirus evolution. *Proc Natl Acad Sci USA*, 105, 20362–20367.

Gilbert, C., Meik, J.M., Dashevsky, D., Card, D.C., Castoe, T.A., and Schaack, S. (2014). Endogenous hepadnaviruses, bornaviruses and circoviruses in snakes. *Proc Biol Sci*, 281, 20141122.

Gire, S.K., Goba, A., Andersen, K.G., Sealfon, R.S., Park, D.J., Kanneh, L., Jalloh, S., Momoh, M., Fullah, M., Dudas, G., Wohl, S., Moses, L.M., Yozwiak, N.L., Winnicki, S., Matranga, C.B., Malboeuf, C.M., Qu, J., Gladden, A.D., Schaffner, S.F., Yang, X., Jiang, P.P., Nekoui, M., Colubri, A., Coomber, M.R., Fonnie, M., Moigboi, A., Gbakie, M., Kamara, F.K., Tucker, V., Konuwa, E., Saffa, S., Sellu, J., Jalloh, A.A., Kovoma, A., Koninga, J., Mustapha, I., Kargbo, K., Foday, M., Yillah, M., Kanneh, F., Robert, W., Massally, J.L., Chapman, S.B., Bochicchio, J., Murphy, C., Nusbaum, C., Young, S., Birren, B.W., Grant, D.S., Scheiffelin, J.S., Lander, E.S., Happi, C., Gevao, S.M., Gnirke, A., Rambaut, A., Garry, R.F., Khan, S.H., and Sabeti, P.C. (2014). Genomic surveillance elucidates Ebola virus origin and transmission during the 2014 outbreak. *Science*, 345, 1369–1372.

Hahn, B.H., Shaw, G.M., De Cock, K.M., and Sharp, P.M. (2000). AIDS as a zoonosis: Scientific and public health implications. *Science*, 287, 607–614.

Han, G.Z. and Worobey, M. (2015). A primitive endogenous lentivirus in a colugo: Insights into the early evolution of lentiviruses. *Mol Biol Evol*, 32, 211–215.

Han, H.J., Wen, H.L., Zhou, C.M., Chen, F.F., Luo, L.M., Liu, J.W., and Yu, X.J. (2015). Bats as reservoirs of severe emerging infectious diseases. *Virus Res*, 205, 1–6.

Harrach, B., Tarján, Z.L., and Benkő, M. (2019). Adenoviruses across the animal kingdom: A walk. *FEBS Lett*, 593, 3660–3673.

Hayward, A., Cornwallis, C.K., and Jern, P. (2015). Pan-vertebrate comparative genomics unmasks retrovirus macroevolution. *Proc Natl Acad Sci USA*, 112, 464–469.

Hedskog, C., Parhy, B., Chang, S., Zeuzem, S., Moreno, C., Shafran, S.D., Borgia, S.M., Asselah, T., Alric, L., Abergel, A., Chen, J.J., Collier, J., Kapoor, D., Hyland, R.H., Simmonds, P., Mo, H., and Svarovskaia, E.S. (2019). Identification of 19 novel Hepatitis C virus subtypes – Further expanding HCV classification. *Open Forum Infect Dis*, 6(3).

Hirsch, V.M., Olmsted, R.A., Murphey-Corb, M., Purcell, R.H., and Johnson, P.R. (1989). An African primate lentivirus (SIVsm) closely related to HIV-2. *Nature*, 339, 389–392.

Holmes, E.C. (2003). Error thresholds and the constraints to RNA virus evolution. *Trends Microbiol*, 11, 543–546.

Holmes, E.C. and Duchêne, S. (2019). Can sequence phylogenies safely infer the origin of the global virome? *MBio*, 10(2).

Holmes, E.C., Dudas, G., Rambaut, A., and Andersen, K.G. (2016). The evolution of Ebola virus: Insights from the 2013–2016 epidemic. *Nature*, 538, 193–200.

Holzerlandt, R., Orengo, C., Kellam, P., and Alba, M.M. (2002). Identification of new herpesvirus gene homologs in the human genome. *Genome Res*, 12, 1739–1748.

Hughes, A.L. and Friedman, R. (2003). Genome-wide survey for genes horizontally transferred from cellular organisms to baculoviruses. *Mol Biol Evol*, 20, 979–987.

Hughes, A.L. and Friedman, R. (2005). Poxvirus genome evolution by gene gain and loss. *Mol Phylogenet Evol*, 35, 186–195.

Hughes, A.L., Irausquin, S., and Friedman, R. (2010). The evolutionary biology of poxviruses. *Infect Genet Evol*, 10, 50–59.

Jackowiak, P., Kuls, K., Budzko, L., Mania, A., Figlerowicz, M., and Figlerowicz, M. (2014). Phylogeny and molecular evolution of the Hepatitis C virus. *Infect Genet Evol*, 21, 67–82.

Kaelber, J.T., Demogines, A., Harbison, C.E., Allison, A.B., Goodman, L.B., Ortega, A.N., Sawyer, S.L., and Parrish, C.R. (2012). Evolutionary reconstructions of the transferrin receptor of Caniforms supports canine parvovirus being a re-emerged and not a novel pathogen in dogs. *PLoS Pathog*, 8, e1002666.

Kafetzopoulou, L.E., Pullan, S.T., Lemey, P., Suchard, M.A., Ehichioya, D.U., Pahlmann, M., Thielebein, A., Hinzmann, J., Oestereich, L., Wozniak, D.M., Efthymiadis, K., Schachten, D., Koenig, F., Matjeschk, J., Lorenzen, S., Lumley, S., Ighodalo, Y., Adomeh, D.I., Olokor, T., Omomoh, E., Omiunu, R., Agbukor, J., Ebo, B., Aiyepada, J., Ebhodaghe, P., Osiemi, B., Ehikhametalor, S., Akhilomen, P., Airende, M., Esumeh, R., Muoebonam, E., Giwa, R., Ekanem, A., Igenegbale, G., Odigie, G., Okonofua, G., Enigbe, R., Oyakhilome, J., Yerumoh, E.O., Odia, I., Aire, C., Okonofua, M., Atafo, R., Tobin, E., Asogun, D., Akpede, N., Okokhere, P.O., Rafiu, M.O., Iraoyah, K.O., Iruolagbe, C.O., Akhideno, P., Erameh, C., Akpede, G., Isibor, E., Naidoo, D., Hewson, R., Hiscox, J.A., Vipond, R., Carroll, M.W., Ihekweazu, C., Formenty, P., Okogbenin, S., Ogbaini-Emovon, E., Gunther, S., and Duraffour, S. (2019). Metagenomic sequencing at the epicenter of the Nigeria 2018 Lassa fever outbreak. *Science*, 363, 74–77.

Kahila Bar-Gal, G., Kim, M.J., Klein, A., Shin, D.H., Oh, C.S., Kim, J.W., Kim, T.H., Kim, S.B., Grant, P.R., Pappo, O., Spigelman, M., and Shouval, D. (2012). Tracing Hepatitis B virus to the 16th century in a Korean mummy. *Hepatology*, 56, 1671–1680.

Kapoor, A., Simmonds, P., Gerold, G., Qaisar, N., Jain, K., Henriquez, J.A., Firth, C., Hirschberg, D.L., Rice, C.M., Shields, S., and Lipkin, W.I. (2011). Characterization of a canine homolog of Hepatitis C virus. *Proc Natl Acad Sci USA*, 108, 11608–11613.

Kapoor, A., Simmonds, P., Scheel, T.K., Hjelle, B., Cullen, J.M., Burbelo, P.D., Chauhan, L.V., Duraisamy, R., Sanchez Leon, M., Jain, K., Vandegrift, K.J., Calisher, C.H., Rice, C.M., and Lipkin, W.I. (2013). Identification of rodent homologs of Hepatitis C virus and pegiviruses. *MBio*, 4, e00216-13.

Katzourakis, A., Tristem, M., Pybus, O.G., and Gifford, R.J. (2007). Discovery and analysis of the first endogenous lentivirus. *Proc Natl Acad Sci USA*, 104, 6261–6265.

Kazlauskas, D., Varsani, A., Koonin, E.V., and Krupovic, M. (2019). Multiple origins of prokaryotic and eukaryotic single-stranded DNA viruses from bacterial and archaeal plasmids. *Nat Commun*, 10(3425).

Keeling, M.J. and Grenfell, B.T. (1997). Disease extinction and community size: Modeling the persistence of measles. *Science*, 275, 65–67.

Kerr, P.J. (2012). Myxomatosis in Australia and Europe: A model for emerging infectious diseases. *Antiviral Res*, 93, 387–415.

Kindhauser, M.K., Allen, T., Frank, V., Santhana, R.S., and Dye, C. (2016). Zika: The origin and spread of a mosquito-borne virus. *Bull World Health Organ*, 94, 675–686C.

Kock, R.A. (2015). Vertebrate reservoirs and secondary epidemiological cycles of vector-borne diseases. *Rev Sci Tech*, 34, 151–163.

Koonin, E.V. and Dolja, V.V. (2014). Virus world as an evolutionary network of viruses and capsidless selfish elements. *Microbiol Mol Biol Rev*, 78, 278–303.

Koonin, E.V., Wolf, Y.I., Nagasaki, K., and Dolja, V.V. (2008). The Big Bang of picorna-like virus evolution antedates the radiation of eukaryotic supergroups. *Nat Rev Microbiol*, 6, 925–939.

Koonin, E.V., Krupovic, M., and Yutin, N. (2015a). Evolution of double-stranded DNA viruses of eukaryotes: From bacteriophages to transposons to giant viruses. *Ann N Y Acad Sci*, 1341, 10–24.

Koonin, E.V., Dolja, V.V., and Krupovic, M. (2015b). Origins and evolution of viruses of eukaryotes: The ultimate modularity. *Virology*, 479–480, 2–25.

Korber, B., Muldoon, M., Theiler, J., Gao, F., Gupta, R., Lapedes, A., Hahn, B.H., Wolinsky, S., and Bhattacharya, T. (2000). Timing the ancestor of the HIV-1 pandemic strains. *Science*, 288, 1789–1796.

Krause-Kyora, B., Susat, J., Key, F.M., Kuhnert, D., Bosse, E., Immel, A., Rinne, C., Kornell, S.C., Yepes, D., Franzenburg, S., Heyne, H.O., Meier, T., Losch, S., Meller, H., Friederich, S., Nicklisch, N., Alt, K.W., Schreiber, S., Tholey, A., Herbig, A., Nebel, A., and Krause, J. (2018). Neolithic and Medieval virus genomes reveal complex evolution of Hepatitis B. *eLife, 7, e36666.*

Kreuder Johnson, C., Hitchens, P.L., Smiley Evans, T., Goldstein, T., Thomas, K., Clements, A., Joly, D.O., Wolfe, N.D., Daszak, P., Karesh, W.B., and Mazet, J.K. (2015). Spillover and pandemic properties of zoonotic viruses with high host plasticity. *Sci Rep*, 5, 14830.

Kruger, D.H., Schonrich, G., and Klempa, B. (2011). Human pathogenic hantaviruses and prevention of infection. *Hum Vaccin*, 7, 685–693.

Krupovic, M. (2013). Networks of evolutionary interactions underlying the polyphyletic origin of ssDNA viruses. *Curr Opin Virol*, 3, 578–586.

Krupovic, M. and Koonin, E.V. (2015). Polintons: A hotbed of eukaryotic virus, transposon and plasmid evolution. *Nat Rev Microbiol*, 13, 105–115.

Krupovic, M. and Koonin, E.V. (2017). Multiple origins of viral capsid proteins from cellular ancestors. *Proc Natl Acad Sci USA*, 114, E2401–E2410.

Krupovic, M., Blomberg, J., Coffin, J.M., Dasgupta, I., Fan, H., Geering, A.D., Gifford, R., Harrach, B., Hull, R., Johnson, W., Kreuze, J.F., Lindemann, D., Llorens, C., Lockhart, B., Mayer, J., Muller, E., Olszewski, N.E., Pappu, H.R., Pooggin, M.M., Richert-Poggeler, K.R., Sabanadzovic, S., Sanfacon, H., Schoelz, J.E., Seal, S., Stavolone, L., Stoye, J.P., Teycheney, P.Y., Tristem, M., Koonin, E.V., and Kuhn, J.H. (2018). Ortervirales: New virus order unifying five families of reverse-transcribing viruses. *J Virol*, 92(12), e00515-18.

Kuhn, J.H., Andersen, K.G., Baize, S., Bao, Y., Bavari, S., Berthet, N., Blinkova, O., Brister, J.R., Clawson, A.N., Fair, J., Gabriel, M., Garry, R.F., Gire, S.K., Goba, A., Gonzalez, J.P., Gunther, S., Happi, C.T., Jahrling, P.B., Kapetshi, J., Kobinger, G., Kugelman, J.R., Leroy, E.M., Maganga, G.D., Mbala, P.K., Moses, L.M., Muyembe-Tamfum, J.J., N'Faly, M., Nichol, S.T., Omilabu, S.A., Palacios, G., Park, D.J., Paweska, J.T., Radoshitzky, S.R., Rossi, C.A., Sabeti, P.C., Schieffelin, J.S., Schoepp, R.J., Sealfon, R., Swanepoel, R., Towner, J.S., Wada, J., Wauquier, N., Yozwiak, N.L., and Formenty, P. (2014). Nomenclature- and database-compatible names for the two Ebola virus variants that emerged in Guinea and the Democratic Republic of the Congo in 2014. *Viruses*, 6, 4760–4799.

Kuhn, J.H., Wolf, Y.I., Krupovic, M., Zhang, Y.Z., Maes, P., Dolja, V.V., and Koonin, E.V. (2019). Classify viruses – the gain is worth the pain. *Nature*, 566, 318–320.

Kuzmin, I.V., Niezgoda, M., Franka, R., Agwanda, B., Markotter, W., Breiman, R.F., Shieh, W.J., Zaki, S.R., and Rupprecht, C.E. (2010). Marburg virus in fruit bat, Kenya. *Emerg Infect Dis*, 16, 352–354.

Lahm, S.A., Kombila, M., Swanepoel, R., and Barnes, R.F. (2007). Morbidity and mortality of wild animals in relation to outbreaks of Ebola haemorrhagic fever in Gabon, 1994–2003. *Trans R Soc Trop Med Hyg*, 101, 64–78.

Lauber, C., Seitz, S., Mattei, S., Suh, A., Beck, J., Herstein, J., Borold, J., Salzburger, W., Kaderali, L., Briggs, J.A.G., and Bartenschlager, R. (2017). Deciphering the origin and evolution of Hepatitis B viruses by means of a family of non-enveloped fish viruses. *Cell Host Microbe*, 22, 387–399.e6.

Lemey, P., Pybus, O.G., Wang, B., Saksena, N.K., Salemi, M., and Vandamme, A.M. (2003). Tracing the origin and history of the HIV-2 epidemic. *Proc Natl Acad Sci USA*, 100, 6588–6592.

Lemey, P., Pybus, O.G., Rambaut, A., Drummond, A.J., Robertson, D.L., Roques, P., Worobey, M., and Vandamme, A.M. (2004). The molecular population genetics of HIV-1 group O. *Genetics*, 167, 1059–1068.

Leroy, E.M., Kumulungui, B., Pourrut, X., Rouquet, P., Hassanin, A., Yaba, P., Delicat, A., Paweska, J.T., Gonzalez, J.P., and Swanepoel, R. (2005). Fruit bats as reservoirs of Ebola virus. *Nature*, 438, 575–576.

Leroy, E.M., Epelboin, A., Mondonge, V., Pourrut, X., Gonzalez, J.P., Muyembe-Tamfum, J.J., and Formenty, P. (2009). Human Ebola outbreak resulting from direct exposure to fruit bats in Luebo, Democratic Republic of Congo, 2007. *Vector Borne Zoonotic Dis*, 9, 723–728.

Liu, W., Pan, S., Yang, H., Bai, W., Shen, Z., Liu, J., and Xie, Y. (2012). The first full-length endogenous hepadnaviruses: Identification and analysis. *J Virol*, 86, 9510–9513.

Long, S.Y., Latimer, E.M., and Hayward, G.S. (2016). Review of elephant endotheliotropic herpesviruses and acute hemorrhagic disease. *ILAR J*, 56, 283–296.

Ludwig, B., Kraus, F.B., Allwinn, R., Doerr, H.W., and Preiser, W. (2003). Viral zoonoses – A threat under control? *Intervirology*, 46, 71–78.

Luis, A.D., Hayman, D.T., O'Shea, T.J., Cryan, P.M., Gilbert, A.T., Pulliam, J.R., Mills, J.N., Timonin, M.E., Willis, C.K., Cunningham, A.A., Fooks, A.R., Rupprecht, C.E., Wood, J.L., and Webb, C.T. (2013). A comparison of bats and rodents as reservoirs of zoonotic viruses: Are bats special?. *Proc Biol Sci*, 280, 20122753.

Ma, W., Kahn, R.E., and Richt, J.A. (2008). The pig as a mixing vessel for influenza viruses: Human and veterinary implications. *J Mol Genet Med*, 3, 158–166.

Macnamara, F.N. (1954). Zika virus: A report on three cases of human infection during an epidemic of jaundice in Nigeria. *Trans R Soc Trop Med Hyg*, 48, 139–145.

Magiorkinis, E.N., Magiorkinis, G.N., Paraskevis, D.N., and Hatzakis, A.E. (2005). Re-analysis of a human Hepatitis B virus (HBV) isolate from an East African wild born Pan troglodytes schweinfurthii: Evidence for interspecies recombination between HBV infecting chimpanzee and human. *Gene*, 349, 165–171.

Malvy, D., McElroy, A.K., de Clerck, H., Gunther, S., and van Griensven, J. (2019). Ebola virus disease. *Lancet*, 393, 936–948.

Martin, D.P., Biagini, P., Lefeuvre, P., Golden, M., Roumagnac, P., and Varsani, A. (2011). Recombination in eukaryotic single stranded DNA viruses. *Viruses*, 3, 1699–1738.

Mayer, S.V., Tesh, R.B., and Vasilakis, N. (2017). The emergence of arthropod-borne viral diseases: A global prospective on dengue, chikungunya and zika fevers. *Acta Trop*, 166, 155–163.

McDonald, S.M., Nelson, M.I., Turner, P.E., and Patton, J.T. (2016). Reassortment in segmented RNA viruses: Mechanisms and outcomes. *Nat Rev Microbiol*, 14, 448–460.

McGeoch, D.J. and Cook, S. (1994). Molecular phylogeny of the alphaherpesvirinae subfamily and a proposed evolutionary timescale. *J Mol Biol*, 238, 9–22.

McGeoch, D.J., Cook, S., Dolan, A., Jamieson, F.E., and Telford, E.A. (1995). Molecular phylogeny and evolutionary timescale for the family of mammalian herpesviruses. *J Mol Biol*, 247, 443–458.

McGeoch, D.J., Dolan, A., and Ralph, A.C. (2000). Toward a comprehensive phylogeny for mammalian and avian herpesviruses. *J Virol*, 74, 10401–10406.

McGeoch, D.J., Gatherer, D., and Dolan, A. (2005). On phylogenetic relationships among major lineages of the Gammaherpesvirinae. *J Gen Virol*, 86, 307–316.

McMullan, L.K., Flint, M., Chakrabarti, A., Guerrero, L., Lo, M.K., Porter, D., Nichol, S.T., Spiropoulou, C.F., and Albarino, C. (2019). Characterisation of infectious Ebola virus from the ongoing outbreak to guide response activities in the Democratic Republic of the Congo: A phylogenetic and *in vitro* analysis. *Lancet Infect Dis*, 19, 1023–1032.

Messina, J.P., Humphreys, I., Flaxman, A., Brown, A., Cooke, G.S., Pybus, O.G., and Barnes, E. (2015). Global distribution and prevalence of Hepatitis C virus genotypes. *Hepatology*, 61, 77–87.

Moens, U., Krumbholz, A., Ehlers, B., Zell, R., Johne, R., Calvignac-Spencer, S., and Lauber, C. (2017). Biology, evolution, and medical importance of polyomaviruses: An update. *Infect Genet Evol*, 54, 18–38.

Mokili, J.L., Rohwer, F., and Dutilh, B.E. (2012). Metagenomics and future perspectives in virus discovery. *Curr Opin Virol*, 2, 63–77.

Moreira, D. and Brochier-Armanet, C. (2008). Giant viruses, giant chimeras: The multiple evolutionary histories of Mimivirus genes. *BMC Evol Biol*, 8(12).

Morse, S.S., Mazet, J.A., Woolhouse, M.E.J., Parrish, C.R., Carroll, D., Karesh, W.B., Zambrana-Torrelio, C., Lipkin, W.I., and Daszak, P. (2012). Prediction and prevention of the next pandemic zoonosis. *Lancet*, 380, 1956–1965.

Moss, B. (2001). Poxviridae: The viruses and their replication. In *Fields Virology, Volume 2*, 4th edition, Fields, B.N., Knipe, D.M., and Howley, P.M. (eds). Lippincott-Raven, Philadelphia, 2849–2883.

Muhlemann, B., Jones, T.C., Damgaard, P.B., Allentoft, M.E., Shevnina, I., Logvin, A., Usmanova, E., Panyushkina, I.P., Boldgiv, B., Bazartseren, T., Tashbaeva, K., Merz, V., Lau, N., Smrcka, V., Voyakin, D., Kitov, E., Epimakhov, A., Pokutta, D., Vicze, M., Price, T.D., Moiseyev, V., Hansen, A.J., Orlando, L., Rasmussen, S., Sikora, M., Vinner, L., Osterhaus, A.D.M.E., Smith, D.J., Glebe, D., Fouchier, R.A.M., Drosten, C., Sjogren, K.G., Kristiansen, K., and Willerslev, E. (2018). Ancient Hepatitis B viruses from the Bronze Age to the Medieval period. *Nature*, 557, 418–423.

Muylkens, B., Thiry, J., Kirten, P., Schynts, F., and Thiry, E. (2007). Bovine herpesvirus 1 infection and infectious bovine rhinotracheitis. *Vet Res*, 38, 181–209.

Nambulli, S., Sharp, C.R., Acciardo, A.S., Drexler, J.F., and Duprex, W.P. (2016). Mapping the evolutionary trajectories of morbilliviruses: What, where and whither. *Curr Opin Virol*, 16, 95–105.

Nassal, M. (2008). Hepatitis B viruses: Reverse transcription a different way. *Virus Res*, 134, 235–249.

Olival, K.J., Hosseini, P.R., Zambrana-Torrelio, C., Ross, N., Bogich, T.L., and Daszak, P. (2017). Host and viral traits predict zoonotic spillover from mammals. *Nature*, 546, 646–650.

Osterholm, M.T., Moore, K.A., Kelley, N.S., Brosseau, L.M., Wong, G., Murphy, F.A., Peters, C.J., LeDuc, J.W., Russell, P.K., Van Herp, M., Kapetshi, J., Muyembe, J.J., Ilunga, B.K., Strong, J.E., Grolla, A., Wolz, A., Kargbo, B., Kargbo, D.K., Sanders, D.A., and Kobinger, G.P. (2015). Transmission of Ebola viruses: What we know and what we do not know. *MBio*, 6, e00137-15.

Parrish, C.R., Holmes, E.C., Morens, D.M., Park, E.C., Burke, D.S., Calisher, C.H., Laughlin, C.A., Saif, L.J., and Daszak, P. (2008). Cross-species virus transmission and the emergence of new epidemic diseases. *Microbiol Mol Biol Rev*, 72, 457–470.

Paweska, J.T., Blumberg, L.H., Liebenberg, C., Hewlett, R.H., Grobbelaar, A.A., Leman, P.A., Croft, J.E., Nel, L.H., Nutt, L., and Swanepoel, R. (2006). Fatal human infection with rabies-related Duvenhage virus, South Africa. *Emerg Infect Dis*, 12, 1965–1967.

Peeters, M., D'Arc, M., and Delaporte, E. (2014). Origin and diversity of human retroviruses. *AIDS Rev*, 16, 23–34.

Perez-Losada, M., Arenas, M., Galan, J.C., Palero, F., and Gonzalez-Candelas, F. (2015). Recombination in viruses: Mechanisms, methods of study, and evolutionary consequences. *Infect Genet Evol*, 30, 296–307.

Periasamy, P., Hutchinson, P.E., Chen, J., Bonne, I., Shahul Hameed, S.S., Selvam, P., Hey, Y.Y., Fink, K., Irving, A.T., Dutertre, C.A., Baker, M., Crameri, G., Wang, L.F., and Alonso, S. (2019). Studies on B Cells in the fruit-eating black flying fox (Pteropus alecto). *Front Immunol*, 10, 489.

Petrova, V.N. and Russell, C.A. (2018). The evolution of seasonal influenza viruses. *Nat Rev Microbiol*, 16, 47–60.

Philippe, N., Legendre, M., Doutre, G., Coute, Y., Poirot, O., Lescot, M., Arslan, D., Seltzer, V., Bertaux, L., Bruley, C., Garin, J., Claverie, J.M., and Abergel, C. (2013). Pandoraviruses: Amoeba viruses with genomes up to 2.5 Mb reaching that of parasitic eukaryotes. *Science*, 341, 281–286.

Pigott, D.M., Golding, N., Mylne, A., Huang, Z., Henry, A.J., Weiss, D.J., Brady, O.J., Kraemer, M.U., Smith, D.L., Moyes, C.L., Bhatt, S., Gething, P.W., Horby, P.W., Bogoch, I.I., Brownstein, J.S., Mekaru, S.R., Tatem, A.J., Khan, K., and Hay, S.I. (2014). Mapping the zoonotic niche of Ebola virus disease in Africa. *eLife*, 3, e04395.

Pimenoff, V.N., de Oliveira, C.M., and Bravo, I.G. (2017). Transmission between archaic and modern human ancestors during the evolution of the oncogenic human papillomavirus 16. *Mol Biol Evol*, 34, 4–19.

Pinsent, A., Fraser, C., Ferguson, N.M., and Riley, S. (2016). A systematic review of reported reassortant viral lineages of influenza A. *BMC Infect Dis*, 16(3).

Pinto-Santini, D.M., Stenbak, C.R., and Linial, M.L. (2017). Foamy virus zoonotic infections. *Retrovirology*, 14(1), 55.

Plantier, J.C., Leoz, M., Dickerson, J.E., De Oliveira, F., Cordonnier, F., Lemee, V., Damond, F., Robertson, D.L., and Simon, F. (2009). A new human immunodeficiency virus derived from gorillas. *Nat Med*, 15, 871–872.

Polat, M., Takeshima, S.N., and Aida, Y. (2017). Epidemiology and genetic diversity of bovine leukemia virus. *Virol J*, 14(1), 209.

Pontremoli, C., Forni, D., and Sironi, M. (2019). Arenavirus genomics: Novel insights into viral diversity, origin, and evolution. *Curr Opin Virol*, 34, 18–28.

Pybus, O.G. and Theze, J. (2016). Hepacivirus cross-species transmission and the origins of the Hepatitis C virus. *Curr Opin Virol*, 16, 1–7.

Pybus, O.G., Charleston, M.A., Gupta, S., Rambaut, A., Holmes, E.C., and Harvey, P.H. (2001). The epidemic behavior of the Hepatitis C virus. *Science*, 292, 2323–2325.

Quan, P.L., Firth, C., Conte, J.M., Williams, S.H., Zambrana-Torrelio, C.M., Anthony, S.J., Ellison, J.A., Gilbert, A.T., Kuzmin, I.V., Niezgoda, M., Osinubi, M.O., Recuenco, S., Markotter, W., Breiman, R.F., Kalemba, L., Malekani, J., Lindblade, K.A., Rostal, M.K., Ojeda-Flores, R., Suzan, G., Davis, L.B., Blau, D.M., Ogunkoya, A.B., Alvarez Castillo, D.A., Moran, D., Ngam, S., Akaibe, D., Agwanda, B., Briese, T., Epstein, J.H., Daszak, P., Rupprecht, C.E., Holmes, E.C., and Lipkin, W.I. (2013). Bats are a major natural reservoir for hepaciviruses and pegiviruses. *Proc Natl Acad Sci USA*, 110, 8194–8199.

Rajao, D.S., Vincent, A.L., and Perez, D.R. (2019). Adaptation of human influenza viruses to swine. *Front Vet Sci*, 5, 347.

Raoult, D., Audic, S., Robert, C., Abergel, C., Renesto, P., Ogata, H., La Scola, B., Suzan, M., and Claverie, J.M. (2004). The 1.2-megabase genome sequence of Mimivirus. *Science*, 306, 1344–1350.

Rasche, A., Souza, E.F.C.D., and Drexler, J.F. (2016). Bat hepadnaviruses and the origins of primate Hepatitis B viruses. *Curr Opin Virol*, 16, 86–94.

Rector, A. and Van Ranst, M. (2013). Animal papillomaviruses. *Virology*, 445, 213–223.

Rixon, F.J. and Schmid, M.F. (2014). Structural similarities in DNA packaging and delivery apparatuses in Herpesvirus and dsDNA bacteriophages. *Curr Opin Virol*, 5, 105–110.

Ruckert, C. and Ebel, G.D. (2018). How do virus-mosquito interactions lead to viral emergence? *Trends Parasitol*, 34, 310–321.

Salemi, M., Strimmer, K., Hall, W.W., Duffy, M., Delaporte, E., Mboup, S., Peeters, M., and Vandamme, A.M. (2001). Dating the common ancestor of SIVcpz and HIV-1 group M and the origin of HIV-1 subtypes using a new method to uncover clock-like molecular evolution. *FASEB J*, 15, 276–278.

Sandbulte, M.R., Spickler, A.R., Zaabel, P.K., and Roth, J.A. (2015). Optimal use of vaccines for control of influenza A virus in swine. *Vaccines (Basel)*, 3, 22–73.

Sanjuan, R. and Domingo-Calap, P. (2016). Mechanisms of viral mutation. *Cell Mol Life Sci*, 73, 4433–4448.

Santiago, M.L., Range, F., Keele, B.F., Li, Y., Bailes, E., Bibollet-Ruche, F., Fruteau, C., Noe, R., Peeters, M., Brookfield, J.F., Shaw, G.M., Sharp, P.M., and Hahn, B.H. (2005). Simian immunodeficiency virus infection in free-ranging sooty mangabeys (Cercocebus atys atys) from the Tai Forest, Cote d'Ivoire: Implications for the origin of epidemic human immunodeficiency virus type 2. *J Virol*, 79, 12515–12527.

Sauter, D., Hue, S., Petit, S.J., Plantier, J.C., Towers, G.J., Kirchhoff, F., and Gupta, R.K. (2011). HIV-1 Group P is unable to antagonize human tetherin by Vpu, Env or Nef. *Retrovirology*, 8(103).

Scheel, T.K., Galli, A., Li, Y.P., Mikkelsen, L.S., Gottwein, J.M., and Bukh, J. (2013). Productive homologous and non-homologous recombination of Hepatitis C virus in cell culture. *PLoS Pathog*, 9, e1003228.

Scheel, T.K., Simmonds, P., and Kapoor, A. (2015). Surveying the global virome: Identification and characterization of HCV-related animal hepaciviruses. *Antiviral Res*, 115, 83–93.

Scott, T.W. and Takken, W. (2012). Feeding strategies of anthropophilic mosquitoes result in increased risk of pathogen transmission. *Trends Parasitol*, 28, 114–121.

Selvarajan Sigamani, S., Zhao, H., Kamau, Y.N., Baines, J.D., and Tang, L. (2013). The structure of the herpes simplex virus DNA-packaging terminase pUL15 nuclease domain suggests an evolutionary lineage among eukaryotic and prokaryotic viruses. *J Virol*, 87, 7140–7148.

Shao, W., Li, X., Goraya, M.U., Wang, S., and Chen, J.L. (2017). Evolution of influenza A virus by mutation and re-assortment. *Int J Mol Sci*, 18(8), 1650.

Sharp, P.M. and Hahn, B.H. (2010). The evolution of HIV-1 and the origin of AIDS. *Philos Trans R Soc Lond B Biol Sci*, 365, 2487–2494.

Sharp, P.M. and Hahn, B.H. (2011). Origins of HIV and the AIDS pandemic. *Cold Spring Harb Perspect Med*, 1, a006841.

Sharp, P.M. and Simmonds, P. (2011). Evaluating the evidence for virus/host co-evolution. *Curr Opin Virol*, 1, 436–441.

Shchelkunov, S.N., Marennikova, S.S., and Moyer, R.W. (2006). *Orthopoxviruses Pathogenic for Humans*. Springer Science & Business Media, Berlin/Heidelberg.

Shi, M., Lin, X.D., Tian, J.H., Chen, L.J., Chen, X., Li, C.X., Qin, X.C., Li, J., Cao, J.P., Eden, J.S., Buchmann, J., Wang, W., Xu, J., Holmes, E.C., and Zhang, Y.Z. (2016). Redefining the invertebrate RNA virosphere. *Nature*, 540(7634), 539–543.

Shi, M., Lin, X.D., Chen, X., Tian, J.H., Chen, L.J., Li, K., Wang, W., Eden, J.S., Shen, J.J., Liu, L., Holmes, E.C., and Zhang, Y.Z. (2018). The evolutionary history of vertebrate RNA viruses. *Nature*, 556, 197–202.

Shil, P., Chavan, S., and Cherian, S. (2011). Molecular basis of antigenic drift in Influenza A/H3N2 strains (1968–2007) in the light of antigenantibody interactions. *Bioinformation*, 6, 266–270.

Siddle, K.J., Eromon, P., Barnes, K.G., Mehta, S., Oguzie, J.U., Odia, I., Schaffner, S.F., Winnicki, S.M., Shah, R.R., Qu, J., Wohl, S., Brehio, P., Iruolagbe, C., Aiyepada, J., Uyigue, E., Akhilomen, P., Okonofua, G., Ye, S., Kayode, T., Ajogbasile, F., Uwanibe, J., Gaye, A., Momoh, M., Chak, B., Kotliar, D., Carter, A., Gladden-Young, A., Freije, C.A., Omoregie, O., Osiemi, B., Muoebonam, E.B., Airende, M., Enigbe, R., Ebo, B., Nosamiefan, I., Oluniyi, P., Nekoui, M., Ogbaini-Emovon, E., Garry, R.F., Andersen, K.G., Park, D.J., Yozwiak, N.L., Akpede, G., Ihekweazu, C., Tomori, O., Okogbenin, S., Folarin, O.A., Okokhere, P.O., MacInnis, B.L., Sabeti, P.C., and Happi, C.T. (2018). Genomic analysis of Lassa virus during an increase in cases in Nigeria in 2018. *N Engl J Med*, 379, 1745–1753.

Simmonds, P. (2013). The origin of Hepatitis C virus. *Curr Top Microbiol Immunol*, 369, 1–15.

Simmonds, P. and Midgley, S. (2005). Recombination in the genesis and evolution of Hepatitis B virus genotypes. *J Virol*, 79, 15467–15476.

Simmonds, P., Adams, M.J., Benko, M., Breitbart, M., Brister, J.R., Carstens, E.B., Davison, A.J., Delwart, E., Gorbalenya, A.E., Harrach, B., Hull, R., King, A.M., Koonin, E.V., Krupovic, M., Kuhn, J.H., Lefkowitz, E.J., Nibert, M.L., Orton, R., Roossinck, M.J., Sabanadzovic, S., Sullivan, M.B., Suttle, C.A., Tesh, R.B., van der Vlugt, R.A., Varsani, A., and Zerbini, F.M. (2017). Consensus statement: Virus taxonomy in the age of metagenomics. *Nat Rev Microbiol*, 15, 161–168.

Simon-Loriere, E. and Holmes, E.C. (2011). Why do RNA viruses recombine? *Nat Rev Microbiol*, 9, 617–626.

Simon-Loriere, E., Holmes, E.C., and Pagan, I. (2013). The effect of gene overlapping on the rate of RNA virus evolution. *Mol Biol Evol*, 30, 1916–1928.

Smith, G.J., Vijaykrishna, D., Bahl, J., Lycett, S.J., Worobey, M., Pybus, O.G., Ma, S.K., Cheung, C.L., Raghwani, J., Bhatt, S., Peiris, J.S., Guan, Y., and Rambaut, A. (2009). Origins and evolutionary genomics of the 2009 swine-origin H1N1 influenza A epidemic. *Nature*, 459, 1122–1125.

Söderlund-Venermo, M. (2019). Emerging human parvoviruses: The rocky road to fame. *Annu Rev Virol*, 6, 71–91.

Sood, R., Hemadri, D., and Bhatia, S. (2013). Sheep associated malignant catarrhal fever: An emerging disease of bovids in India. *Indian J Virol*, 24, 321–331.

Steel, J. and Lowen, A.C. (2014). Influenza A virus reassortment. *Curr Top Microbiol Immunol*, 385, 377–401.

Stephenson, I. and Democratis, J. (2006). Influenza: Current threat from avian influenza. *Br Med Bull*, 75–76, 63–80.

Suh, A., Weber, C.C., Kehlmaier, C., Braun, E.L., Green, R.E., Fritz, U., Ray, D.A., and Ellegren, H. (2014). Early mesozoic coexistence of amniotes and hepadnaviridae. *PLoS Genet*, 10, e1004559.

Switzer, W.M., Salemi, M., Qari, S.H., Jia, H., Gray, R.R., Katzourakis, A., Marriott, S.J., Pryor, K.N., Wolfe, N.D., Burke, D.S., Folks, T.M., and Heneine, W. (2009). Ancient, independent evolution and distinct molecular features of the novel human T-lymphotropic virus type 4. *Retrovirology*, 6, 9.

Tate, J.E., Burton, A.H., Boschi-Pinto, C., Steele, A.D., Duque, J., and Parashar, U.D., WHO-coordinated Global Rotavirus Surveillance Network (2012). 2008 estimate of worldwide rotavirus-associated mortality in children younger than 5 years before the introduction of universal Rotavirus vaccination programmes: A systematic review and meta-analysis. *Lancet Infect Dis*, 12, 136–141.

Teeling, E.C., Springer, M.S., Madsen, O., Bates, P., O'Brien, S.J., and Murphy, W.J. (2005). A molecular phylogeny for bats illuminates biogeography and the fossil record. *Science*, 307, 580–584.

Timen, A. and Koopmans, M. (July 11, 2008). Marburg hemorrhagic fever – The Netherlands ex Uganda. Archive number 20080711.2115 [Online]. Available: http://www.promedmail.org.

Uerpmann, H. and Uerpmann, M. (2002). The appearance of the domestic camel in south-east Arabia. *J Oman Stud*, 12, 235–260.

Vallari, A., Holzmayer, V., Harris, B., Yamaguchi, J., Ngansop, C., Makamche, F., Mbanya, D., Kaptue, L., Ndembi, N., Gurtler, L., Devare, S., and Brennan, C.A. (2011). Confirmation of putative HIV-1 Group P in Cameroon. *J Virol*, 85, 1403–1407.

Van Doorslaer, K. (2013). Evolution of the papillomaviridae. *Virology*, 445, 11–20.

Vandamme, A.M., Salemi, M., Van Brussel, M., Liu, H.F., Van Laethem, K., Van Ranst, M., Michels, L., Desmyter, J., and Goubau, P. (1998). African origin of human T-lymphotropic virus type 2 (HTLV-2) supported by a potential new HTLV-2d subtype in Congolese Bambuti Efe Pygmies. *J Virol*, 72, 4327–4340.

Vartanian, J.P., Pineau, P., Henry, M., Hamilton, W.D., Muller, M.N., Wrangham, R.W., and Wain-Hobson, S. (2002). Identification of a Hepatitis B virus genome in wild chimpanzees (Pan troglodytes schweinfurthi) from East Africa indicates a wide geographical dispersion among equatorial African primates. *J Virol*, 76, 11155–11158.

Vijaykrishna, D., Poon, L.L., Zhu, H.C., Ma, S.K., Li, O.T., Cheung, C.L., Smith, G.J., Peiris, J.S., and Guan, Y. (2010). Reassortment of pandemic H1N1/2009 influenza A virus in swine. *Science*, 328, 1529.

Walker, P.J., Siddell, S.G., Lefkowitz, E.J., Mushegian, A.R., Dempsey, D.M., Dutilh, B.E., Harrach, B., Harrison, R.L., Hendrickson, R.C., Junglen, S., Knowles, N.J., Kropinski, A.M., Krupovic, M., Kuhn, J.H., Nibert, M., Rubino, L., Sabanadzovic, S., Simmonds, P., Varsani, A., Zerbini, F.M., and Davison, A.J. (2019). Changes to virus taxonomy and the International Code of Virus Classification and Nomenclature ratified by the International Committee on Taxonomy of Viruses (2019). *Arch Virol*, 164, 2417–2429.

Walsh, P.D., Abernethy, K.A., Bermejo, M., Beyers, R., De Wachter, P., Akou, M.E., Huijbregts, B., Mambounga, D.I., Toham, A.K., Kilbourn, A.M., Lahm, S.A., Latour, S., Maisels, F., Mbina, C., Mihindou, Y., Obiang, S.N., Effa, E.N., Starkey, M.P., Telfer, P., Thibault, M., Tutin, C.E., White, L.J., and Wilkie, D.S. (2003). Catastrophic ape decline in western equatorial Africa. *Nature*, 422, 611–614.

Walter, S., Rasche, A., Moreira-Soto, A., Pfaender, S., Bletsa, M., Corman, V.M., Aguilar-Setien, A., Garcia-Lacy, F., Hans, A., Todt, D., Schuler, G., Shnaiderman-Torban, A., Steinman, A., Roncoroni, C., Veneziano, V., Rusenova, N., Sandev, N., Rusenov, A., Zapryanova, D., Garcia-Bocanegra, I., Jores, J., Carluccio, A., Veronesi, M.C., Cavalleri, J.M., Drosten, C., Lemey, P., Steinmann, E., and Drexler, J.F. (2016). Differential infection patterns and recent evolutionary origins of equine hepaciviruses in donkeys. *J Virol*, 91, e01711-16.

Wang, L.F. and Eaton, B.T. (2007). Bats, civets and the emergence of SARS. *Curr Top Microbiol Immunol*, 315, 325–344.

Weaver, S.C., Salas, R., Rico-Hesse, R., Ludwig, G.V., Oberste, M.S., Boshell, J., and Tesh, R.B. (1996). Re-emergence of epidemic Venezuelan equine encephalomyelitis in South America. VEE Study Group. *Lancet*, 348 (9025), 436–440.

Webster, R.G. and Govorkova, E.A. (2014). Continuing challenges in influenza. *Ann N Y Acad Sci*, 1323, 115–139.

Wertheim, J.O. and Worobey, M. (2009). Dating the age of the SIV lineages that gave rise to HIV-1 and HIV-2. *PLoS Comput Biol*, 5, e1000377.

Wolf, Y.I., Kazlauskas, D., Iranzo, J., Lucia-Sanz, A., Kuhn, J.H., Krupovic, M., Dolja, V.V., and Koonin, E.V. (2018). Origins and evolution of the global RNA virome. *MBio*, 9(6), e02329-18.

Wolf, Y.I., Kazlauskas, D., Iranzo, J., Lucia-Sanz, A., Kuhn, J.H., Krupovic, M., Dolja, V.V., and Koonin, E.V. (2019). Reply to Holmes and Duchêne, "can sequence phylogenies safely infer the origin of the global virome?": Deep phylogenetic analysis of RNA viruses is highly challenging but not meaningless. *MBio*, 10(2), e00542-19.

Wolfe, N.D., Daszak, P., Kilpatrick, A.M., and Burke, D.S. (2005a). Bushmeat hunting, deforestation, and prediction of zoonoses emergence. *Emerg Infect Dis*, 11, 1822–1827.

Wolfe, N.D., Heneine, W., Carr, J.K., Garcia, A.D., Shanmugam, V., Tamoufe, U., Torimiro, J.N., Prosser, A.T., Lebreton, M., Mpoudi-Ngole, E., McCutchan, F.E., Birx, D.L., Folks, T.M., Burke, D.S., and Switzer, W.M. (2005b). Emergence of unique primate T-lymphotropic viruses among central African bushmeat hunters. *Proc Natl Acad Sci USA*, 102, 7994–7999.

Woodford, N. and Ellington, M.J. (2007). The emergence of antibiotic resistance by mutation. *Clin Microbiol Infect*, 13, 5–18.

Woolhouse, M.E.J. and Adair, K. (2013). The diversity of human RNA viruses. *Future Virol*, 8, 159–171.

Woolhouse, M.E.J., Taylor, L.H., and Haydon, D.T. (2001). Population biology of multihost pathogens. *Science*, 292, 1109–1112.

Woolhouse, M.E.J., Haydon, D.T., and Antia, R. (2005). Emerging pathogens: The epidemiology and evolution of species jumps. *Trends Ecol Evol*, 20, 238–244.

Woolhouse, M.E.J., Adair, K., and Brierley, L. (2013). RNA viruses: A case study of the biology of emerging infectious diseases. *Microbiol Spectr*, 1(1).

Xu, X., Zhao, H., Gong, Z., and Han, G.Z. (2018). Endogenous retroviruses of non-avian/mammalian vertebrates illuminate diversity and deep history of retroviruses. *PLoS Pathog*, 14, e1007072.

Zeller, M. and Andersen, K.G. (2018). Backbone of RNA viruses uncovered. *Nature*, 556, 182–183.

Zhang, G., Cowled, C., Shi, Z., Huang, Z., Bishop-Lilly, K.A., Fang, X., Wynne, J.W., Xiong, Z., Baker, M.L., Zhao, W., Tachedjian, M., Zhu, Y., Zhou, P., Jiang, X., Ng, J., Yang, L., Wu, L., Xiao, J., Feng, Y., Chen, Y., Sun, X., Zhang, Y., Marsh, G.A., Crameri, G., Broder, C.C., Frey, K.G., Wang, L.F., and Wang, J. (2013). Comparative analysis of bat genomes provides insight into the evolution of flight and immunity. *Science*, 339, 456–460.

List of Authors

Michelle M. ARNOLD
Department of Microbiology
and Immunology
Louisiana State University
Health Sciences Center
Shreveport
USA

Megan T. BALDRIDGE
Washington University
School of Medicine
St. Louis
USA

Rachele CAGLIANI
IRCCS Eugenio Medea
Bosisio Parini
Italy

Lindsey M. COSTANTINI
North Carolina Central University
Chapel Hill
USA

Blossom DAMANIA
North Carolina Central University
Chapel Hill
USA

Albie VAN DIJK
Department of Biochemistry
North-West University
Potchefstroom
South Africa

Bria F. DUNLAP
Washington University
School of Medicine
St. Louis
USA

Rachel FEARNS
Boston University
School of Medicine
USA

Stephanie M. KARST
University of Florida
College of Medicine
Gainesville
USA

Carolina B. LÓPEZ
Department of Pathobiology
University of Pennsylvania
Philadelphia
USA

Susana LÓPEZ
Department of Developmental Genetics
Instituto de Biotecnología UNAM
Cuernavaca
Mexico

Alessandra MOZZI
IRCCS Eugenio Medea
Bosisio Parini
Italy

Rachel NETZBAND
University at Albany – SUNY
Albany, NY
USA

Norma A. ORDAZ
Department of Plant Pathology,
University of California, Davis
USA

Cara T. PAGER
University at Albany – SUNY
Albany, NY
USA

Matthew PHILLIPS
University of Florida
College of Medicine
Gainesville
USA

Chiara PONTREMOLI
IRCCS Eugenio Medea
Bosisio Parini
Italy

Swapna Priya RAJARAPU
Department of Entomology
and Plant Pathology
North Carolina State University
Raleigh
USA

Félix AUGUSTO REY
Structural Virology Unit
Institut Pasteur
and
CNRS UMR 3569
Paris
France

Dorith ROTENBERG
Department of Entomology
and Plant Pathology
North Carolina State University
Raleigh
USA

María-Carla SALEH
Viruses and RNAi Unit
Institut Pasteur
and
CNRS UMR 3569
Paris
France

Manuela SIRONI
IRCCS Eugenio Medea
Bosisio Parini
Italy

Diane E. ULLMAN
Department of Entomology
and Nematology
University of California, Davis
USA

Marilyne UZEST
BGPI
University of Montpellier
and
INRA
and
CIRAD
and
Montpellier SupAgro
France

Anna E. WHITFIELD
Department of Entomology
and Plant Pathology
North Carolina State University
Raleigh
USA

Index

V, Z

Printed and bound by CPI Group (UK) Ltd, Croydon, CR0 4YY

13/09/2026

14961045-0006